STUDY GUIDE

DonnaJean Fredeen

CHEMISTRY

Second Edition

McMURRY • FAY

Main groups

Transition metals

Main groups

1 1A	2 2A	3 3B	4 4B	5 5B	6 6B	7 7B	8	9 8B	10	11 1B	12 2B	13 3A	14 4A	15 5A	16 6A	17 7A	18 8A
1 H 1.00794																	2 He 4.00260
3 Li 6.941	4 Be 9.01218											5 B 10.81	6 C 12.011	7 N 14.0067	8 O 15.9994	9 F 18.998403	10 Ne 20.1797
11 Na 22.98977	12 Mg 24.305											13 Al 26.98154	14 Si 28.0855	15 P 30.97376	16 S 32.066	17 Cl 35.453	18 Ar 39.948
19 K 39.0983	20 Ca 40.078	21 Sc 44.9559	22 Ti 47.88	23 V 50.9415	24 Cr 51.996	25 Mn 54.9380	26 Fe 55.847	27 Co 58.9332	28 Ni 58.69	29 Cu 63.546	30 Zn 65.39	31 Ga 69.72	32 Ge 72.61	33 As 74.9216	34 Se 78.96	35 Br 79.904	36 Kr 83.80
37 Rb 85.4678	38 Sr 87.62	39 Y 88.9059	40 Zr 91.224	41 Nb 92.9064	42 Mo 95.94	43 Tc (98)	44 Ru 101.07	45 Rh 102.9055	46 Pd 106.42	47 Ag 107.8682	48 Cd 112.41	49 In 114.82	50 Sn 118.710	51 Sb 121.757	52 Te 127.60	53 I 126.9045	54 Xe 131.29
55 Cs 132.9054	56 Ba 137.33	57 *La 138.9055	72 Hf 178.49	73 Ta 180.9479	74 W 183.85	75 Re 186.207	76 Os 190.2	77 Ir 192.22	78 Pt 195.08	79 Au 196.9665	80 Hg 200.59	81 Tl 204.383	82 Pb 207.2	83 Bi 208.9804	84 Po (209)	85 At (210)	86 Rn (222)
87 Fr (223)	88 Ra 226.0254	89 †Ac 227.0278	104 Rf (261)	105 Db (262)	106 Sg (263)	107 Bh (262)	108 Hs (265)	109 Mt (266)	110 (269)	111 (272)	112 (277)						

*Lanthanide series

58 Ce 140.12	59 Pr 140.9077	60 Nd 144.24	61 Pm (145)	62 Sm 150.36	63 Eu 151.96	64 Gd 157.25	65 Tb 158.9254	66 Dy 162.50	67 Ho 164.9304	68 Er 167.26	69 Tm 168.9342	70 Yb 173.04	71 Lu 174.967
90 Th 232.0381	91 Pa 231.0359	92 U 238.0289	93 Np 237.048	94 Pu (244)	95 Am (243)	96 Cm (247)	97 Bk (247)	98 Cf (251)	99 Es (252)	100 Fm (257)	101 Md (258)	102 No (259)	103 Lr (260)

†Actinide series

STUDY GUIDE

DonnaJean Fredeen
Southern Connecticut State University

CHEMISTRY
SECOND EDITION

John McMurry
Cornell University

Robert C. Fay
Cornell University

PRENTICE HALL, Upper Saddle River, NJ 07458

Senior Editor: John Challice
Production Editor: Carole Suraci
Supplement Cover Designer: Liz Nemeth
Special Projects Manager: Barbara A. Murray
Supplement Cover Manager: Paul Gourhan
Manufacturing Buyer: Ben Smith
Associate Editor: Mary Hornby

© 1998 by Prentice-Hall, Inc.
Simon & Schuster / A Viacom Company
Upper Saddle River, NJ 07458

Printed in the United States of America

10 9 8 7 6 5 4 3 2 1

ISBN 0-13-757436-3

Prentice-Hall International (UK) Limited, *London*
Prentice-Hall of Australia Pty. Limited, *Sydney*
Prentice-Hall Canada, Inc., *London*
Prentice-Hall Hispanoamericana, S.A., *Mexico*
Prentice-Hall of India Private Limited, *New Delhi*
Prentice-Hall of Japan, Inc., *Tokyo*
Simon & Schuster Asia Pte. Ltd., *Singapore*
Editora Prentice-Hall do Brazil, Ltda., *Rio de Janeiro*

STUDENT STUDY GUIDE

TABLE OF CONTENTS

PREFACE

To the student:

This study guide was written specifically to assist the student using CHEMISTRY by McMurry & Fay and presents, in outline form, the major concepts, theories, facts and applications found in the text. The second edition of the Study Guide has been expanded to include more examples within the outlines and more questions and problems in the Self Tests.

Every chapter is keyed to the main text, and is presented in four sections:
- Listed Learning Goals – alerting the student to key concepts that will be covered in each chapter.
- Chapter Overview – a brief summary of major material that will be covered in that chapter and placed in its context.
- Chapter Outline – highlighting the major concepts and theories in the text. The Learning Goals are clearly marked and expounded upon within these chapter outlines. Throughout each outline, the detailed examples are included showing applications of major chemical concepts. The outlines also provide the student with an initial set of notes which can then be tailored and annotated with lecture material.
- Self Tests and Solutions - modeled on the text problems and linked to the Chapter Learning Goals, this section is intended to test the student's knowledge of the material covered in the chapters. Successful completion of these problems indicates the student has mastered the learning goals for the chapters. The student who uses this section as a mock exam will discover which topics have been mastered and which topics need more attention.

My efforts in writing this Study Guide began nearly ten years ago when I first put together a "*Concepts and Definitions*" manual for my students. Since that time, the size and amount of work needed to create this book has dramatically increased. However, I still have only one goal in mind: to help you, the student, understand and appreciate the field of chemistry. It takes hard work. But if you take the time to read and work the problems in your text, use the outline in this study guide and work the problems you find here, you will not only begin to appreciate the role that chemistry plays in your life, you will also do well in your course. (You may even find that you like the subject!)

I would like to take this opportunity to acknowledge the following people: Sandi Hakanson whose belief in me from the very beginning made this all possible; Mary Hornby, who has kindly and patiently pushed me along; Robert Snyder, who helped with some of the problems; Bob Fay, who so very carefully checked the manuscript for errors; and Carole Suraci, who caught all those unsightly typos. Mother, Daddy, Ken, Elissa, and Andrew – thank you. To Ken, this golf season is all yours; and yes, Elissa and Andrew, you may now play on Mommy's computer.

Finally, I would like to dedicate this Study Guide to two men who were instrumental in shaping me as a chemist; Mr. Jerry Watts and Mr. Sam Meador, my high school chemistry teachers. Mr. Watts, it all began in your class watching a candle burn. Mr. Meador, with your encouragement, that flickering flame grew into a fire. If I can inspire one student in the way that you inspired me, I will truly be successful. Thank you!

DonnaJean A. Fredeen
Southern Connecticut State University

CHAPTER 1

CHEMISTRY: MATTER AND MEASUREMENT

Chapter Learning Goals

1⊠ Give the symbol and name of elements mentioned in this chapter.
2⊠ Identify the group number and period to which an element belongs.
3⊠ Identify the regions of the periodic table.
4⊠ List the seven basic SI units of measure, and give the numerical equivalent of the common metric prefixes used with these units.
5⊠ Express numbers in scientific notation. (See appendix A in your text.)
6⊠ Interconvert Fahrenheit, Celsius, and Kelvin temperatures.
7⊠ Determine the number of significant digits in a measured quantity.
8⊠ State the result of a calculation involving measured quantities to the correct number of significant digits.
9⊠ Use dimensional analysis to solve metric-English conversion problems.
10⊠ Identify properties as extensive or intensive.
11⊠ Perform calculations using density.

Chapter in Brief

Chemistry is the study of the composition, properties, and transformations of matter and of chemical laws which are responsible for the changes that take place in nature. In this chapter, you will begin your study of chemistry by learning the names and symbols of the elements on the periodic table. You will also begin to see that the periodic table of the elements is the most important organizing principle of chemistry. To understand chemistry, you must have an understanding of the measurements that we make. You will be introduced to the International System of Units seven base units which, along with other derived units, suffice for all scientific measurements. It is necessary to know the precision of the measurements you make, which requires the use of significant figures. You will learn how to determine the number of significant figures in a measurement and the rules for carrying significant figures through a calculation. Finally, you will be introduced to the method of dimensional-analysis and how to use this method to convert one unit into an equivalent unit.

Approaching Chemistry: Experimentation

    A. Scientific Method - systematic approach to research.
        1. Experimentation.
        2. Hypothesis - an interpretation that explains the results of many experiments.
        3. Theory - consistent explanation of known observations; logical interpretations of experimental results.

1⊠ **Chemistry and the Elements**
    A. Element - a fundamental substance that can't be chemically changed or broken down into anything simpler.
        1. Chemical symbol - used to represent specific elements
            a. capitalize the first letter; second letter is lower case
    B. Periodic Table - a tabular organization of all 112 elements.

2⊠ **Elements and the Periodic Table**
    A. Periods – seven horizontal rows in the periodic table.

B. Groups - 18 vertical columns in the periodic table.

C. The elements in a given group have similar chemical properties.

3⊠ **Some Characteristics of the Elements**

A. The periodic table of the elements is the most important organizing principle of chemistry. (See inside cover.)

1. Regular progression in the size of the seven periods.

a. reflects a similar regularity in atomic structure

2. Main Group (or Representative) Elements - Groups 1A - 8A; (two larger groups on the left and the six larger groups on the right of the table).

3. Transition-metal Elements - Groups 1B - 8B; (the 10 smaller groups in the middle of the table).

4. Inner transition-metal (or Rare Earth) - the 14 groups shown separately at the bottom of the table.

B. Similarities of Groups.

1. Group 1A - Alkali metals; lustrous, silvery metals; react rapidly with water to form highly alkaline products.

2. Group 2A - Alkaline earth metals; lustrous, silvery metals; less reactive than alkali metals.

3. Group 7A - Halogens; corrosive, nonmetallic elements; salt formers.

4. Group 8A - Noble gases; gases with low reactivity.

C. Three major classes of elements in the periodic table.

1. Metals - largest category of elements; found on the left side of the periodic table.

a. solids (except mercury)

b. malleable

c. ductile - can be drawn into thin wires without breaking

d. conduct heat and electricity

2. Nonmetals - found on the right side of the periodic table.

a. gases, liquids or solids

b. brightly colored

c. brittle solids

d. poor conductors of heat and electricity

3. Semimetals (metalloids) - elements adjacent to the zigzag boundary between metals and nonmetals.

a. properties fall between metals and nonmetals

b. brittle

c. poor conductors of heat and electricity

4⊠ **Experimentation and Measurement**

A. International System (SI) of Units - Seven base units (and units derived from them) that suffice for all scientific measurements. (See Table 1.2 in text, page 10)

B. Common prefixes used to modify SI units

1. mega (M); factor = 10^6.

2. kilo (k); factor = 10^3.

3. deci (d); factor = 10^{-1}.

4. centi (c); factor = 10^{-2}.

5. milli (m); factor = 10^{-3}.

6. micro (μ); factor = 10^{-6}.

7. nano (n); factor = 10^{-9}.

5⊠ C. Scientific Notation - an exponential format for numbers that are either very large or very small (see Appendix A in your text).

D. All measurements of physical quantities contain both a number and a unit label.

Measuring Mass

A. Mass (SI unit = kg) - the amount of matter in an object.
B. Matter - term used to describe anything physically real.
C. Weight - the pull of gravity on an object by the earth or other celestial body.
D. The mass of an object can be measured on a balance by comparing the weight of the object to the weight of a reference standard of known mass.

Measuring Length

A. Meter - standard unit of length in the SI system.
   1. Distance traveled by light in a vacuum in 1/299,792,458th of a second.

Derived Units: Measuring Volume

A. Derived quantities - quantities expressed in terms of one or more of the seven base units. (See Table 1.4 in text, page 13).
B. Volume - the amount of space occupied by an object.
   1. Measured in SI units by the cubic meter (m^3).
   2. Commonly used measurements:
      a. cubic decimeter (dm^3) = metric liter (L)
      b. cubic centimeter (cm^3) = metric milliliter (mL)

6⊠ ### Measuring Temperature

A. Common unit - Celsius degree (°C).
B. Scientific unit - Kelvin.
C. Celsius and Kelvin scales have 100 degrees between the freezing point and boiling point of water.

Temperature in K = temperature in °C + 273.15

Temperature in °C = temperature in K - 273.15

D. Fahrenheit scale has 180° between the freezing point and boiling point of water.
   1. 180° F encompasses the same range as 100° C.

$$a. \quad 1°C \times \frac{180°F}{100°C} = \frac{9}{5} \ °F$$

E. To convert from Celsius to Fahrenheit - do a size correction followed by a zero-point correction.
F. To convert from Fahrenheit to Celsius - do a zero-point correction followed by a size correction.

$$°F = \left(\frac{9}{5} \times °C\right) + 32 \qquad\qquad °C = \frac{5}{9} \times \left(°F - 32\right)$$

EXAMPLE:

Titanium, an important structural metal used in aircraft, has a melting point of 1725° C. Express this temperature in °F and Kelvin.

SOLUTION: This is a good time to use the "thinking approach." A Fahrenheit degree is smaller than a Celsius degree. Just think about the melting (freezing) point of water. Water melts at 0° on the Celsius scale and at 32° on the Fahrenheit scale. Therefore, when converting from Celsius to Fahrenheit, the temperature should be higher. Now, apply the formula given above and see if

3

your answer compares well with your thought process.

$$^\circ F = \left(\frac{9}{5} \times 1725\right) + 32 = 3137\ ^\circ F$$

Your calculated value for the temperature on the Fahrenheit scale agrees well with the thinking approach we used. The same logic can be used in calculating the value of the melting point on the Kelvin scale. The freezing point of water is 273.15° higher on the Kelvin scale, so the melting point of titanium also should be 273.15° higher.

$$K = 1725 + 273.15 = 1998.15\ K$$

7⊠   **Accuracy, Precision, and Significant Figures in Measurement**
   A. Accuracy - how close a measurement is to the true value.
   B. Precision - how well a number of independent measurements agree with one another.
      1. To indicate the precision of a measurement use all the digits known with certainty, plus one additional estimated digit.
      2. Total number of digits in the measurement = the number of significant figures.
   C. Determining the number of significant figures - see rules, page 18 in text.

EXAMPLE:

12.502 g - 5 significant figures (rule 1)

0.005 25 mL - 3 significant figures (rule 2); If you write the number as 5.25×10^{-3} Ml, it is easier to see the number of significant figures

5.30 m - 3 significant figures (rule 3)

79,300 s - uncertain (rule 4); If the number is 7.9300×10^{4}, there are five significant figures. If the number is 7.930×10^{4}, there are four significant figures.   If the number is 7.93×10^{4}, there are three significant figures.

   D. Using scientific notation can be very helpful in determining the number of significant figures (see above example).
   E. Exact numbers have an infinite number of significant figures.

8⊠   **Rounding Numbers**
   A. In carrying out a multiplication or division, the answer can't have more significant figures than either of the original numbers.

EXAMPLE:

five significant
figures   *three significant figures*

$$\frac{55.678\,g}{32.5\,mL} = 1.71\ g/mL$$

three significant
figures

B. In carrying out an addition or subtraction, the answer can't have more digits to the right of the decimal point than either of the original numbers.

EXAMPLE:

$$24.7835 \text{ g} \quad \textit{ends 4 places past decimal point}$$
$$- \ 0.45 \ \ \text{g} \quad \textit{ends 2 places past decimal point}$$
$$24.33 \ \ \ \text{g} \quad \textit{ends 2 places past decimal point}$$

C. Rules for rounding off numbers - see pages 19 and 20 in text.
  1. In doing calculations, use all figures, significant or not, and then round off the final answer.

EXAMPLE:

If it takes 0.75 hours to drive a distance of 26.8 miles, what is the average speed of the car in miles per hour?

SOLUTION: Average speed $= \dfrac{26.8 \text{ mi}}{0.75 \text{ h}} = 35.73333 \qquad$ mi/h

Decide how many significant figures should be in your answer. The denominator only has two significant figures; therefore, the answer must also have only two significant figures. Round off your answer. The first digit to be dropped is greater than 5. 35.733 33 mi/h becomes 36 mi/h.

EXAMPLE:

Round off the numbers to three significant figures: (a) 7.835 g   (b) 4.265 g.

SOLUTION: (a) Because the last digit is a 5 with nothing following, the answer is 7.84 since the digit to the left of the 5 is odd.

  (b) Because the last digit is a 5 with nothing following, the answer is 4.26 since the digit to the left of the 5 is even.

9⌦   **Calculations: Converting from One Unit to Another**
  A. Dimensional-analysis method - a quantity described in one unit is converted into an equivalent quantity described in a different unit by using a conversion factor to express the specific relationship between units.

Starting quantity x conversion factor = equivalent quantity

  1. Units are treated like numbers and can be multiplied and divided.
  2. Set up an equation so that all unwanted units cancel, leaving only the desired units.
    a.   the right answer is obtained only if the equation is set up so that the unwanted units cancel
  3. Start with the information given.
  4. Use a conversion factor that allows the unit in the information given to be cancelled.
  5. Continue to use conversion factors to cancel the previous unit until the desired unit is left.

EXAMPLE:

What would be the weight in kilograms of a person weighing 175.5 lb?

SOLUTION: Given that 2.205 lb = 1 kg, set up the equation knowing that you want to convert 175.5 lb to kg and that the unit lb should cancel. (Use the thinking approach to rationalize that 1 kg is larger than a pound; therefore, your answer in kg should be less than the 175.5 lb you started with.)

$$175.5 \text{ lb} \times \frac{1 \text{ kg}}{2.205 \text{ lb}} = 79.59 \text{ kg}$$

Warning: It's easy to get the "right" answer using dimensional-analysis without really understanding what you're doing.

EXAMPLE:

An international exchange student rents a car while studying in Germany. The speedometer in this car is calibrated in km/hr. Once out on the Autobahn, he wants to test the car's performance at 80 mi/hr. How fast must he drive in km/hr?

SOLUTION: Using the conversion table on the back cover, we learn that 1 mi = 1.6093 km. Set up the equation beginning with the information given (80 mi/hr) and write the conversion factor so that the unit mi cancels.

$$\frac{80 \text{ mi}}{\text{hr}} \times \frac{1.6093 \text{ km}}{1 \text{ mi}} = 130 \ \frac{\text{km}}{\text{hr}}$$

EXAMPLE:

Calculate the amount of time it would take an athlete to run 3200 m if she normally runs a mile in 4 min, 38 s.

SOLUTION: To determine the amount of time it takes for our athlete to run 3200 m, we should first convert the distance to miles. Once we have converted 3200 m to miles, we can then calculate the amount of time it will take for our athlete to run that distance. However, it will be easier to calculate the amount of time to run 3200 m if we first convert 4 min, 38 s to s.

$$4 \text{ min} \times \frac{60 \text{ s}}{1 \text{ min}} = 240 \text{ s}; \quad 240 \text{ s} + 38 \text{ s} = 278 \text{ s}$$

Using the conversion table found on the back cover, we find that the conversion given for distance is between a mile and kilometer. Therefore, we should first convert meters to kilometers.

$$3200 \text{ m} \times \frac{1 \text{ km}}{1000 \text{ m}} \times \frac{1 \text{ mi}}{1.6093 \text{ km}} \times \frac{278 \text{ s}}{1 \text{ mi}} = 553 \text{ s}$$

Knowing that there are 60 s in 1min, we can convert our time to 9.2 min which is equivalent to 9 min, 12 s.

6

Properties of Matter: Density

10⊠ A. Property - any characteristic that can be used to describe or identify matter.
    1. ...ysical properties - characteristics that can be determined without changing the chemical makeup of the sample.
    2. Chemical properties - properties that do change the chemical makeup of the sample.
    3. Intensive properties - properties that do not depend on the size of the sample.
    4. Extensive properties - properties that depend on the size of the sample.

11⊠ B. Density - intensive property that relates mass to volume.
    1. Expressed in units of g/cm^3 or g/mL.
    2. Temperature dependent property.

EXAMPLE:

A student determined that a metal cylinder having a mass of 57.893 g has a volume of 38.32 cm^3. What is the density of this metal cylinder?

SOLUTION: Density $= \dfrac{57.893\ g}{38.34\ mL} = 1.511$ g/mL

Self-Test

This section is intended to test your knowledge of the material covered in this chapter. Think through these problems and make certain you understand what is going on. Ask yourself if your answer makes sense. Many of these questions are linked to the chapter learning goals. Therefore, successful completion of these problems indicates you have mastered the learning goals for this chapter. You will receive the greatest benefit from this section if you use it as a mock exam. You will then discover which topics you have mastered and which topics you need to study in more detail.

True/False

1. The symbol for the element cobalt is CO.

2. The element sodium belongs in the group referred to as the alkaline earth metals.

3. The element bromine is a nonmetal.

4. Semimetals have properties somewhere between those of metals and nonmetals, and therefore are excellent conductors of electricity.

5. The SI base unit for length is the meter.

6. The unit most commonly used in the laboratory for volume is the m^3.

7. The number 0.000 252 contains three significant figures.

8. 1×10^{-6} g is equal to 1 μg.

9. The symbol for iron comes from its Latin root ferrum.

10. The number 5,000,000 can be expressed in scientific notation as 5×10^6.

Multiple Choice

1. The symbol for antimony is
   a. At
   b. Ar
   c. As
   d. Sb

2. The symbol Mg represents the element
   a. Manganese
   b. Magnesium
   c. Mercury
   d. Molybdenum

3. The symbol for boron is:
   a. B
   b. Br
   c. Ba
   d. Bk

4. The definition of mass is:
   a. term used to describe anything physically real.
   b. the pull of gravity on an object by the earth or other celestial body.
   c. the amount of matter in an object.

5. The following data were collected by a student during a laboratory exercise:

   25.78 mL, 25.82 mL, 25.65 mL.

   The true volume in this experiment is 30.00 mL. These numbers represent data that is:
   a. very precise and accurate.
   b. very precise but not accurate.
   c. very accurate but not precise.
   d. neither precise nor accurate.

6. If 5.078 93 were divided by 0.0789, the answer would contain:
   a. 6 significant figures.
   b. 4 significant figures.
   c. 5 significant figures.
   d. 3 significant figures.

7. The element oxygen is found in which region of the periodic table?
   a. metal
   b. semimetal
   c. halogens
   d. nonmetal

8. The transition metals:
    a. are the six larger groups on the right of the periodic table.
    b. are the 10 smaller groups in the middle of the periodic table.
    c. are the 14 groups shown separately at the bottom of the table.

9. A physical property is one that:
    a. changes the chemical makeup of the sample.
    b. depends on the size of the sample.
    c. can be determined without changing the chemical makeup of the sample.
    d. does not depend on the size of the sample.

10. The SI unit that is used for temperature is:
    a. °F
    b. °C
    c. °K
    d. K

11. The element bromine is found in the group known as the
    a. alkali metals
    b. alkaline earth metals
    c. chalcogens
    d. halogens

12. The element europium, Eu, is a(n)
    a. alkali metal
    b. lanthanide element
    c. transition metal
    d. actinide element

13. The symbol for potassium is
    a. K
    b. Pt
    c. Po
    d. P

14. The SI prefix, micro, corresponds to the multiplier
    a. 10^{-3}
    b. 10^{-6}
    c. 10^{-9}
    d. 10^{-12}

15. The Fahrenheit degree is
    a. smaller than a Celsius degree
    b. smaller than a Kelvin
    c. larger than a Celsius degree
    d. the same as a Celsius degree

Chapter 1 – Chemistry: Matter and Measurement

Matching

Hypothesis

a. The amount of space occupied by an object.

Theory

b. property that does not depend on the size of the sample.

Element

c. how close a measurement is to the true value.

Main Group Elements

d. how well a number of independent measurements agree with one another.

Matter

e. an interpretation that explains the results of many experiments.

Volume

f. a fundamental substance that can't be chemically changed or broken down into anything simpler.

Precision

g. a method in which a quantity described in one unit is converted into an equivalent quantity described in a different unit by using a conversion factor to express the specific relationship between units.

Accuracy

h. a consistent explanation of known observations; logical interpretations of experimental results.

Dimensional Analysis

i. term used to describe anything physically real.

Intensive property

j. Groups 1A - 8A on the periodic table.

Fill-in-the-Blank

1. The symbol for the element potassium is _____.

2. The element Ti can be found in the _____ period and the group _____.

3. The element Ge is called _____ and is a _____.

4. Collectively, the elements in group 8A are called the _____.

5. The SI unit for pressure is the _____.

6. The definition and SI unit for mass is _____

    _____.

7. All measurements of physical quantities contain both a _____.

8. One nanosecond is equal to _____ seconds.

9. One meter is equal to _____ centimeters.

10. One gram is equal to _____ kilograms.

11. To indicate the precision of a measurement, use all of the digits known with certainty plus _____

    _____.

12. In carrying out a multiplication or division, the answer can't have more significant figures than

    _____.

13. Density is an _____ property that relates _____.

14. The three steps involved in the scientific method are _____.

15. Group 1A metals react _____ with water to form highly _____

    products. These metals are called the _____ metals.

16. Chemistry is defined as _____

    _____

    _____.

17. _____ describe the changes that take place in nature.

18. The system of units used by chemists is the _____ and

    contains _____ fundamental units.

19. The columns in the periodic table are referred to as a _____, and the rows are

    referred to as a _____.

20. Elements within a group have similar _____.

Problems

1. 3.89 μg = _____ mg = _____ ng

2. Perform the following calculations and report your answer with the correct number of significant
    figures:

    a. $\dfrac{7.984}{3.2}$ =   b. 7.53 x 23.945 62 =   c. 3.268 + 4 =   d. 34.21 - 0.039 =

3. The land area of Australia is 2,941,526 mi^2. Round off this quantity to four significant figures; to
    two significant figures. Express your answers in scientific notation.

4. How many inches are there in 35.67 cm?

5. The boiling point of ethanol is 78.5°C. Convert this temperature to °F and K.

6. A student going home for the weekend traveled 700 km at a speed of 103 km/h. How many miles did she travel? What was her average speed in miles/h?

7. A student determines the mass of a metal cylinder to be 53.487 g. She then places the metal cylinder in a graduated cylinder containing 25.34 mL of water. The water level in the cylinder rose to 59.72 mL. What is the density of the metal cylinder?

8. What would the density of the metal cylinder in the preceding problem be in lb/gal?

9. The density of Si is 2.33 g/cm^3. What is the mass of a piece of silicon having a volume of 5.78 cm^3?

10. What are the names and symbols of the elements that are vertical and horizontal neighbors of arsenic?

11. Write the names and symbols of all elements that occupy the same column as oxygen in the periodic table. What name is given to this group?

12. Many people look forward to the year 2000 with anticipation since they consider this to be the beginning of the next millenium (1×10^3 years). How many hours are in a millenium if a year is defined to contain exactly 365.24 days?

13. How many minutes does it take light from the sun to reach Earth? (The distance from the sun to the Earth is 93 million mi; the speed of light = 3.00×10^8 m/s.)

14. The density of titanium is 4.55 g/ml. Calculate the mass of a solid cylinder of titanium 2.75 cm in diameter and 5.05 cm long.

15. A pycnometer is a device used to make precise density determinations. An empty pycnometer weighs 36.20 g when empty and 50.27 g when filled with water at 20 °C. The density of water at 20 °C is 0.9982 g/mL. When 12.05 g of iron is placed in the pycnometer, a total mass of 60.79 g is obtained when the pycnometer is again filled with water. What is the density of iron in g/mL?

Solutions

True-False

1. F. The symbol for the element cobalt is Co.

2. F. Sodium belongs in the group referred to as the alkali metals.

3. T

4. F. Semimetals do have properties somewhere between those of metals and nonmetals, but they are poor conductors of electricity.

5. T

6. F. The unit most commonly used in the laboratory to measure volume is either the liter (dm^3) or milliliter (cm^3).

7. T

8. T

9. T

10. T

Multiple Choice

1. d
2. b
3. a
4. c
5. b
6. d
7. d
8. b
9. c
10. d
11. d
12. b
13. a
14. b
15. c

Matching

Hypothesis - e
Theory - h
Element - f
Main group elements - j
Matter - i
Volume - a
Precision - d
Accuracy - c
Dimensional Analysis - g
Intensive property - b

Fill-in-the-Blank

1. K
2. 4th period, 4B
3. germanium, semimetal
4. noble gases
5. pascal
6. the amount of matter in an object; kilogram
7. number and unit label
8. 1×10^{-9}
9. 100
10. 0.001
11. one additional estimated digit

12. either of the original numbers
13. intensive, the mass of an object to its volume
14. experimentation, hypothesis, theory.
15. violently; alkaline; alkali
16. the study of the composition, properties, and transformations of matter and of chemical laws which are responsible for the changes that take place in nature.
17. Chemical laws
18. International System of Measurements (SI); 7
19. group; period
20. properties

Problems

1. $3.89 \; \mu g \times \dfrac{1 \, g}{1 \times 10^6 \; \mu g} \times \dfrac{1000 \; mg}{1 \, g} = 3.89 \times 10^{-3} \; mg$

    $3.89 \; \mu g \times \dfrac{1 \, g}{1 \times 10^6 \; \mu g} \times \dfrac{1 \times 10^9 \; ng}{1 \, g} = 3.89 \times 10^{3} \; ng$

2. $\dfrac{7.984}{3.2} = 2.5$ (The least amount of significant figures, two, is found in the denominator.)

    $7.53 \times 23.945 \; 62 = 180$ (The least amount of significant figures, 3, is found in the number 7.53.)

    $3.268 + 4 = 7$ (There are no digits to the right of the decimal point in the number 4.)

    $34.21 - 0.039 = 34.17$ (There are only two digits to the right of the decimal point in 34.21.)

3. 2.942×10^{6} ; The first digit removed is 5 which is followed by more nonzero digits; therefore, you round up one digit.

    2.9×10^{6} ; The first digit removed is less than 5; therefore all of the digits to the right of the nine are simply dropped.

4. The conversion factor needed to solve this problem is 1 in = 2.54 cm

    $35.67 \; cm \times \dfrac{1 \, in}{2.54 \; cm} = 14.04 \; in$

5. $^\circ F = \left(\dfrac{9}{5} \times 78.5 \right) + 32 = 173 \; ^\circ F$

    $K = 78.5 + 273.15 = 351.6 \; K$

6. The conversion factor needed to solve this problem is 1 km = 0.6214 mi.

    $700 \; km \times \dfrac{0.6214 \; mi}{1 \; km} = 435 \; mi$

$$103 \ \frac{km}{hr} \times \frac{0.6214 \ mi}{1 \ km} = 64.0 \ \frac{mi}{hr}$$

7. volume of metal cylinder $= 59.72 \ mL - 25.34 \ mL = 34.38 \ mL$

$$density = \frac{53.487 \ g}{34.38 \ mL} = 1.556 \ g \, / \, mL$$

8. The conversion factors needed to solve this problem are:
   $1 \ lb = 453.6 \ g$; $1 \ qt = 946 \ mL$; $1 \ gal = 4 \ qt$.

   This problem can be solved in a single step.

   $$1.556 \ \frac{g}{mL} \times \frac{1 \ lb}{453.6 \ g} \times \frac{946 \ mL}{1 \ qt} \times \frac{4 \ qt}{1 \ gal} = 12.99 \ \frac{lb}{gal}$$

9. mass $= 2.33 \ \dfrac{g}{mL} \times 5.78 \ mL = 13.5 \ g$

10. Neighbors vertical to As are P (phosphorous), Sb (antimony). Neighbors horizontal to As are Ge (germanium) and Se (selenium).

11. O (oxygen), S (sulfur), Se (selenium), Te (tellurium), Po (polonium). This group is referred to as the chalcogens.

12. $1 \times 10^3 \ yr \times \dfrac{365.24 \ days}{1 \ yr} \times \dfrac{24 \ hr}{1 \ day} = 9 \times 10^6 \ hr$

13. 93 million miles $= 9.3 \times 10^7 \ mi$. However, the speed of light is expressed in terms of m/s. Once we convert miles to meters, we can calculate time in seconds and convert to minutes.

   $$9.3 \times 10^7 \ mi \times \frac{1.6093 \ km}{1 \ mi} \times \frac{1000 \ m}{1 \ km} \times \frac{1 \ s}{3.00 \times 10^8 \ m} \times \frac{1 \ min}{60 \ s} = 8.3 \ min$$

14. The volume of a cylinder is calculated with the formula $\pi r^2 h$. Therefore, the volume of the titanium cylinder is

   $$Volume = \pi \left(\frac{2.75}{2}\right)^2 5.05 = 30.0 \ cm^3$$

   $Mass = Density \times Volume$; $Mass = 4.55 \dfrac{g}{cm^3} \times 30.0 \ cm^3 = 137 \ g$

15. The first part of this problem can be used to determine the total volume of the pycnometer. This volume is calculated from the mass of water in the pycnometer and the density of water.

   Mass of water $= 50.27 \ g - 36.20 \ g = 14.07 \ g$

$$\text{Volume of water} = \frac{\text{mass of water}}{\text{density of water}} = \frac{14.07 g}{0.9982 \; {}^{g}\!/_{mL}} = 14.10 \text{ mL}$$

This volume represents the volume of the pycnometer. Knowing the volume of the pycnometer, we can calculate the density of iron with the information given in the second part of the problem. To determine the density of iron, we need the volume of iron. (We are given the mass of iron.) We can calculate the volume of iron from the volume of the pycnometer and the volume of water. The volume of water is calculated from the mass of water. The mass of water in the pycnometer with the iron is calculated by subtracting the mass of the iron and the mass of the empty pycnometer from the mass of the pycnometer with iron and water.

Mass of water = 60.79 g - 12.05 g - 36.20 g = 12.54 g water

The volume of water can now be calculated from the mass and density of water.

$$\text{Volume of water} = \frac{12.54 \text{ g}}{0.9982 \; {}^{g}\!/_{mL}} = 12.56 \text{ mL}$$

Volume of iron = Volume of pycnometer - Volume of water = 14.10 mL - 12.56 mL = 1.54 mL

$$\text{Density of iron} = \frac{12.05 \text{ g}}{1.54 \text{ mL}} = 7.82 \; {}^{g}\!/_{mL}$$

CHAPTER 2

ATOMS, MOLECULES, AND IONS

Chapter Learning Goals

1⊠ Explain the law of a) definite proportions; b) multiple proportions; c) conservation of mass.
2⊠ For two different compounds comprised of the same two elements, show that the law of multiple proportions is obeyed.
3⊠ Explain what information about the atom was revealed by the experiments of a) Thomson, b) Millikan, c) Rutherford.
4⊠ Describe the atom in terms of the composition, mass, and volume of the nucleus relative to the mass and volume occupied by the electrons.
5⊠ Given the symbol for an isotope of an atom or ion, determine the number of protons, neutrons, and electrons.
6⊠ Given the mass and natural abundance of all isotopes of a given element, calculate the average atomic weight of that element.
7⊠ For any element, calculate a) the mass in grams of a single atom and b) the number of atoms in a given number of grams.
8⊠ State the difference between a) compounds and mixtures, b) heterogeneous and homogeneous mixtures, c) elements and atoms, d) atoms and molecules.
9⊠ Identify which substances are ionic and which are molecular.
10⊠ Identify which substances are acids and which are bases.
11⊠ Give the formulas and names of a) common polyatomic ions, b) ionic compounds, c) binary molecular compounds, d) common acids.

Chapter in Brief

To have an understanding of substances that we call atoms, molecules and ions, you first need to understand the fundamental laws that govern chemistry. You then will begin to look at the structure of the atom and learn about the three subatomic particles: protons, neutrons, and electrons. Knowing the atomic number and atomic mass of an atom allows you to determine the number of protons (= electrons) and neutrons for an element. An atom's mass is measured using the atomic mass unit. You will see how the total mass of an atom in amu, or the atomic weight, can be used to calculate the mass of a single atom. You will also learn the differences between compounds and mixtures, molecules and ions, and acids and bases. Finally, you will begin applying the rules for naming inorganic compounds.

1⊠ **Conservation of Mass and the Law of Definite Proportions**
   A. Element – a substance that cannot be further broken down.
   B. Law of Mass Conservation – Mass is neither created nor destroyed in chemical reactions.
       1. Cornerstone of chemical science.
   C. Law of Definite Proportions – Different samples of a pure chemical substance always contain the same proportion of elements by mass.
       1. Elements do not combine chemically in random proportion.

Dalton's Atomic Theory and the Law of Multiple Proportions
   A. Dalton's Atomic Theory
       1. Elements are made of tiny particles called atoms.
       2. Each element is characterized by the mass of its atoms.
          a. atoms of the same element have the same mass.

        b. atoms of different elements have different masses.

     3. Chemical combination of elements to make different substances occurs when atoms join together in small, whole-number ratios.

     4. Chemical reactions only rearrange the way that atoms are combined; the atoms themselves are not changed.

1⊠   B. Law of Multiple Proportions - If two elements combine in different ways to form different substances, the mass ratios are small, whole-number multiples of each other.

2⊠   ***EXAMPLE:***

Carbon monoxide and carbon dioxide are both gases that contain only carbon and oxygen. A sample of carbon monoxide contains 1.00 g of C and 1.33 g of O. A sample of carbon dioxide contains 1.00 g of C and 2.67 g of O. Show that the two substances obey the law of multiple proportions.

SOLUTION: Find the C:O mass ratio in each compound.

Carbon monoxide: C:O mass ratio $= \dfrac{1.00 \text{g C}}{1.33 \text{ g O}} = 0.752$

Carbon dioxide: C:O mass ratio $= \dfrac{1.00 \text{ g C}}{2.67 \text{ g O}} = 0.375$

The two C:O mass ratios are clearly small, whole-number multiples of each other.

$$\frac{\text{C:O mass ratio in carbon monoxide}}{\text{C:O mass ratio in carbon dioxide}} = \frac{0.752}{0.375} = 2.00$$

3⊠ **The Structure of Atoms: Electrons**

     A. Thomson: Cathode rays consist of tiny negatively charged particles called electrons.

       1. Electrons are emitted from electrodes made of many different metals.

         a. different metals contain electrons

       2. Cathode rays can be deflected by bringing either a magnet or an electrically charged plate near the tube. This deflection depends on:

         a. the strength of the deflecting magnetic or electric field

         b. the size of the negative charge on the electron

         c. the mass of the electron

       3. Charge-to-mass ratio, e/m of the electron $= 1.758\ 819 \times 10^8$ C/g.

     B. Millikan: Determined that the charge on a drop of oil (see Fig. 2.4 in your text, page 43) was always a small, whole-number multiple of e.

       1. $e = 1.602\ 177 \times 10^{-19}$ C.

       2. Knowing the values for e/m and e for an electron, m can be calculated.

         a. $m = 9.109\ 390 \times 10^{-28}$ g

3⊠ **The Structure of Atoms: Rutherford's Nuclear Model**

     A. Alpha (α) particles.

       1. 7000 times more massive than electrons.

       2. Have a positive charge that is twice the magnitude of but opposite in sign to the charge on an electron.

       3. Beam of α particles - most pass through a thin metal foil but a few are deflected at large angles.

     B. Nucleus - a tiny central core in an atom where the mass of the atom is concentrated.

       1. Contains the atom's positive charges.

4⊠ **The Structure of Atoms: Protons and Neutrons**
    A. Nucleus is composed of two kinds of particles.
        1. Protons - have a mass of $1.672\ 623 \times 10^{-24}$ g and are positively charged.
        2. Neutrons - almost identical in mass to protons but carry no charge.
    B. Elements differ from one another according to the number of protons their atoms contain.
        1. Atomic number (Z) = the number of protons in an atom = the number of electrons in an atom.
        2. Mass number (A) = number of protons + number of neutrons in an atom.
5⊠   C. Isotopes - atoms with identical atomic numbers but different mass numbers.
        1. Mass number written as left superscript.
        2. Atomic number written as left subscript. (The atomic number is sometimes left off since all atoms of an element always contain the same number of protons.)
        3. The number of neutrons in an isotope can be calculated from A - Z.
        4. Number of neutrons in an atom has little effect on the chemical properties of the atom.

EXAMPLE:
Determine the number of protons, neutrons, and electrons in the following isotopes:

$$^{199}_{80}\text{Hg} \qquad ^{195}_{78}\text{Pt} \qquad ^{84}_{38}\text{Sr}$$

SOLUTION: The subscript is the atomic number (Z), and the superscript is the mass number(A). Z = the number of protons = the number of electrons, and A-Z = the number of neutrons.

$^{199}_{80}\text{Hg}$: Z = 80; # of protons = # of electrons = 80

A-Z = 199-80 = 119; # of neutrons = 119

$^{195}_{78}\text{Pt}$: Z = 78; # of protons = # of electrons = 78

A-Z = 195 - 78 = 117; # of neutrons = 117

$^{84}_{38}\text{Sr}$: Z = 38; # of protons = # of electrons = 38

A-Z = 84 - 38 = 46; # of neutrons = 46

Atomic Weight
7⊠   A. Atomic mass unit (amu) - exactly 1/12th the mass of an atom of $^{12}_{6}\text{C}$.
        1. 1 amu = 1.6605×10^{-24} g.
    B. Atomic weight - the mass of an atom in amu (numerically close to the atom's mass number).
        1. Atomic weight values are weighted averages for the naturally occurring mixtures of different isotopes.
6⊠   2. Atomic weight of an element = Σ(mass of each isotope × the fraction of the isotope).
            a. Σ is used for the term "the sum of."

EXAMPLE:
Calculate the mass in grams of a single atom of iridium.

SOLUTION: From the mass given in the periodic table we know that 1 atom of Ir has a mass of 192.2 amu. We also know that 1 amu = 1.6605×10^{-24} g. The final unit we want to end up with is g/atom of Ir. Therefore, we should set up our problem in the following manner:

$$\frac{192.2 \text{ amu}}{1 \text{ atom Ir}} \times \frac{1.6605 \times 10^{-24} \text{ g}}{1 \text{ amu}} = 3.191 \times 10^{-22} \text{ g}$$

EXAMPLE:
How many atoms of cadmium are present in 25.2 g?

SOLUTION: $25.2 \text{ g} \times \dfrac{1 \text{ amu}}{1.6605 \times 10^{-24} \text{ g}} \times \dfrac{1 \text{ atom}}{112.4 \text{ amu}} = 1.35 \times 10^{23}$

EXAMPLE:
Magnesium has three naturally occurring isotopes: ^{24}Mg with an abundance of 78.99% and an atomic weight of 23.985 amu; ^{25}Mg with an abundance of 10.00% and an atomic weight of 24.986 amu; and ^{26}Mg with an abundance of 11.01% and an atomic weight of 25.983 amu. Calculate the average atomic weight of Mg.

SOLUTION: $(0.7899 \times 23.985 \text{amu}) + (0.1000 \times 24.986 \text{amu}) + (0.1101 \times 25.983 \text{amu}) = 24.31 \text{ amu}$

8⊠ Compounds and Mixtures

A. Chemical reactions - atoms from two or more different elements combine, creating new materials with properties completely unlike those of their constituent elements.
  1. Chemical equation – format used to write a chemical reaction.
     a. reactants - starting substances in chemical reaction; written on the left side of the arrow
     b. products - new materials formed; written on the right side of the arrow
     c. an arrow is placed between the reactants and products to indicate a transformation
B. Compounds - new materials of constant composition created from chemical reactions.
  1. Formed when atoms undergo chemical combination in a specific manner.
  2. Composition indicated by a chemical formula
     a. lists the symbols of the individual constituent elements.
     b. the number of each atom is given by a subscript
C. Mixtures - formed when two or more substances are blended together in some random proportion.
  1. Individual substances do not undergo a chemical change.
  2. Homogeneous mixture (solution) - uniform properties throughout.
  3. Heterogeneous mixture - regions with differing compositions.

9⊠ Molecules, Ions, and Chemical Bonds

A. Chemical bonds - formed by the electrons in the atoms; join the atoms together in compounds.
  1. Covalent bonds - occur between two nonmetals.
  2. Ionic bonds - occur between a metal and a nonmetal.
  3. Metallic bonds – occur between two metals.

8⊠ B. Covalent bond - two atoms share electrons.
  1. Molecule - unit of matter that results when two or more atoms are joined by covalent bonds.
     a. structural formula - shows the specific connections between atoms
        i. contains more information than the chemical formula

      b. some elements exist as molecules
        i. H_2, O_2, N_2, F_2, Cl_2, Br_2, I_2
    C. Ionic Bonds - a complete transfer of one or more electrons from one atom to another.
      1. Ions - charged particles resulting from the loss or gain of electrons.
        a. cation - positively charged particle resulting from the loss of one or more electrons
        b. anion - negatively charged particle resulting from the gain of one or more electrons
        c. polyatomic ions - charged, covalently bonded groups of atoms
          i. charged molecules - consist of specific numbers and kinds of atoms joined together by covalent bonds in a definite way
      2. Ionic solids – cations and anions are packed together in a regular manner so that the charges cancel.

10⊠ Acids and Bases

    A. Two important ions: H^+ and OH^-
      1. Fundamental to the concept of acids and bases.
    B. Acid - a substance that provides H^+ ions when dissolved in water.
      1. Produces H^+ along with the corresponding anion.
    C. Base - a substance that provides OH^- ions when dissolved in water.
      1. Produces OH^- along with the corresponding cation.

11⊠ Naming Binary Ionic Compounds

    A. Binary Ionic Compounds - ionic compounds (contains a cation and an anion) containing only two elements.
      1. Identify the cation first; then the anion.
        a. cation takes the same name as the element (Remember: metals form cations.)
        b. anions take the first part of its name from the element and adds the ending "*ide*" (Remember: nonmetals form anions.)
      2. Common main group and transition metal ions – see Table 2.2 and 2.3 in text (pages 58 – 59).
        a. elements within a group often form similar kinds of ions
        b. main-group metal cations: charge = group number
        c. main-group nonmetal anions: charge = group number - 8
        d. some metals from more than one kind of cation
          i. when naming these ions, the charge is indicated by a Roman numeral in parenthesis
          ii. use for transition metal complexes
    B. Electrical Neutrality - cations and anions combine in such a manner that the overall charge on a compound is equal to zero.
    C. Formulas for ionic compounds always contain the smallest whole number ratio of cation to anion.

EXAMPLE:
Name the following compounds:
CaS   CuO   CsI $SnCl_4$

SOLUTION:
CaS: calcium sulfide   no Roman numeral needed: calcium is a group 2A metal and forms only Ca^{2+}
CuO: copper(II) oxide   the charge on oxygen is a -2. To have electrical neutrality the charge on copper, which is a transition metal, must be +2. The Roman numeral (II) is present because copper is

a transition metal.

CsI: cesium iodide   cesium is a group 1A metal and forms only Cs^+

$SnCl_4$:   tin(IV) chloride   there are 4 Cl^- ions with a total charge of -4. To have electrical neutrality, the charge on tin must be +4. Tin can form more than one kind of ion; therefore the charge should be indicated with a Roman numeral.

EXAMPLE:

Write formulas for the following compounds:

barium iodide   cobalt(II) phosphide   lead(II) bromide

SOLUTION:

BaI_2 barium is a group 2A metal and only forms Ba^{2+}. To balance the charge, you need 2 I^- ions.

Co_3P_2 cobalt(II) has a +2 charge, while phosphorous has a -3 charge. To have electrical neutrality, you need 3 Co^{2+} and 2 P^{3-}. $[(3 \times +2) + (2 \times -3)] = 0$.

$PbBr_2$ lead(II) has a +2 charge, while bromine has a -1 charge. To have electrical neutrality, you need 1 Pb^{2+} and 2 Br^-.

11 ⊠   **Naming Binary Molecular Compounds**

    A. Binary molecular compounds - molecular compounds (nonmetals) containing only two elements.
        1. One of the elements is more cationlike.
            a. takes the name of the element.
        2. One of the elements is more anionlike.
            a. takes an *"ide"* ending
        3. Character depends upon the relative positions of the two elements in the periodic table.
            a. more cationlike - farther left in the periodic table
            b. more anionlike - farther right in the periodic table
    B. To specify the numbers of each element present, use numerical prefixes. (See Table 2.4 in text, page 61.)
        1. the *mono* prefix is not used for the atom named first
    C. When naming binary molecular compounds which contain hydrogen, it is necessary to indicate whether the molecule is in the gaseous or aqueous (in water) state.
        1. If the molecule is a gas, use the above rules.
            a. When writing the formula, always indicate the molecule is a gas with (*g*)
        2. If the molecule is in aqueous solution, name the compound as a binary acid (see below).

EXAMPLE:

Name the following compounds:

SF_4 PCl_5 P_4O_{10} SiO_2

SOLUTION:

SF_4:   sulfur tetrafluoride

PCl_5:   phosphorus pentachloride

P_4O_{10}:   tetraphosphorus decaoxide

SiO_2:   silicon dioxide

EXAMPLE:
Write formulas for the following compounds:

dihydrogen sulfide   xenon tetrafluoride   dinitrogen trioxide

SOLUTION:

dihydrogen sulfide: H_2S (g)   if naming as a binary molecule, the compound is
                                                in a gaseous state which needs to be indicated.

xenon tetrafluoride: XeF_4

dinitrogen trioxide: N_2O_3

11⊠   Naming Compounds with Polyatomic Ions
A. Follow the rules for naming binary ionic compounds.
   1. Identify the cation.
   2. Identify the anion.
      a. learn the names, formulas, and charge numbers of the most common polyatomic anions
        (See Table 2.5 in text, page 63)
      b. most names end with the suffix "*ite*" or "*ate*"
      c. oxoanions - an atom of the same element is combined with different numbers of
        oxygen atoms
      d. several pairs of ions are related by the presence or absence of a hydrogen atom
B. Naming oxoanions.
   1. Learn the name and formula of the ion whose name ends with "ate" (including the charge
      on the anion).
   2. Add one O; add prefix "per".
   3. Remove one O; change "ate" to "ite".
   4. Remove two O; add prefix "hypo" and change "*ate*" to "*ite*".

EXAMPLE:
Name the following compounds:
$AlPO_3$ $KMnO_4$ $Co_2(CO_3)_3$ $CuBrO_3$

SOLUTION:

$AlPO_3$:   aluminum phosphite

$KMnO_4$:   potassium permanganate

$Co_2(CO_3)_3$:   cobalt(III) carbonate. The carbonate anion has a -2 charge. The total
                            charge on the carbonate anion is -6 (3 × -2). Therefore, the total charge on
                            the cobalt cation must be a +6. Each cobalt cation would then have a
                            charge of +3. (+6/2). The charge on the cobalt is indicated with Roman
                            numerals because cobalt is a transition metal.

$CuBrO_3$: copper(I) bromate. The bromate anion has a -1 charge. Therefore, the charge on copper must be a +1. The charge on the copper is indicated with Roman numerals because copper is a transition metal.

EXAMPLE:
Write formulas for the following compounds:

potassium sulfate   silver(I) phosphate   iron (III) periodate

SOLUTION:

potassium sulfate: K_2SO_4 The sulfate anion has a -2 charge. Potassium has a +1 charge, so 2 K^+ are needed.

silver(I) phosphate: Ag_3PO_4 The phosphate anion has a -3 charge. Silver has a +1 as indicated by the Roman numeral. 3 Ag^+ are needed for every PO_4^{3-} ion.

iron(III) periodate: $Fe(IO_4)_3$ The periodate anion has a -1 charge. Iron has a +3 charge as indicated by the Roman numeral. 3 IO_4^- are needed for every Fe^{3+}.

11⊠   **Naming Acids**
   A. Oxoacids - contain oxygen in addition to hydrogen and another element.
      1. Yields H^+ and a polyatomic oxoanion when dissolved in H_2O.
         a. See Table 2.6 in text (page 64)
      2. Names are related to the names of the corresponding oxoanions.
         a. replace the suffix "*ite*" with "*ous acid*"
         b. replace the suffix "*ate*" with "*ic acid*"
   B. Binary Acids - contain hydrogen and one other element or a polyatomic anion that is not an oxoanion.
      1. Aqueous solutions.
      2. Use the prefix "*hydro*" followed by the first part of the name of the other element, the suffix "*ic*" and acid.
      3. When writing the formula, always indicate that it is an aqueous solution by (*aq*).

EXAMPLE:
Name the following acids:
H_3PO_3 H_2S (*aq*) H_2CO_3

SOLUTION:

H_3PO_3:   This compound is an acid that contains the phosphite anion. It is named phosphorous acid.

H_2S (*aq*)   This compound is a binary compound containing hydrogen in aqueous solution. It is named hydrosulfuric acid.

H_2CO_3 This compound is an acid that contains the carbonate anion. It is named carbonic acid.

Self-Test

This section is intended to test your knowledge of the material covered in this chapter. Think through these problems and make certain you understand what is going on. Ask yourself if your answer makes sense. Many of these questions are linked to the chapter learning goals. Therefore, successful completion of these problems indicates you have mastered the learning goals for this chapter. You will receive the greatest benefit from this section if you use it as a mock exam. You will then discover which topics you have mastered and which topics you need to study in more detail.

True-False

1. Different samples of a pure chemical substance always contain the same proportion of elements by mass.

2. In a chemical reaction, the atoms undergo a change.

3. Cathode rays consist of tiny positively charged particles.

4. The nucleus is composed of two kinds of particles: protons and electrons.

5. Isotopes are atoms with identical atomic numbers but different mass numbers.

6. The mass number, A, represents the number of protons and neutrons in an atom.

7. A cation is a positively charged particle resulting from the gain of a proton.

8. An anion is a negatively charged particle resulting from the gain of an electron.

9. The correct name for FeS is iron sulfide.

10. The correct formula for dinitrogen pentoxide is N_2O_5.

Multiple Choice

1. The nucleus of an atom contains
    a. the protons, neutrons, and electrons of the atom.
    b. the protons of the atom.
    c. the neutrons of the atom.
    d. the protons and neutrons of the atom.

2. Elements differ from one another according to the number of
    a. protons.
    b. neutrons.
    c. isotopes.
    d. amu.

3. The isotope $_{27}^{60}\text{Co}$ contains
    a. 27 neutrons.
    b. 60 protons.
    c. 33 electrons.
    d. 33 neutrons.

4. Chemical bonds are formed by the interaction of
   a. the protons in an atom.
   b. the neutrons in the atom.
   c. the electrons in the atom.

5. A covalent bond is the result of
   a. a transfer of electrons between atoms
   b. the interaction of a cation and anion.
   c. the sharing of electrons between atoms.

6. A homogeneous mixture is a mixture
   a. formed when atoms undergo chemical combination in a specific manner.
   b. with uniform properties throughout.
   c. that has regions of differing composition.

7. The correct name for HCl (*aq*) is
   a. hydrogen chloride.
   b. hydrochloric acid.
   c. chloric acid.
   d. hypochloric acid.

8. The correct formula for calcium phosphide is
   a. Ca_2P_3.
   b. C_3P_2.
   c. Ca_3P_2.

9. The correct name of $Co(NO_3)_2$ is
   a. cobalt nitrate.
   b. cobalt(II) nitrate.
   c. cobalt(III) nitrate.
   d. cobaltic nitrate.

10. The correct formula for bromous acid is:
    a. HBr (*aq*).
    b. $HBrO_3$.
    c. HBrO.
    d. $HBrO_2$.

Fill-in-the-Blank
1. If two elements combine in different ways to form different substances, the mass ratios are

   _____.

2. In comparison with the electron, alpha (α) particles are _____

   _____.

3. The atomic number of an atom (Z) is the number of _____ in the atom.

4. The number of electrons in a neutral atom is equal to the number of _____ in the atom.

5. The number of _____ in an atom has little effect on the chemical properties of the atom.

6. The mass of an atom in amu is referred to as the _____.

7. Covalent bonds occur between two _____ .

8. Ionic bonds occur between a _____ .

9. A substance that provides H^+ ions in water is called an _____ .

10. In a chemical formula, the number of each atom is given as a _____ .

11. Elements are made of tiny particles called _____ .

12. Most substances on earth are _____, formed when atoms of two or more elements combine in a _____ .

13. The atoms in a compound are held together by one of two fundamental kinds of _____

_____ .

14. _____ form when one atom completely transfers one or more electrons to another atom, resulting in the formation of _____ .

15. According to the nuclear model of an atom, protons and neutrons are clustered into a dense core called the _____ .

Matching

Al_2O_3	a. mercury(II) oxide
Cations	b. Atoms with identical atomic numbers but different mass numbers.
FeO	c. phosphorus trichloride
Anions	d. fundamental atomic particle that is negatively charged.
NH_4Cl	e. calcium carbonate
Bases	f. fundamental atomic particle that is neutral.
$CaCO_3$	g. sulfur tetrafluoride
isotopes	h. substances that yield OH^- ions when dissolved in water.
SF_4	i. aluminum oxide
Protons	j. positively charged ions
TiF_3	k. chromium(III) hydroxide
Neutrons	l. negatively charged ions.

PCl_3 m. barium sulfite

Electrons   n. fundamental atomic particle that is
                                                 positively charged.

$BaSO_3$ o. iron(II) oxide

HgO   p. ammonium chloride

$Cr(OH)_3$ q. titanium(III) fluoride

Problems

1. Phosphorus forms two compounds with chlorine. In one compound, 0.75 g of phosphorus
   combines with 2.58 g of chlorine. In another compound, 1.35 g of phosphorus combines with 7.73
   g of chlorine. Show that these two substances obey the law of multiple proportions.

2. Nitrogen forms a number of compounds with oxygen. It is found that in one compound, 0.681 g of
   N is combined with 0.778 g of O; in another compound, 0.560 g of N is combined with 1.28 g of
   O. Prove that these data support the law of multiple proportions.

3. Europium, Eu, has two naturally occurring isotopes: $^{151}_{63}Eu$ with an abundance of 47.82% and an
   atomic weight of 150.9 amu, and $^{153}_{63}Eu$ with an abundance of 52.18% and an atomic weight of
   152.9 amu. Calculate the average atomic weight of europium.

4. The four isotopes of lead along with their abundances and atomic weights are: ^{204}Pb, atomic
   weight = 203.973 amu, abundance = 1.48%; ^{206}Pb, atomic weight = 205.9745 amu, abundance =
   23.6%; ^{207}Pb, atomic weight = 206.9759 amu, abundance = 22.6%, and ^{208}Pb, atomic weight =
   207.9766 amu, abundance = 52.3%. Calculate the average atomic weight of lead.

5. How many protons, neutrons, and electrons are present in the following elements?

   a. $^{18}_{8}O$ b. $^{57}_{26}Fe^{2+}$ c. $^{197}_{79}Au$ d. $^{131}_{53}I^-$

6. What is the identity of the element X in the following ions?

   a. X^{3+}, a cation that has 13 protons.

   b. X^{2-}, an anion that has 34 protons.

7. What anion will the following acids produce when dissolved in water?

   a. HNO_3 b. $HClO_4$ c. H_2SO_4 d. H_2S

8. What cation will the following bases produce when dissolved in water?

   a. KOH   b. $Ca(OH)_2$ c. $Mg(OH)_2$ d. $Li(OH)$

9. What is the mass (in grams) of 150 atoms of iron?

10. How many atoms are there in 332 g of gold?

11. If the average mass of a single iodine atom is 2.107×10^{-22} g, what is the mass in grams of 6.02×10^{23} atoms? How does this answer compare to the atomic mass of iodine?

12. A 15.51 g quantity of iron was allowed to react with 25.00 mL of nitric acid that had a density of 1.40 g/mL. The reaction yielded H_2 gas and a solution of iron (II) sulfate. The density of the gas was determined to be 0.0899 g/L and the mass of the solution was found to be 49.958 g. How many liters of H_2 were formed? What chemical law did you use to work this problem?

13. Name the following compounds:

   a. LiF   b. $GaCl_3$ c. Ca_3N_2 d. V_2O_5

   e. $Sc(NO_3)_3$ f. Mn_2O_7 g. $FeBr_3$ h. KIO_3

   i. $K_2Cr_2O_7$ j. $Sn(NO_2)_4$ k. N_2O_5 l. HBr (g)

   m. $HClO_2$ n. HBr (aq) o. H_2SO_3 p. N_2O

14. Write formulas for the following compounds:

   a. rubidium iodide   b. magnesium phosphide   c. cobalt(II) chloride

   d. copper(I) oxide   e. ammonium thiosulfate   f. chromium(III) nitride

   g. hydrogen cyanide gas   h. sulfur trioxide   i. hydrofluoric acid

   j. calcium hydride   k. bromic acid   l. sodium sulfate

   m. sodium acetate   n. potassium permanganate   o. hypochlorous acid

   p. potassium hydrogen carbonate

Solutions

True-False

1. T

2. F. In a chemical reaction, the atoms only rearrange the way they are combined; the atoms themselves are not changed.

3. F. Cathode rays consist of tiny, negatively charged particles.

4. F. The nucleus is composed of two kinds of particles: protons and neutrons.

5. T.

6. T

7. F. A cation is a positively charged particle resulting from the loss of an electron.

8. T.

9. F. The correct name for FeS is iron(II) sulfide.

10. T.

Multiple Choice

1. d
2. a
3. d
4. c
5. c
6. b
7. b
8. c
9. b
10. d

Fill-in-the-Blank

1. small, whole number multiples of each other.

2. 7000 times more massive than an electron and have a positive charge twice the magnitude of the charge on the electron.

3. protons

4. protons

5. neutrons

6. atomic weight

7. nonmetal atoms

8. cation and anion

9. acid

10. subscript

11. atoms

12. compounds; chemical reaction

13. chemical bonds

14. Ionic solids; ions

15. nucleus

Matching

Al_2O_3 – i
Cations - j
FeO – o
Anions - l
NH_4Cl – p
Bases - h
$CaCO_3$ – e
Isotopes - b
SF_4 – g
Protons - n
TiF_3 – q
Neutrons - f
PCl_3 – c
Electrons - d
$BaSO_3$ - m
HgO - a
$Cr(OH)_3$ - k

Problems

1. first compound: Cl:P mass ratio $= \dfrac{2.58 \text{ g Cl}}{0.75 \text{ g P}} = 3.44$

   second compound: Cl:P mass ratio $= \dfrac{7.73 \text{ g Cl}}{1.35 \text{ g P}} = 5.73$

   $\dfrac{\text{Cl:P mass ratio in 1st compound}}{\text{Cl:P mass ratio in 2nd compound}} = \dfrac{3.44}{5.73} = 0.6 = \dfrac{3}{5}$

2. first compound: N:O ratio $= \dfrac{0.681 \text{ g N}}{0.778 \text{ g O}} = 0.875$

   second compound: N:O ratio $= \dfrac{0.560 \text{ g N}}{1.28 \text{ g O}} = 0.438$

   $\dfrac{\text{N:O ratio in 1st compound}}{\text{N:O ratio in 2nd compound}} = \dfrac{0.875}{0.438} = 2.0$

3. avg. atomic weight $= (0.4782 \times 150.9 \text{amu}) + (0.5218 \times 152.9 \text{amu}) = 151.9 \text{amu}$

4. avg. atomic weight
   $= (203.973 \text{ amu} \times 0.0148) + (205.9745 \text{ amu} \times 0.236) + (206.9759 \text{ amu} \times 0.226) + (207.9766 \text{ amu} \times 0.523) = 207 \text{ amu}$

5. a) p = 8, n = 10, e = 8
   b) p = 26, n = 31, e = 24
   c) p = 79, n = 118, e = 79
   d) p = 53, n = 78, 3 = 54

6. a) Al^{3+} b) Se^{2-}

7. a) NO_3^- b) ClO_4^- c) HSO_4^- d) HS^-

8. a) K^+ b) Ca^{2+} c) Mg^{2+} d) Li^+

9. $150 \text{ atoms} \times \dfrac{55.85 \text{ amu}}{1 \text{ atom}} \times \dfrac{1.6605 \times 10^{-24} \text{ g}}{1 \text{ amu}} = 1.39 \times 10^{-20} \text{ g}$

10. $332 \text{ g Au} \times \dfrac{1 \text{ amu}}{1.6605 \times 10^{-24} \text{ g}} \times \dfrac{1 \text{ atom Au}}{197.0 \text{ amu}} = 1.01 \times 10^{24} \text{ atoms Au}$

11. $\left(2.107 \times 10^{-22} \dfrac{\text{g}}{\text{atom}}\right) \times \left(6.02 \times 10^{23} \text{ atoms}\right) = 126.841 \text{ g}$

    This answer is very close to the atomic mass of iodine.

12. The chemical law needed to work this problem is the Law of Conservation of Mass. According to this law, "Matter cannot be created nor destroyed, only changed from one form to another," In other words, the total mass of the reactants must equal the mass of the products. In this chemical reaction, the mass of the reactants is the mass of the iron and the nitric acid. The mass of the nitric acid is calculated from the volume and density:

    $25.00 \text{ mL} \times 1.40 \dfrac{\text{g}}{\text{mL}} = 35.0 \text{ g}$

    Mass of reactants = 15.51 g Fe + 35.0 g HNO_3 = 50.51 g reactants

    The mass of products also must equal 50.51 g. Knowing the mass of products, we can now calculate the mass of H_2 gas and determine the volume of H_2 from its density.

    50.51 g products = mass of H_2 + 49.958 g $Fe(NO_3)_2$; mass of H_2 = 0.55 g H_2

    $\text{Volume of } H_2 = 0.55 \text{ g } H_2 \times \dfrac{1 \text{ L}}{0.0899 \text{ g } H_2} = 6.1 \text{ L } H_2$

13. a. lithium fluoride   b. gallium chloride   c. calcium nitride

    d. vanadium(V) oxide   e. scandium(III) nitrate   f. manganese(VII) oxide

    g. iron(III) bromide   h. potassium iodate   i. potassium dichromate

    j. tin(IV) nitrite   k. dinitrogen pentoxide   l. hydrogen bromide

    m. chlorous acid   n. hydrobromic acid   o. sulfurous acid

    p. dinitrogen oxide

14. a. RbI

    b. Mg_3P_2

    c. $CoCl_2$

    d. Cu_2O

    e. $(NH_4)_2S_2O_3$

    f. CrN

    g. HCN (*g*)

    h. SO_3

    i. HF (*aq*)

    j. CaH_2

    k. $HBrO_3$

    l. $(Na)_2SO_4$

    m. $NaC_2H_3O_2$

    n. $KMnO_4$

    o. HClO

    p. $KHCO_3$

CHAPTER 3

FORMULAS, EQUATIONS, AND MOLES

Chapter Learning Goals

1⊠ For simple chemical reactions, write and balance chemical equations.
2⊠ Calculate molar mass.
3⊠ Interconvert grams, moles, and numbers of formula units.
4⊠ Determine the number of moles and grams of one reactant needed to react with a given number of moles and grams of another reactant and the number of moles and grams of product(s) that result from the reaction.
5⊠ Calculate percent yield.
6⊠ Calculate number of grams of products produced from a given number of grams of reactants when the theoretical yield is less than 100%.
7⊠ Identify the limiting and excess reagents in a reaction mixture.
8⊠ Determine the number of grams of excess reagent remaining at the end of a reaction and the number of grams of product(s) produced.
9⊠ Describe how to prepare a solution of known molarity by 1) dissolving a solid in a solvent; b) by diluting a more concentrated solution.
10⊠ Interconvert solution molarity, solution volume, solute moles, and solute grams.
11⊠ Determine the volume of one reactant needed to react with a given volume of a second reactant.
12⊠ Determine the percent composition and empirical formula of a compound.
13⊠ Understand how to use combustion analysis to obtain the empirical formula of a compound containing carbon, hydrogen, and one other element.
14⊠ From empirical formula and molar mass, determine the molecular formula of a compound.

Chapter in Brief

The central concern of chemistry is the change of one substance into another. As a consequence, chemical reactions are at the heart of the science. In this chapter, you begin your study of chemical reactions by learning how to balance chemical equations. Using balanced chemical equations along with the concepts of the mole and stoichiometry you will learn how to calculate the amounts of reactants needed and the theoretical amount of products produced in a reaction. These calculations are carried out using molar masses for gram ↔ mole conversions or molarities for mole ↔ volume ↔ molarity conversions. You will also learn how to determine an empirical formula from the percent composition of a compound and vice versa. Finally, you will see how combustion analysis, a type of elemental analysis, along with mass spectrometry can be used to determine both the empirical and molecular formula of a compound.

1⊠ **Balancing Chemical Equations**
A. Balanced equations - the numbers and kinds of atoms on both sides of the arrow are the same.
   1. All chemical equations must be balanced.
      a. obeys law of mass conservation
B. Formula unit - one unit (an atom, ion, or molecule) corresponding to a given formula.
C. Balancing process.
   1. Write the unbalanced equation using the correct chemical formula for all reactants and products.
   2. Use coefficients, the number placed before the formula, to indicate how many formula units are required to balance the equation.

34

a. only the coefficients can be changed: the formulas must stay the same
3. Reduce the coefficients to their smallest whole-number values.
4. Check your answer.

EXAMPLE:
Dinitrogen pentoxide reacts with water to produce nitric acid. Write a balanced equation for this reaction.

SOLUTION:
Step 1: Write the unbalanced equation.

$$N_2O_5 + H_2O \rightarrow HNO_3$$

Step 2: Use coefficients to balance the equation. Think about one element at a time. (It usually helps to save oxygen for last.) Notice on the reactant side that there are two N's but only one N on the product side. Begin by placing a 2 in front of HNO_3.

$$N_2O_5 + H_2O \rightarrow 2 HNO_3$$

You now have two H's on both the reactant and product side. You also have a total of six O's on both the reactant and product side. (Remember to multiply the coefficient in front of a formula by the subscript of an element to get the total number of atoms of the element; for the oxygen in HNO_3, the total number = 2(coefficient) × 3(subscript) = 6).

Step 3: The coefficients are already reduced to their smallest whole number ratio.

Step 4: Check your answer.

Reactant side	Product side
2 N	2 N
6 O	6 O
2 H	2 H

EXAMPLE:
Write a balanced equation for the combustion reaction of pentane (C_5H_{12}).

SOLUTION:
The term combustion reaction is used to indicate reaction with oxygen. When hydrocarbons (compounds containing primarily C and H) undergo combustion reaction, carbon dioxide and water are produced.

Step 1: Write the unbalanced equation:

$$C_5H_{12} + O_2 \rightarrow CO_2 + H_2O$$

Step 2: Use coefficients to balance the equation. Begin with carbon. There are five C's on the reactant side, but only one C on the product side. Begin by placing a 5 in front of CO_2.

$$C_5H_{12} + O_2 \rightarrow 5 CO_2 + H_2O$$

There are 12 H's on the reactant side, but only two H's on the product side. Place a 6 in front of H_2O.

$$C_5H_{12} + O_2 \rightarrow 5\,CO_2 + 6\,H_2O$$

Now balance the oxygens. There are two O's on the reactant side and 16 O's on the product side. Place an 8 in front of O_2.

$$C_5H_{12} + 8\,O_2 \rightarrow 5\,CO_2 + 6\,H_2O$$

Step 3: Reduce the coefficients to their smallest whole-number ratio.

The ratio is 1:8:5:6, which is the smallest whole-number ratio.

Step 4: Check your answer.

Reactant side	Product side
5 C	5 C
12 H	12 H
16 O	16 O

Chemical Symbols on a Different Level

 A. Symbols represent both a microscopic and macroscopic level.
  1. Microscopic level - chemical symbols represent the behavior of individual atoms and molecules.
  2. Macroscopic level - formulas and equations represent the large-scale behavior of atoms and molecules that gives rise to observable properties.
   a. deal with macroscopic behavior in the laboratory

Avogadro's Number and the Mole

 A. Mole (mol) = 6.022×10^{23} particles.
  1. Avogadro's number; abbreviated N_A
  2. 1 mol HNO_3 = 6.022×10^{23} molecules of HNO_3; 1 mol electrons = 6.022×10^{23} electrons
  3. Importance of mole - provides a relationship between numbers of molecules and masses of molecules.
   a. must convert a number ratio into a mass ratio
    i. number ratios determined from the coefficients in the balanced equation
    ii. mass ratios are determined from the molecular weights of the substances
 B. Molecular weight - (mol. wt., MW) - sum of the atomic weights of all atoms in a molecule.
  1. Formula weight - (form. wt.) - sum of atomic weights of all atoms in one formula unit of any substance.
 C. Molar mass of a substance - one mole of any substance has a mass equal to its molecular or formula weight in grams.
  1. Mass of one mole of a substance.
  2. Mass of 6.022×10^{23} molecules, ions, atoms, or compounds.
  3. Molecular weight of substance in grams.
  4. Serves as a conversion factor between numbers of formula units and mass.
  5. Always specify the formula of the particles.
 D. The coefficients in a balanced chemical equation indicate the number of moles of each substance needed for the reaction.
  1. Use molar masses to calculate reactant masses.

EXAMPLE:
What is the formula weight of ammonium dihydrogen phosphate? What is its molar mass?

SOLUTION:
The formula for ammonium dihydrogen phosphate is $NH_4H_2PO_4$

The elements present along with their atomic weights (rounded off to one decimal place) are:

N (14.0 amu) H (1.0 amu) P (31.0 amu) O (16.0 amu)

Multiply the atomic weight of each element by the number of times that element appears in the formula and add together the results.

N (1×14.0 amu)	= 14.0 amu
H (6×1.0 amu)	= 6.0 amu
P (1×31.0 amu)	= 31.0 amu
O (4×16.0 amu)	= 64.0 amu
	= 115.0 amu

The molar mass is the formula weight in grams: 115.0 g/mol

EXAMPLE:
The balanced equation for the combustion of pentane is:

$$C_5H_{12} + 8\,O_2 \rightarrow 5\,CO_2 + 6\,H_2O$$

How many moles of oxygen will react with a) 1 mol C_5H_{12} and b) 0.5 mol C_5H_{12}?

SOLUTION: From the balanced equation we know that 8 moles of O_2 will react with 1 mole of C_5H_{12} since the coefficients represent the number of moles of each substance. The ratio 8 moles O_2:1 mole C_5H_{12} represents a conversion factor which can be used to calculate the number of moles of O_2 that will react with 0.5 moles C_5H_{12}.

$$0.5 \text{ mol } C_5H_{12} \times \frac{8 \text{ mol } O_2}{1 \text{ mol } C_5H_{12}} = 4 \text{ mol } O_2$$

Stoichiometry: Chemical Arithmetic

    A. Stoichiometry - refers to the conversion between moles and grams of reactants and products in a chemical equation.

3⊠   B. To convert between grams, moles and formula units.

        1. Use the molar mass of the substance as a conversion factor.

EXAMPLE:
How many moles of PCl_3 are present in 8.0 g PCl_3?

SOLUTION: First calculate the molar mass (in g/mol) of PCl_3.

Molar mass PCl_3 = 31.0 g/mol + (3 x 35.5 g/mol) = 137.5 g/mol

This molar mass also represents a conversion factor of $\dfrac{137.5 \text{ g } PCl_3}{1 \text{ mol } PCl_3}$ or $\dfrac{1 \text{ mol } PCl_3}{137.5 \text{ g } PCl_3}$.

To determine the moles in 8.0 g of PCl_3

$$8.0 \text{ g } PCl_3 \times \frac{1 \text{ mol } PCl_3}{137.5 \text{ g } PCl_3} = 0.058 \text{ mol } PCl_3$$

4⊠

C. Calculating the grams of reactants needed and products produced knowing the grams and moles of another reactant.
   1. Beginning with the known grams of reactant, convert to moles.
   2. Using the balanced chemical equation, convert between moles of substances.
   3. Knowing moles, convert to grams.

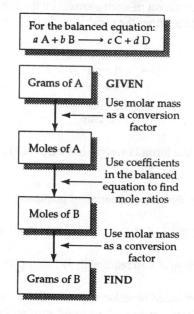

For the balanced equation:
$$a\,A + b\,B \longrightarrow c\,C + d\,D$$

Grams of A — GIVEN

Use molar mass as a conversion factor

Moles of A

Use coefficients in the balanced equation to find mole ratios

Moles of B

Use molar mass as a conversion factor

Grams of B — FIND

EXAMPLE:

How many grams of water will react with 8.0 g PCl_3 in the following balanced chemical equation?

$$PCl_3 + 3\,H_2O \rightarrow H_3PO_3 + 3\,HCl$$

SOLUTION: From the above example, we know that 8.0 g PCl_3 = 0.058 mol PCl_3. From the balanced equation, we know that 3 moles of water will react with 1 mole of PCl_3. We can use this mole:mole ratio to calculate the number of grams of water needed to react with 8.0 g PCl_3.

$$0.058 \text{ mol } PCl_3 \times \frac{3 \text{ mol } H_2O}{1 \text{ mol } PCl_3} \times \frac{18.0 \text{ g } H_2O}{1 \text{ mol } H_2O} = 3.1 \text{ g } H_2O$$

Notice, we first converted from moles of PCl_3 to moles of water, then from moles of water to grams of water.

EXAMPLE:

How many grams of phosphorous acid will be produced in the above example?

SOLUTION: Again, we start knowing that 8.0 g PCl_3 = 0.058 mol PCl_3 and that 1 mole PCl_3 will react with 1 mole H_3PO_3. We will first convert from moles of PCl_3 to moles of H_3PO_3; then from moles of H_3PO_3 to grams of H_3PO_3.

$$0.058 \text{ mol PCl}_3 \times \frac{1 \text{ mol H}_3\text{PO}_3}{1 \text{ mol PCl}_3} \times \frac{82 \text{ g H}_3\text{PO}_3}{1 \text{ mol H}_3\text{PO}_3} = 4.8 \text{ g H}_3\text{PO}_3$$

Yields of Chemical Reactions

    A. Actual yield - amount of product actually formed.

       1. Less than the amount predicted by theory.

    B. Theoretical yield - amount of product calculated to form if all of the reactants are converted to products.

5⊠   C. Percent yield $= \dfrac{\text{Actual yield}}{\text{Theorectical yield}} \times 100\%$

6⊠   ***EXAMPLE:***

Aspirin is produced from salicylic acid ($C_7H_6O_3$) and acetic anhydride ($C_4H_6O_3$) according to the following balanced equation:

$$\underset{\substack{\text{salicylic} \\ \text{acid}}}{C_7H_6O_3} + \underset{\substack{\text{acetic} \\ \text{anhydride}}}{C_4H_6O_3} \rightarrow \underset{\text{aspirin}}{C_9H_8O_4} + \underset{\substack{\text{acetic} \\ \text{acid}}}{C_2H_4O_2}$$

How many grams of aspirin would you obtain from 4.5 g salicylic acid if the percent yield of the reaction is 75%?

SOLUTION: We can use the following steps to solve this problem:

1. Calculate the molar mass of both aspirin and salicylic acid.

Molar mass salicylic acid = (7 x 12 g/mol) + (6 x 1.0 g/mol) + (3 x 16.0 g/mol) = 138.0 g/mol

Molar mass aspirin = (9 x 12.0 g/mol) + (8 x 1.0 g/mol) + (4 x 16.0 g/mol) = 180.0 g/mol

2. To determine the actual yield, we must first determine the theoretical yield. To do this, we must first convert grams of salicylic acid to moles of salicylic acid.

$$4.5 \text{ g C}_7\text{H}_6\text{O}_3 \times \frac{1 \text{ mol C}_7\text{H}_6\text{O}_3}{138 \text{ g C}_7\text{H}_6\text{O}_3} = 0.033 \text{ mol C}_7\text{H}_6\text{O}_3$$

3. To determine the theoretical yield of aspirin, we must convert moles of salicylic acid to moles of aspirin and then convert moles of aspirin to grams of aspirin.

$$0.033 \text{ mol C}_7\text{H}_6\text{O}_3 \times \frac{1 \text{ mol C}_9\text{H}_8\text{O}_4}{1 \text{ mol C}_7\text{H}_6\text{O}_3} \times 180.0 \text{ g C}_7\text{H}_6\text{O}_3 = 5.9 \text{ g C}_9\text{H}_8\text{O}_4$$

4. To determine the actual yield of aspirin, we substitute the theoretical yield and percent yield into the equation for percent yield.

$$75\% = 100\% \times \frac{\text{actual yield}}{5.9 \text{ g aspirin}}$$

g aspirin = 0.75 x 5.9 g aspirin = 4.4 g aspirin

Reactions with Limiting Amounts of Reactants

A. Many reactions are carried out using an excess of one reactant.

B. Excess reactant - reactant that is present in more than the amount that is needed according to the stoichiometry.

   1. Only the amount required by stoichiometry reacts.
   2. Excess reactant acts as a spectator and takes no role in the reaction.

C. Limiting reactant - the reactant present in the limiting amount.

   1. This determines the extent to which a chemical reaction takes place.

7,8⊠   D. To determine if a limiting amount of one reactant is present and calculate the amount of nonlimiting reactant consumed.

   1. Find how many moles of each reactant are present.
   2. Look at the coefficients in the balanced equation to see the required mole ratio of the two reactants and compare to the number of moles calculated in the first step.
   3. Subtract the number of moles of excess reactant consumed from the number of moles of excess reactant present. Convert to grams.
      a. this is grams of nonlimiting reactant in excess
   4. To determine the amount of product formed, use the number of moles of limiting reactant and the balanced equation to first convert to moles of product then grams of product.

EXAMPLE:

Boron sulfide, B_2S_3, reacts violently with water to form dissolved boric acid, H_3BO_3, and hydrogen sulfide gas. Assume that 5.88 g of B_2S_3 is mixed with 7.85 g of water. Which reactant is limiting and which reactant is in excess? How many grams of the excess reactant are consumed? How many grams of boric acid are produced?

SOLUTION: First, write a balanced chemical equation. (**Always make sure the chemical equation is balanced before beginning any stoichiometry problem!**)

$$B_2S_3 + 6 H_2O \rightarrow 2 H_3BO_3 + 3 H_2S$$

Determine the number of moles of each reactant present.

$$5.88 \text{ g } B_2S_3 \times \frac{1 \text{ mol } B_2S_3}{117.6 \text{ g } B_2S_3} = 0.0500 \text{ mol } B_2S_3$$

$$7.85 \text{ g } H_2O \times \frac{1 \text{ mol } H_2O}{18.0 \text{ g } H_2O} = 0.436 \text{ mol } H_2O$$

The required mole ratio is 6 moles of water for every 1 mole of B_2S_3 . Therefore, the required number of moles of water to react with 0.0500 moles of B_2S_3 is 6 x 0.0500 = 0.300. Since we have more than enough moles of water to react with B_2S_3, water is the excess reactant and B_2S_3 is the limiting reactant.

Moles of B_2S_3 consumed: 0.0500 mol B_2S_3
Moles of H_2O consumed: 0.300 mol H_2O
Moles of H_2O not consumed: 0.436 mol H_2O - 0.300 mol H_2O = 0.136 mol H_2O

To determine the grams of boric acid, use the balanced chemical equation to convert moles of B_2S_3 to moles of boric acid; then convert from moles of boric acid to grams of boric acid.

$$0.0500 \text{ mol } B_2S_3 \times \frac{2 \text{ mol } H_3BO_3}{1 \text{ mol } B_2S_3} \times \frac{61.8 \text{ g } H_3BO_3}{1 \text{ mol } H_3BO_3} = 6.18 \text{ g } H_3BO_3$$

Concentrations of Reactants in Solution: Molarity

A. Most chemical reactions are carried out in solution.
   1. Reactants must have considerable mobility for a reaction to occur.
B. Solute - a substance dissolved in a solution.
C. Concentration - describes the relative amount of solute to a solvent or solution.

10⊠   D. Molarity (M) - the number of moles of a solute dissolved in each liter of solution.
   1. Dissolve the solute in enough solution to give a final solution volume of 1.00 L.
   2. Can be used as a conversion factor.

$$\text{Molarity} = \frac{\text{Moles of solute}}{\text{Volume of solution (L)}}$$

$$\text{Moles of solute} = \text{Molarity} \times \text{Volume of solution}$$

$$\text{Volume of solution} = \frac{\text{Moles of solute}}{\text{Molarity}}$$

9⊠   **EXAMPLE:**

What is the molarity of a solution prepared by dissolving 16.8 g NaOH and diluting to a final volume of 2.5 L?

SOLUTION: First, find the number of moles of NaOH in 16.8 g NaOH (molar mass = 40.0 g/mol)

$$16.8 \text{ g NaOH} \times \frac{1 \text{ mol NaOH}}{40.0 \text{ g NaOH}} = 0.420 \text{ mol NaOH}$$

Next, divide the number of moles of NaOH by the volume (in liters) of solution.

$$\frac{0.420 \text{ mol NaOH}}{2.5 \text{ L}} = 0.17 \text{ M}$$

10⊠   **EXAMPLE:**

How many moles of $K_2Cr_2O_7$ are present in 250 mL of a 0.025 M solution of $K_2Cr_2O_7$?

SOLUTION: You can calculate the number of moles of $K_2Cr_2O_7$ by multiplying the molarity of the solution by the volume (in liters).

$$\frac{0.025 \text{ mol } K_2Cr_2O_7}{1 \text{ L}} \times 0.250 \text{ L} = 0.0063 \text{ mol } K_2Cr_2O_7$$

10⊠   **EXAMPLE:**

A particular reaction requires that you use 36.7 g of H_2SO_4. How many mL of a 5.00 M solution of H_2SO_4 are required to provide 36.7 g?

SOLUTION: First, determine the number of moles of H_2SO_4 in 36.7 g H_2SO

$$36.7 \text{ g } H_2SO_4 \times \frac{1 \text{ mol } H_2SO_4}{98.0 \text{ g } H_2SO_4} = 0.374 \text{ mol } H_2SO_4$$

The volume of solution can now be calculated knowing both the moles of solute and the molarity of the solution.

$$0.374 \text{ mol } H_2SO_4 \times \frac{1 \text{ L } H_2SO_4}{5.00 \text{ mol } H_2SO_4} = 74.8 \text{ mL}$$

10⊠ **EXAMPLE:**

How many grams of $Pb(NO_3)_2$ are needed to prepare 250 mL of a 0.10 M solution?

SOLUTION: First, determine the number of moles of $Pb(NO_3)_2$ found in 250 mL of a 0.10 M solution.

$$0.250 \text{ L} \times \frac{0.10 \text{ mol } Pb(NO_3)_2}{1 \text{ L}} = 0.025 \text{ mol } Pb(NO_3)_2$$

The grams of $Pb(NO_3)_2$ can now be calculated knowing the moles of $Pb(NO_3)_2$.

$$0.025 \text{ mol } Pb(NO_3)_2 \times \frac{331.2 \text{ g } Pb(NO_3)_2}{1 \text{ mol } Pb(NO_3)_2} = 8.3 \text{ g } Pb(NO_3)_2$$

9⊠ **Diluting Concentrated Solutions**

A. Dilution - the number of moles of solute remains the same; only the volume is changed by adding more solvent.

$$\text{Moles of solute } = M_i\left(\frac{\text{mol}}{\text{L}}\right) \times V_i(\text{L}) = M_f\left(\frac{\text{mol}}{\text{L}}\right) \times V_f(\text{L})$$

This equation should only be used for problems which deal with the dilution of a solution!

EXAMPLE:

What would the final concentration be if 25 mL of 12.0 M HCl were diluted to a final volume of 500 mL?

SOLUTION: We know that $M_i \times V_i = M_f \times V_f$. Rearranging this equation gives $M_f = \dfrac{M_i \times V_i}{V_f}$.

The final concentration of the above solution would be:

$$\frac{12.0 \text{ M} \times 0.025 \text{ L}}{0.500 \text{ L}} = 0.60 \text{ M}$$

11⊠ **Solution Stoichiometry**

A. Knowing the molarity of the solutions is critical to carry out stoichiometry calculations.
   1. Can be used to calculate either the number of moles (if volume is known) or volume (if the number of moles is known).

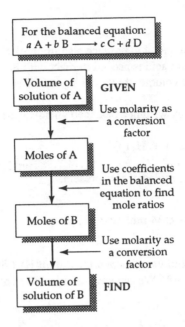

EXAMPLE:

Barium chloride will react with sodium phosphate to produce barium phosphate (a solid) and sodium chloride. How many mL of 0.025 M barium chloride are needed to react with 50 mL of 0.015 M sodium phosphate? How many grams of barium phosphate will be produced?

SOLUTION: First, write the balanced equation for the reaction.

$$3\ BaCl_2\ +\ 2\ Na_3PO_4\ \rightarrow\ Ba_3(PO_4)_2\ +\ 6\ NaCl$$

Next, calculate the moles of Na_3PO_4 present.

$$0.050\ L \times \frac{0.015\ mol\ Na_3PO_4}{1\ L} = 7.5 \times 10^{-4}\ mol\ Na_3PO_4$$

Using the balanced equation, calculate the number of moles of $BaCl_2$ needed to react with 7.5×10^{-4} moles Na_3PO_4.

$$7.5 \times 10^{-4}\ mol\ Na_3PO_4 \times \frac{3\ mol\ BaCl_2}{2\ mol\ Na_3PO_4} = 1.1 \times 10^{-3}\ mol\ BaCl_2$$

The volume of the $BaCl_2$ solution can now be calculated knowing both the molarity and the number of moles.

$$\frac{1.1 \times 10^{-3}\ mol\ BaCl_2}{0.025\ mol\ BaCl_2\ /\ L} = 0.044\ L = 44\ mL$$

To calculate the grams of barium phosphate produced, we can either begin with moles of sodium phosphate or moles of barium chloride.

$$7.3 \times 10^{-4}\ mol\ Na_3PO_4 \times \frac{1\ mol\ Ba_3(PO_4)_2}{2\ mol\ Na_3PO_4} \times \frac{601.9\ g\ Ba_3(PO_4)_2}{1\ mol\ Ba_3(PO_4)_2} = 0.22\ g\ Ba_3(PO_4)_2$$

43

EXAMPLE:

Hydrochloric acid reacts with zinc metal to produce a solution of zinc chloride and hydrogen gas. What volume of 12 M hydrochloric acid is needed to react with 8.75 g of zinc? If the density of hydrogen gas is 0.0899 g/L, what volume of hydrogen gas will be produced?

SOLUTION: Once again, we begin by writing a balanced chemical equation.

$$Zn\ (s)\ +\ 2\ HCl\ (aq)\ \rightarrow\ ZnCl_2\ (aq)\ +\ H_2\ (g)$$

Next, we determine the number of moles of zinc present.

$$mol\ Zn = 8.75\ g\ Zn \times \frac{1\ mol\ Zn}{65.4\ g\ Zn} = 0.134\ mol\ Zn$$

From the balanced equation, we know that there are 2 mol of HCl for every 1 mol Zn. Therefore, the number of moles of HCl is 0.268. We can now calculate the volume of 12 M HCl needed for this reaction.

$$Volume\ HCl = 0.268\ mol\ HCl \times \frac{1\ L\ soln}{12\ mol\ HCl} = 0.0223\ L = 22.3\ mL\ soln$$

To calculate the volume of H_2 gas produced, we first need to calculate the grams of H_2 gas produced. This is easily accomplished by beginning with the moles of zinc present.

$$0.134\ mol\ Zn \times \frac{1\ mol\ H_2}{1\ mol\ Zn} \times \frac{2.00\ g\ H_2}{1\ mol\ H_2} = 0.268\ g\ H_2$$

$$Volume\ of\ H_2 = 0.268\ g\ H_2 \times \frac{1\ L\ H_2}{0.0899\ g\ H_2} = 2.98\ L\ H_2$$

Titration

A. Procedure for determining the concentration of a solution.
 1. Carefully measured volume of one solution reacts with a second solution of known concentration and volume.

EXAMPLE:

A student finds that 38.4 mL of 0.215 M hydrochloric acid is required to neutralize a 20.0 mL sample of barium hydroxide. What is the molarity of $Ba(OH)_2$?

SOLUTION: First, we need to write a balanced equation for the neutralization reaction.

$$2\ HCl\ (aq)\ +\ Ba(OH)_2\ (aq)\ \rightarrow\ BaCl_2\ (aq)\ +\ H_2O\ (l)$$

From the information given, we can calculate the number of moles of hydrochloric used to reach the equivalence point of the titration.

$$mol\ HCl = 0.215\ \frac{mol\ HCl}{L\ soln} \times 0.0384\ L\ soln = 8.26 \times 10^{-3}\ mol\ HCl$$

From the balanced equation, we know that we have 2 mol HCl for every 1 mol of $Ba(OH)_2$. Therefore, the number of moles of $Ba(OH)_2$ present is 4.13×10^{-3}. Knowing the moles of

44

$Ba(OH)_2$ present allows us to calculate the molarity of $Ba(OH)_2$.

$$Molarity = \frac{4.13 \times 10^{-3} \text{ mol } Ba(OH)_2}{0.020 \text{ L soln}} = 0.207 \text{ M}$$

12⊠   **Percent Composition and Empirical Formulas**
A. Composition of a compound - the identity and amount of elements present in a compound.
B. Percent composition - the mass percent of each element present in a compound.
1. Calculate the chemical formula from percent composition.
a. find the relative numbers of moles of each element in the compound
i. use molar masses of the elements as conversion factors
b. find the ratio of the numbers of moles by dividing the larger number of moles by the smaller number
c. multiply the subscripts by small integers in a trial-and-error procedure until whole numbers are found

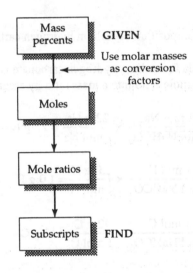

C. Empirical formula - gives only the ratios of atoms in a compound.
D. Molecular formula - gives the actual numbers of atoms in a molecule.
1. May be the same as the empirical formula.
2. May be a multiple of the empirical formula.

a. Multiple $= \dfrac{\text{Molecular weight}}{\text{Empirical formula weight}}$

E. Can derive % composition from a chemical formula.

EXAMPLE:
The percent composition of a solid is known to be 68.4% Ba, 10.3% P, and 21.3% O. What is the empirical formula of the compound?

SOLUTION: Assuming a 100 g sample gives us 68.4 g Ba, 10.3 g P, and 21.3 g O, convert these masses to numbers of moles.

$$68.4 \text{ g Ba} \times \frac{1 \text{ mol Ba}}{137.3 \text{ g Ba}} = 0.498 \text{ mol Ba}$$

$$10.3 \text{ g P} \times \frac{1 \text{ mol P}}{31.0 \text{ g P}} = 0.332 \text{ mol P}$$

$$21.3 \text{ g O} \times \frac{1 \text{ mol O}}{16 \text{ g O}} = 1.33 \text{ mol O}$$

Knowing the relative numbers of moles, find the ratio by dividing the two larger numbers by the smaller number.

$$\frac{1.33}{0.332} = 4 \qquad \frac{0.498}{0.332} = 1.5$$

The O:Ba:P ratio of 4:1.5:1 gives an empirical formula of $Ba_{1.5}PO_4$. However, we must multiply the subscripts by a small integer (in this case 2) to find whole numbers for the formula. $Ba_3P_2O_8$, or $Ba_3(PO_4)_2$.

EXAMPLE:

What is the percent composition of sodium hydrogen carbonate?

SOLUTION: The formula for sodium hydrogen carbonate is $NaHCO_3$. The Na:H:C:O mole ratio is 1:1:1:3. Convert this mole ratio into a mass ratio by assuming there is a 1 mole sample present.

$$1 \text{ mol NaHCO}_3 \times \frac{1 \text{ mol Na}}{1 \text{ mol NaHCO}_3} \times \frac{23 \text{ g Na}}{1 \text{ mol Na}} = 23 \text{ g Na}$$

$$1 \text{ mol NaHCO}_3 \times \frac{1 \text{ mol H}}{1 \text{ mol NaHCO}_3} \times \frac{1.0 \text{ g H}}{1 \text{ mol H}} = 1.0 \text{ g H}$$

$$1 \text{ mol NaHCO}_3 \times \frac{1 \text{ mol C}}{1 \text{ mol NaHCO}_3} \times \frac{12 \text{ g C}}{1 \text{ mol C}} = 12 \text{ g C}$$

$$1 \text{ mol NaHCO}_3 \times \frac{3 \text{ mol O}}{1 \text{ mol NaHCO}_3} \times \frac{16 \text{ g O}}{1 \text{ g O}} = 48 \text{ g O}$$

To determine the percent composition, divide the mass of each element present by the total mass of the compound and multiply by 100.

Total mass of 1 mole of $NaHCO_3$ = 84 g

$$\% \text{ Na} = \frac{23 \text{ g Na}}{84 \text{ g}} \times 100\% = 27\%$$

$$\% \text{ H} = \frac{1.0 \text{ g H}}{84 \text{ g}} \times 100\% = 1.2\%$$

$$\% \text{ C} = \frac{12 \text{ g C}}{84 \text{ g}} \times 100\% = 14\%$$

$$\% \text{ O} = \frac{48 \text{ g O}}{84 \text{ g}} \times 100\% = 57\%$$

13⊠ **Determining Empirical Formulas: Elemental Analysis**

   A. Combustion analysis - a compound of unknown composition is burned with oxygen to produce the volatile combustion products CO_2 and H_2O.

      1. CO_2 and H_2O are separated by a gas chromatograph and their masses are determined.

      2. Used to determine the empirical formula of a compound.

   B. To determine the empirical formula of a compound from a combustion reaction:

      1. Carry out gram-to-mole conversions to find the molar amounts of C and H in CO_2 and H_2O.

      2. Carry out mole-to-gram conversions to find the number of grams of C and H in the original sample.

      3. Subtract the masses of C and H from the mass of the starting sample to determine the number of grams of the substance unaccounted for.

      4. Determine the number of moles of the remaining substance.

      5. Find the ratio of the numbers of moles by dividing the larger number of moles by the smaller number of moles.

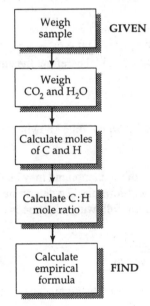

EXAMPLE:

   When 4.50 g of ethyl butyrate, a compound containing C, H, and O, undergoes combustion, 10.24 g of CO_2 and 4.19 g H_2O are produced. Determine the empirical formula of this compound.

SOLUTION: First, find the molar amounts of C and H in CO_2 and H_2O.

$$10.24 \text{ g } CO_2 \times \frac{1 \text{ mol } CO_2}{44.0 \text{ g } CO_2} \times \frac{1 \text{ mol C}}{1 \text{ mol } CO_2} = 0.233 \text{ mol C}$$

$$4.19 \text{ g } H_2O \times \frac{1 \text{ mol } H_2O}{18.0 \text{ g } H_2O} \times \frac{2 \text{ mol H}}{1 \text{ mol } H_2O} = 0.466 \text{ mol H}$$

Next, carry out mole-to-gram conversions to find the number of grams of C and H in the original sample.

$$0.233 \text{ mol C} \times \frac{12.0 \text{ g C}}{1 \text{ mol C}} = 2.79 \text{ g C}$$

$$0.466 \text{ mol H} \times \frac{1.00 \text{ g H}}{1 \text{ mol H}} = 0.466 \text{ g H}$$

Subtract the masses of C and H from the mass of the starting sample to determine the mass of O.

$$4.50 \text{ g} - 2.79 \text{ g} - 0.466 \text{ g} = 1.24 \text{ g}$$

Convert the mass of O to moles of O.

$$1.24 \text{ g O} \times \frac{1 \text{ mol O}}{16.0 \text{ g O}} = 0.0778 \text{ mol O}$$

Find the ratio of the numbers of moles by dividing the larger number of moles by the smaller number of moles.

$$\frac{0.233}{0.0778} = 3 \qquad \frac{0.446}{0.0778} = 6$$

This gives a mole ratio of C:H:O of 3:6:1. Therefore, the empirical formula for ethyl butyrate is C_3H_6O.

14⬦ ***EXAMPLE:***

Determine the molecular formula of ethyl butyrate knowing that the molecular weight of the compound is 116 g/mol.

SOLUTION: The molecular formula of the compound may be the same as the empirical formula or it may be a multiple of the empirical formula. To determine if the molecular formula is a multiple of the empirical formula the following equation is used.

$$\text{Multiple} = \frac{\text{Molecular weight}}{\text{Empirical formula weight}}$$

For ethyl butyrate:

$$\text{Multiple} = \frac{116 \text{ g/mol}}{58 \text{ g/mol}} = 2$$

The subscripts in the empirical formula are multiplied by 2 giving a molecular formula of $C_6H_{12}O_2$.

Self-Test

This section is intended to test your knowledge of the material covered in this chapter. Think through these problems and make certain you understand what is going on. Ask yourself if your answer makes sense. Many of these questions are linked to the chapter learning goals. Therefore, successful completion of these problems indicates you have mastered the learning goals for this chapter. You will receive the greatest benefit from this section if you use it as a mock exam. You will then discover which topics you have mastered and which topics you need to study in more detail.

True/False

1. A balanced chemical equation obeys the law of definite proportions.

2. When balancing a chemical equation, the coefficients or the subscripts of the formula units may be changed.

3. When hydrocarbons undergo combustion, carbon dioxide and water are produced.

4. Chemical symbols represent the behavior of individual atoms and molecules only on the microscopic level.

5. Laboratory work involves the behavior of atoms and molecules on a large-scale (macroscopic) level.

6. The coefficients in a balanced chemical equation indicate the number of moles of each substance needed for the reaction.

7. The excess reactant determines the extent to which a chemical reaction takes place.

8. When diluting a solution, the number of moles of solute changes.

9. The theoretical yield of a reaction is always greater than the actual yield.

10. The empirical formula and molecular formula of a compound are always different.

Matching

Combustion reaction	a. reactant that is present in more than the amount that is actually needed according to the stoichiometry.
Microscopic level	b. gives the ratio of atoms in a compound.
Macroscopic level	c. reaction with oxygen.
Formula weight	d. a procedure for determining the concentration of a solution in which a carefully measured volume of one solution reacts with a second solution of known concentration and volume.
Molecular weight	e. the sum of the atomic weights of all atoms in one formula unit of any substance.

Molar mass

    f. chemical symbols represent the behavior of individual atoms and molecules.

Excess reactant

    g. the mass percent of each element present in a compound.

Limiting reactant

    h. give the actual number of atoms in a molecule.

Molarity

    i. the sum of the atomic weights of all atoms in a molecule.

Titration

    j. the reactant that is present in the limiting amount and determines the extent to which a chemical reaction takes place.

Percent composition

    k. the molecular or formula weight in grams of one mole of any substance.

Empirical formula

    l. the number of moles of solute dissolved in one liter of solution.

Molecular formula

    m. the large-scale behavior of atoms and molecules that gives rise to observable properties.

Fill-in-the-Blank

1.   All chemical equations must be _____.

2.   One mole of any object, atom, molecule, or ion contains _____ of formula units.

3.   Using the relationship between numbers of moles and numbers of grams to carry out chemical calculations is called _____.

4.   The concentration of a substance in solution is usually expressed as _____.

5.   When carrying out a dilution, the _____ of the solution changes, but the _____ stays the same.

6.   In carrying out a chemical reaction, the _____ determines the amount of product formed.

7.   _____ makes it possible to calculate the amount of volume of one solution needed to react with a given volume of another solution.

8.   The empirical formula of a substance can be derived from _____.

9.   In a combustion analysis, the products _____ and _____ are formed.

10. To do actual laboratory work, it is necessary to convert between _____ and _____ of reactants and products.

Problems

1. Balance the following equations:

   a. $Mg_3N_2 + HCl \rightarrow MgCl_2 + NH_4Cl$

   b. calcium carbonate reacts with nitric acid to produce carbon dioxide gas, water, and calcium nitrate.

   c. Octane (C_8H_{18}) undergoes combustion.

   d. $Al + H_2SO_4 \rightarrow Al_2(SO_4)_3 + H_2$

   e. $AgNO_3 + Na_2SO_4 \rightarrow Ag_2SO_4 + NaNO_3$

2. Phosphine (PH_3) is an extremely poisonous gas with a garlic odor and is used in the process for making flame resistant cotton cloth . How many grams of phosphine will be produced when 15.4 g of calcium phosphide reacts with water to produce phosphine and calcium hydroxide?

3. Many metal ions will react with hydrogen sulfide to form precipitates with a characteristic color. This reaction has been widely used in the laboratory to identify the presence of metal ions in an unknown solution. The safest way of producing hydrogen sulfide is by the reaction of water with thioacetamide.

$$\overset{\overset{\displaystyle S}{\|}}{CH_3CNH_2} + H_2O \longrightarrow \overset{\overset{\displaystyle O}{\|}}{CH_3CHNH_2} + H_2S$$

How many grams of water are needed to react with 5.75 g of thioacetamide? How many grams of hydrogen sulfide will be produced?

4. Sodium sulfite is used in the manufacture of paper pulp and as a food additive to prevent decomposition. It is prepared according to the following equation:

$$NaHSO_3 + Na_2CO_3 \rightarrow Na_2SO_3 + CO_2 + H_2O$$

How many grams of sodium sulfite would you obtain from 75.8 g of sodium hydrogen sulfite if the reaction had a % yield of 64%?

5. The largest use of nitric acid is in the reaction with ammonia to produce ammonium nitrate for fertilizer. If 156.5 g of nitric acid reacted with 275.0 g of ammonia, which reactant would be in excess? How many moles of the excess reactant would be left over after the reaction? How many grams of ammonium nitrate would be produced?

6. Describe how you would prepare 500 mL of 3.75 M NaOH.

7. Describe how you would prepare 2.50 L of 0.125 M H_3PO_4 beginning with a concentrated solution that has a molarity of 15.0 M.

8. How many milliliters of 0.150 M HCl are needed to react with 2.45 g of zinc in the following reaction?

$$Zn + 2\,HCl \rightarrow ZnCl_2 + H_2$$

9. 50 mL of 0.025 M $Pb(NO_3)_2$ is reacted with 50 mL of 0.10 M K_2CrO_4 to produce $PbCrO_4$, a bright yellow solid, and KNO_3. Which reactant is in excess? How many moles of excess reactant will be left over at the end of the reaction? How many grams of $PbCrO_4$ will be produced?

10. What is the percent composition of aspirin, $C_9H_8O_4$?

11. What is the empirical formula for ethylene glycol if the percent composition is 38.7% C, 9.7% H and 51.6% O? What is the molecular formula for this compound if the molar mass is 62 g/mol?

12. For the reaction

$$Ag_2CO_3 + HNO_3 \rightarrow CO_2 + AgNO_3 + H_2O$$

How many grams of silver nitrate will be formed when 9.85 g of silver carbonate reacts with 250 mL of 0.250 M HNO_3?

13. 12.78 g of ethyl acetate, a compound containing C, H, and O, undergoes combustion to produce 25.56 g of carbon dioxide and 10.46 g of water. What is the empirical formula of this compound? What is the molecular formula of this compound if the molar mass is 88.0 g/mol?

14. What is the percent composition of calcium phosphite?

15. 4-*n*-hexylresorcinol is used as an antiseptic in throat lozenges and mouth washes. Its percent composition is 74.2% C, 9.3% H, and 16.5% O. What is its empirical formula? What is its molecular formula if the molar mass is 194 g/mol?

Solutions

True/False

1. F. A balanced chemical equation obeys the law of mass conservation.

2. F. When balancing a chemical equation, only the coefficients may be changed; the subscripts of the formula units must stay the same.

3. T

4. F. Chemical symbols represent the behavior of atoms and molecules on both the microscopic and macroscopic level.

5. T

6. T

7. F. The limiting reactant determines the extent to which a chemical reaction takes place.

8. F. When diluting a solution, the number of moles of solute stays the same; only the total volume of solution changes.

9. T

10. F. The molecular formula of a compound may be the same as the empirical formula.

Matching

Combustion reaction - c
Microscopic level - f
Macroscopic level - m
Formula weight - e
Molecular weight - i
Molar mass - k
Excess reactant - a
Limiting reactant - j
Molarity - l
Titration - d
Percent composition - g
Empirical formula - b
Molecular formula - h

Fill-in-the-Blank
1. Balanced
2. Avogadro's number
3. Stoichiometry
4. Molarity
5. Volume; number of moles
6. Limiting reactant
7. Titration
8. Percent composition
9. Carbon dioxide; water
10. Moles and grams

Problems
1. a. $Mg_3N_2 + 8 HCl \rightarrow 3 MgCl_2 + 2 NH_4Cl$

   b. $CaCO_3 + 2 HNO_3 \rightarrow CO_2 + H_2O + Ca(NO_3)_2$

   c. $2 C_8H_{18} + 25 O_2 \rightarrow 16 CO_2 + 18 H_2O$: Remember that combustion refers to the reaction with oxygen and that hydrocarbons produce CO_2 and water when they undergo combustion. To balance this equation, first begin by balancing the C's and H's as shown below.

   $C_8H_{18} + O_2 \rightarrow 8 CO_2 + 9 H_2O$

   Notice that this gives you a total of 25 O's on the product side. Because the oxygen on the reactant side is present in the form of O_2, you would need 25/2 as the coefficient. This would give the following balanced equation:

$$C_8H_{18} + 25/2\ O_2 \rightarrow 8\ CO_2 + 9\ H_2O$$

However, coefficients are not allowed to be fractions. If we multiply all of the coefficients in the equation by 2, we will remove the fraction and end up with the smallest whole number ratio of molecules.

$$2\ C_8H_{18} + 25\ O_2 \rightarrow 16\ CO_2 + 18\ H_2O$$

d. $2\ Al + 3\ H_2SO_4 \rightarrow Al_2(SO_4)_3 + 3\ H_2$

e. $2\ AgNO_3 + Na_2SO_4 \rightarrow Ag_2SO_4 + 2\ NaNO_3$

2. First, write the balanced chemical equation:

$$Ca_3P_2 + 6H_2O \rightarrow 2PH_3 + 3Ca(OH)_2$$

Next, convert grams of Ca_3P_2 to moles of Ca_3P_2

$$15.4\ g\ Ca_3P_2 \times \frac{1\ mol\ Ca_3P_2}{182.3\ g\ Ca_3P_2} = 0.0845\ mol\ Ca_3P_2$$

Knowing moles of Ca_3P_2, we can convert to moles of PH_3 and then grams of PH_3.

$$0.0845\ mol\ Ca_3P_2 \times \frac{2\ mol\ PH_3}{1\ mol\ Ca_3P_2} \times \frac{34\ g\ PH_3}{1\ mol\ PH_3} = 5.75\ g\ PH_3$$

3. First, check that the chemical equation given is balanced. Next, convert grams of thioacetamide to moles.

$$5.75\,g\,CH_3\,CSNH_2 \times \frac{1\,mol\,CH_3\,CSNH_2}{75.0\,g\,CH_3\,CSNH_2} = 0.0767\,mol$$

Knowing moles of thioacetamide, we can now calculate grams of water and grams of hydrogen sulfide.

$$0.0767\,mol\,CH_3\,CSNH_2 \times \frac{1\,mol\,H_2O}{1\,mol\,CH_3\,CSNH_2} \times \frac{18.0\,g\,H_2O}{1\,mol\,H_2O} = 1.38\,g\,H_2O$$

$$0.0767\,mol\,CH_3\,CSNH_2 \times \frac{1\,mol\,H_2S}{1\,mol\,CH_3\,CSNH_2} \times \frac{34.0\,g\,H_2S}{1\,mol\,H_2S} = 2.61\,g\,H_2S$$

4. First, check to see if the chemical equation is balanced. It is not! Balance the equation.

$$2\ NaHSO_3 + Na_2CO_3 \rightarrow 2\ Na_2SO_3 + CO_2 + H_2O$$

Next, calculate the theoretical yield of sodium sulfite by converting grams of sodium hydrogen sulfite to moles of sodium hydrogen sulfite. Next, convert moles of sodium hydrogen sulfite to moles of sodium sulfite and then to grams of sodium sulfite.

$$75.8\,g\,NaHSO_3 \times \frac{1\,mol\,NaHSO_3}{104.0\,g\,NaHSO_3} = 0.729\,mol\,NaHSO_3$$

$$0.729 \text{ mol NaHSO}_3 \times \frac{2 \text{ mol Na}_2\text{SO}_3}{2 \text{ mol NaHSO}_3} \times \frac{126.0 \text{ g Na}_2\text{SO}_3}{1 \text{ mol Na}_2\text{SO}_3} = 91.8 \text{ g Na}_2\text{SO}_3$$

Knowing the percent yield and theoretical yield, we can now calculate the actual yield.

$$64\% = 100\% \times \frac{\text{actual yield}}{91.8 \text{ g Na}_2\text{SO}_3}; \qquad\qquad \text{actual yield} = 59 \text{ g Na}_2\text{SO}_3$$

5. The chemical equation for this problem is:

$$HNO_3 + NH_3 \rightarrow NH_4NO_3$$

This problem is a limiting reactant problem. First, we need to determine how many moles of HNO_3 and NH_3 are present.

$$156.5 \text{ g HNO}_3 \times \frac{1 \text{ mol HNO}_3}{63.0 \text{ g HNO}_3} = 2.48 \text{ mol HNO}_3$$

$$275.0 \text{ g NH}_3 \times \frac{1 \text{ mol NH}_3}{17.0 \text{ g NH}_3} = 16.2 \text{ mol NH}_3$$

The required mole ratio is 1 mole of nitric acid for every 1 mole of ammonia. Therefore, the number of moles of ammonia needed to react with nitric acid is 2.48. Since we have more than enough moles of ammonia to react with nitric acid, ammonia is the excess reactant and nitric acid is the limiting reactant.

Moles of nitric acid consumed: 2.48 mol HNO_3
Moles of ammonia consumed: 2.48 mol NH_3
Moles of ammonia not consumed: 16.2 mol NH_3 - 2.48 mol NH_3 = 13.7 mol NH_3

The amount of NH_4NO_3 produced is determined by the 2.48 mol of HNO_3

.

$$2.48 \text{ mol HNO}_3 \times \frac{1 \text{ mol NH}_4\text{NO}_3}{1 \text{ mol HNO}_3} \times \frac{80.0 \text{ g NH}_4\text{NO}_3}{1 \text{ mol NH}_4\text{NO}_3} = 198 \text{ g NH}_4\text{NO}_3$$

6. First calculate the number of moles of NaOH.

$$0.500 \text{ L} \times \frac{3.75 \text{ mol NaOH}}{1 \text{ L}} = 1.88 \text{ mol NaOH}$$

Next, determine how many grams of NaOH are in 1.88 mol NaOH.

$$1.88 \text{ mol NaOH} \times \frac{40.0 \text{ g NaOH}}{1 \text{ mol NaOH}} = 75.2 \text{ g NaOH}$$

To prepare this solution, you would dissolve 75.2 g of NaOH in water and dilute to a final volume of 500 mL.

7. To prepare this solution, we need to know the volume of the concentrated phosphoric acid. We will use the equation

$$i = \frac{M_f \times V_f}{M_i}.$$

Substituting in the given information leads to

$$i = \frac{0.125 \, M \times 2.50 \, L}{15.0 \, M} = 0.0208 \, L = 20.8 \, mL$$

To prepare this solution, we would **carefully** add 20.8 mL to $\approx$ 1 L of water and dilute to 2.5 L. **CAUTION**: Always add acid to water, not water to acid!

8. First, calculate the number of moles of zinc.

$$2.45 \, g \, Zn \times \frac{1 \, mol \, Zn}{65.4 \, g \, Zn} = 0.0375 \, mol \, Zn$$

The Zn:HCl mole ratio is 1:2. Therefore, the moles of HCl = 0.0750. The volume of HCl needed for this reaction can now be calculated.

$$\frac{0.0750 \, mol \, HCl}{0.150 \, mol \, HCl \, / \, L} = 0.500 \, L = 500 \, mL$$

9. The balanced equation for this reaction is:

$$Pb(NO_3)_2 + K_2CrO_4 \rightarrow PbCrO_4 + 2 \, KNO_3$$

This problem is a limiting reactant problem. First, calculate the number of moles of $Pb(NO_3)_2$ and K_2CrO_4.

$$0.050 \, L \times \frac{0.025 \, mol \, Pb(NO_3)_2}{1 \, L} = 1.25 \times 10^{-3} \, mol \, Pb(NO_3)_2$$

$$0.050 \, L \times \frac{0.10 \, mol \, K_2 \, CrO_4}{1 \, L} = 5.0 \times 10^{-3} \, mol \, K_2 \, CrO_4$$

The $Pb(NO_3)_2$:K_2CrO_4 ratio is 1:1. Therefore, 1.25×10^{-3} moles of K_2CrO_4 are needed to react with 1.25×10^{-3} moles of $Pb(NO_3)_2$. There is more than enough K_2CrO_4 for this reaction. Therefore, the excess reactant is K_2CrO_4, and the limiting reactant is $Pb(NO_3)_2$.

Moles of $Pb(NO_3)_2$ consumed: 1.25×10^{-3} mol $Pb(NO_3)_2$
Moles of K_2CrO_4 consumed: 1.25×10^{-3} mol K_2CrO_4
Moles of K_2CrO_4 not consumed: $(5.0 \times 10^{-3}$ mol $K_2CrO_4) - (1.25 \times 10^{-3}$ mol $K_2CrO_4) = 3.75 \times 10^{-3}$ mol K_2CrO_4

Grams of $PbCrO_4$ produced can now be calculated.

$$1.25 \times 10^{-3} \, mol \, Pb(NO_3)_2 \times \frac{1 \, mol \, PbCrO_4}{1 \, mol \, Pb(NO_3)_2} \times \frac{323.2 \, g \, PbCrO_4}{1 \, mol \, PbCrO_4} = 0.404 \, g \, PbCrO_4$$

10. The C:H:O mole ratio is 9:8:4. Convert this mole ratio into a mass ratio by assuming there is a 1 mole sample present.

$$1 \text{ mol } C_9H_8O_4 \times \frac{9 \text{ mol } C}{1 \text{ mol } C_9H_8O_4} \times \frac{12.0 \text{ g } C}{1 \text{ mol } C} = 108 \text{ g } C$$

$$1 \text{ mol } C_9H_8O_4 \times \frac{8 \text{ mol } H}{1 \text{ mol } C_9H_8O_4} \times \frac{1.00 \text{ g } H}{1 \text{ mol } H} = 8.00 \text{ g } H$$

$$1 \text{ mol } C_9H_8O_4 \times \frac{4 \text{ mol } O}{1 \text{ mol } C_9H_8O_4} \times \frac{16.0 \text{ g } O}{1 \text{ mol } O} = 64.0 \text{ g } O$$

To determine the percent composition, divide the mass of each element present by the total mass of the compound and multiply by 100.

$$\%C = \frac{108 \text{ g } C}{180 \text{ g } C_9H_8O_4} \times 100\% = 60.0\%$$

$$\%H = \frac{8.00 \text{ g } H}{180 \text{ g } C_9H_8O_4} \times 100\% = 4.44\%$$

$$\%O = \frac{64.0 \text{ g } O}{180 \text{ g } C_9H_8O_4} \times 100\% = 35.6\%$$

11. Assuming a 100 g sample gives us 38.7 g C, 9.7 g H, and 51.6 g O. Convert these masses to number of moles.

$$38.7 \text{ g } C \times \frac{1 \text{ mol } C}{12.0 \text{ g } C} = 3.22 \text{ mol } C$$

$$9.7 \text{ g } H \times \frac{1 \text{ mol } H}{1.0 \text{ g } H} = 9.7 \text{ mol } H$$

$$51.6 \text{ g } O \times \frac{1 \text{ mol } O}{16.0 \text{ g } O} = 3.22 \text{ mol } O$$

Knowing the relative number of moles, find the ratio by dividing the larger number by the smaller number.

$\frac{9.7}{3.22} = 3$; The C:H:O ratio of 1:3:1 gives the empirical formula CH_3O.

To determine the molecular formula, you must find the multiplier.

$\frac{62 \text{ g/mol}}{31 \text{ g/mol}} = 2$; Therefore, the molecular formula is $C_2H_6O_2$.

12. Because you have been given information about both reactants in this problem, you must first check to see if one of the reactants is in excess. Begin by writing the balanced chemical equation for the reaction.

$$Ag_2CO_3 + 2 HNO_3 \rightarrow CO_2 + 2 AgNO_3 + H_2O$$

Next, determine the number of moles of both reactants.

$$9.85\,g\,Ag_2\,CO_3 \times \frac{1\,mol\,Ag_2\,CO_3}{275.8\,g\,Ag_2\,CO_3} = 0.0357\,mol\,Ag_2\,CO_3$$

$$0.250\,L\,HNO_3 \times \frac{0.250\,mol\,HNO_3}{1\,L} = 0.0625\,mol\,HNO_3$$

The Ag_2CO_3:HNO_3 mole ratio is 1:2; therefore you need 0.0714 moles of HNO_3 (2 × 0.0357). You only have 0.0625 moles of HNO_3, so it is the limiting reactant and will determine the number of grams of silver nitrate produced.

$$0.0625\,mol\,HNO_3 \times \frac{2\,mol\,AgNO_3}{2\,mol\,HNO_3} \times \frac{169.9\,g\,AgNO_3}{1\,mol\,AgNO_3} = 10.6\,g\,AgNO_3$$

13. Begin by finding the molar amounts of C and H in CO_2 and H_2O.

$$25.56\,g\,CO_2 \times \frac{1\,mol\,CO_2}{44.0\,g\,CO_2} \times \frac{1\,mol\,C}{1\,mol\,CO_2} = 0.581\,mol\,C$$

$$10.46\,g\,H_2O \times \frac{1\,mol\,H_2O}{18.0\,g\,H_2O} \times \frac{2\,mol\,H}{1\,mol\,H_2O} = 1.16\,mol\,H$$

Next, carry out mole-to-gram conversions to find the number of grams of C and H in the original sample.

$$0.581\,mol\,C \times \frac{12\,g\,C}{1\,mol\,C} = 6.97\,g\,C$$

$$1.16\,mol\,H \times \frac{1.0\,g\,H}{1\,mol\,H} = 1.16\,g\,H$$

Subtract the masses of C and H from the mass of the starting sample to determine the mass of O.

$$12.78\,g - 6.97\,g - 1.16\,g = 4.65\,g$$

Convert the mass of oxygen to moles of oxygen.

$$4.65\,g\,O \times \frac{1\,mol\,O}{16.0\,g\,O} = 0.291\,mol\,O$$

Find the ratio of the numbers of moles by dividing the larger numbers of moles by the smaller number of moles.

$$\frac{1.16}{0.291} = 4 \qquad\qquad \frac{0.581}{0.291} = 2$$

This gives a mole ratio of C:H:O of 2:4:1. Therefore, the empirical formula is C_2H_4O. To determine the molecular formula, you must determine the multiplier, $\frac{88.0\,g/mol}{44.0\,g/mol} = 2$; Therefore, the molecular formula for this compound is $C_4H_8O_2$.

14. The formula for calcium phosphite is $Ca_3(PO_3)_2$. The Ca:P:O mole ratio is 3:2:6. Begin by converting this mole ratio into a mass ratio by assuming 1 mole of sample is present.

$$1\,mol\,Ca_3(PO_3)_2 \times \frac{3\,mol\,Ca}{1\,mol\,Ca_3(PO_3)_2} \times \frac{40.1\,g\,Ca}{1\,mol\,Ca} = 120.3\,g\,Ca$$

$$1\,mol\,Ca_3(PO_3)_2 \times \frac{2\,mol\,P}{1\,mol\,Ca_3(PO_3)_2} \times \frac{31.0\,g\,P}{1\,mol\,P} = 62.0\,g\,P$$

$$1\,mol\,Ca_3(PO_3)_2 \times \frac{6\,mol\,O}{1\,mol\,Ca_3(PO_3)_2} \times \frac{16.0\,g\,O}{1\,mol\,O} = 96.0\,g\,O$$

The percent composition is determined by dividing the mass of each element present by the total mass of the compound and multiplying by 100%.

$$\%\,Ca = \frac{120.3\,g\,Ca}{278.3\,g\,Ca_3(PO_3)_2} \times 100\% = 43.2\%$$

$$\%\,P = \frac{62.0\,g\,P}{278.3\,g\,Ca_3(PO_3)_2} \times 100\% = 22.3\%$$

$$\%\,O = \frac{96.0\,g\,O}{278.3\,g\,Ca_3(PO_3)_2} \times 100\% = 34.5\%$$

15. Assuming a 100 g sample gives us 74.2 g C, 9.3 g H, and 16.5 g O. Convert these masses to numbers of moles.

$$74.2\,g\,C \times \frac{1\,mol\,C}{12\,g\,C} = 6.18\,mol\,C$$

$$9.3\,g\,H \times \frac{1\,mol\,H}{1.0\,g\,H} = 9.3\,mol\,H$$

$$16.5\,g\,O \times \frac{1\,mol\,O}{16.0\,g\,O} = 1.03\,mol\,O$$

Knowing the relative numbers of moles, find the ratio by dividing the two larger numbers by the smaller number.

$$\frac{6.18}{1.03} = 6 \qquad \frac{9.3}{1.03} = 9$$

The C:H:O ratio of 6:9:1 gives an empirical equation of C_6H_9O. To determine the molecular formula, find the multiplier.

$$\frac{194\,g/mol}{97\,g/mol} = 2\,;\ \text{Therefore, the molecular formula is } C_{12}H_{18}O_2.$$

CHAPTER 4

REACTIONS IN AQUEOUS SOLUTIONS

Chapter Learning Goals

1⊠ Classify substances as electrolytes or nonelectrolytes.

2⊠ Write molecular, ionic, and net ionic equations for precipitation, acid-base, and redox reactions.

3⊠ State solubility rules, and use them to predict whether a precipitate might form when aqueous salt solutions are mixed.

4⊠ Identify the common strong acids and strong bases.

5⊠ Assign oxidation numbers to each atom in a chemical species.

6⊠ In a redox reaction, identify the species oxidized, the species reduced, the oxidizing agent, and the reducing agent.

7⊠ Using an activity series, predict whether a redox reaction will occur when a metal is placed in contact with a solution containing an ion of a different metal.

8⊠ Balance redox reactions by the oxidation-number method and by the half-reaction method.

9⊠ Determine the concentration of a species using data from a redox titration.

Chapter in Brief

This chapter discusses three different types of chemical reactions in aqueous solution: precipitation reactions, acid-base reactions, and oxidation-reduction reactions. These reactions are often written as net ionic equations. As a consequence, it is necessary to know the difference between strong and weak electrolytes and nonelectrolytes.   You can predict the outcome of each of these reactions if you know the solubility rules, can recognize acids and bases, and know how to assign oxidation numbers to compounds. You are also shown how to balance oxidation-reduction reactions with either the oxidation number method or the method of half reactions. Finally, you will apply the concepts you learned in earlier chapters about stoichiometry to the reactions which are discussed in this chapter.

Some Ways that Chemical Reactions Occur

A. Reactions are driven from reactants to products by some energetic driving force that pushes them along.

B. Driving forces of reactions:

1. Precipitation Reactions - solid precipitate forms and drops out of solution.
Driving force = Removal of material from solution.

2. Acid-Base Neutralization - an acid reacts with a base to produce a salt and water.
Driving force = formation of water.

3. Oxidation-Reduction (Redox) Reaction - transfer of electrons between reactants.
Driving force = decrease in electrical potential.

1⊠ **Electrolytes in Aqueous Solution**

A. Electrolytes - Dissolve in water to produce ionic solutions.

1. Strong Electrolytes - completely dissociate in water.

2. Weak Electrolytes - incompletely dissociate in water.

a. establish an **equilibrium** between the forward and backward reactions

b. water is a weak electrolyte

B. Nonelectrolytes - do not produce ions in aqueous solution.

2⊠   **Aqueous Reactions and Net Ionic Equations**
   A. Molecular Equations - reactants and products are written using their full formulas as if they were molecules.
   B. Ionic Equations - any reactants or products which completely dissociate in water are shown in terms of their free ions.
   C. Spectator Ions - ions that do not undergo a change during the reaction
       1. The specific identity is not important.
       2. Balance the charge.
   D. Net Ionic Equation - the equation for the net change that takes place during the reaction; the spectator ions are not included.

EXAMPLE:
   When solutions of barium chloride and sodium phosphate are mixed, solid barium phosphate and a solution of NaCl are produced. Write molecular, ionic, and net ionic equations for this reaction.

SOLUTION:

   Molecular Eqn: $3\, BaCl_2\,(aq)\ +\ 2\, Na_3(PO_4)\,(aq) \rightarrow Ba_3(PO_4)_2\,(s)\ +\ 6\, NaCl\,(aq)$

   Ionic Eqn: $3\, Ba^{2+}\,(aq) + 6\, Cl^-\,(aq) + 6\, Na^+\,(aq) + 2\, PO_4^{3-}\,(aq) \rightarrow Ba_3(PO_4)_2\,(s) + 6\, Na^+\,(aq) + 6\, Cl^-\,(aq)$

   Spectator Ions: Na^+ and Cl^-

   Net Ionic Equation: $3\, Ba^{2+}\,(aq)\ +\ 2\, PO_4^{3-}\,(aq) \rightarrow Ba_3(PO_4)_2\,(s)$

3⊠   **Precipitation Reactions and Solubility Rules**
   A. Solubility - how much of each compound will dissolve in a given amount of solvent at a given temperature.
       1. Low solubility - precipitate forms.
       2. High solubility - no precipitate will form.
       3. Solubilities can be predicted using the guidelines found on page 121 in the text.
   B. Solubility Guidelines
       1. Use to predict if a precipitate will form in a reaction.
       2. Use to prepare and isolate a specific compound by carrying out a precipitation reaction.

EXAMPLE:
   Write molecular, ionic and net ionic equations for the reaction between:
   $Pb(NO_3)_2$ and $(NH_4)_2SO_4$; $SnCl_2$ and $NaOH$; $NaNO_3$ and $BaCl_2$.

SOLUTION: First, predict which products will be formed by exchanging the cations between the two salts. Then use the solubility guidelines to predict if any of the products are insoluble.
$Pb(NO_3)_2$ and $(NH_4)_2SO_4$:
Molecular:
$Pb(NO_3)_2\,(aq)\ +(NH_4)_2SO_4\,(aq)\ \rightarrow\ Pb(SO_4)\,(s)\ +\ 2\, NH_4NO_3\,(aq)$
Ionic:
$Pb^{2+}\,(aq) + 2\, NO_3^-\,(aq) + 2\, NH_4^+\,(aq) + SO_4^{2-}\,(aq)\ \rightarrow PbSO_4\,(s) + 2\, NH_4^+\,(aq) + 2\, NO_3^-\,(aq)$
Spectator ions: NH_4^+ and NO_3^-
Net ionic:
$Pb^{2+}\,(aq)\ +\ SO_4^{2-}\,(aq)\ \rightarrow\ PbSO_4\,(s)$

$SnCl_2$ and NaOH:

    Molecular:

    $SnCl_2$ (aq) + 2 NaOH(aq) $\rightarrow$ $Sn(OH)_2$ (s) + 2 NaCl (aq)

    Ionic:

    Sn^{2+} (aq) + 2 Cl^- (aq) + 2 Na^+ (aq) + 2 OH^- (aq) $\rightarrow$ $Sn(OH)_2$ (s) + 2 Na^+ (aq) + 2 Cl^- (aq)

    Spectator ions: Na^+ and Cl^-

    Net ionic:

    Sn^{2+} (aq) + 2 OH^- (aq) $\rightarrow$ $Sn(OH)_2$ (s)

$NaNO_3$ and $BaCl_2$:

    Molecular:

    2 $NaNO_3$ (aq) + $BaCl_2$ (aq) $\rightarrow$ $Ba(NO_3)_2$ (aq) + 2 NaCl (aq)

    Ionic:

    2 Na^+ (aq) + 2 NO_3^- (aq) + Ba^{2+} (aq) + 2 Cl^- (aq) $\rightarrow$ Ba^{2+} (aq) + 2 NO_3^- (aq) + 2 Na^+ (aq) + 2 Cl^- (aq)

    Spectator ions: all ions are spectator ions

    No net ionic reaction

Acids, Bases, and Neutralization Reactions

    A. Arrhenius Acid - dissociates in water to produce H^+.

        1. Hydronium Ion - H^+ attaches to a water molecule; represented as H_3O^+.

        2. Strong acids - strong electrolytes.

        3. Weak acids - weak electrolytes.

        4. Polyprotic acids - acids with more than one acidic hydrogen; dissociate in steps.

    B. Arrhenius Base - dissociates in water to produce OH^-.

        1. Strong bases - strong electrolytes; most metal hydroxides.

        2. Weak bases - weak electrolytes.

           a. NH_3 (g) + H_2O (l) $\rightleftarrows$ NH_4^+ (aq) + OH^- (aq)

    C. Neutralization Reaction – a reaction between an acid and base which produces a salt and water.

        1. Salt produced from cation of base and anion of acid.

        2. Net reaction for a strong acid + strong base: H^+ (aq) + OH^- (aq) $\rightarrow$ H_2O

        3. Net reaction for a weak acid + strong base: HA (aq) + OH^- (aq) $\rightarrow$ H_2O + A^- (aq)

    D. Common acid and bases – see Table 4.2 in text (page 124).

EXAMPLE:

    Determine if the following substances are acids or bases: HBr (aq), KOH, $HClO_3$, $Ca(OH)_2$, CH_3COOH, NH_3.

SOLUTION: To determine if a species is either an acid or base according to the Arrhenius definition, we must look for the ability of the substance to produce either H^+ or OH^- in aqueous solution. (You also can memorize Table 4.2 in your text, but we prefer you understand the concept.)

    HBr (aq) – produces H^+ in aqueous solution; acid

    KOH – produces OH^- in aqueous solution; base

    $HClO_3$ – produces H^+ in aqueous solution; acid

    $Ca(OH)_2$ – produces OH^- in aqueous solution; base

    CH_3COOH – produces H^+ in aqueous solution; acid

    NH_3 – produces OH^- in aqueous solution; base

EXAMPLE:
Write the molecular, ionic, and net ionic equations for the reaction between nitric acid and calcium hydroxide.

SOLUTION: Remember that the reaction between an acid and a base produces a salt and water. The cation of the salt comes from the base and the anion of the salt comes from the acid. In this example, the salt produced is $Ca(NO_3)_2$. Therefore the molecular equation for this reaction is:

Molecular:
$$2\ HNO_3\ +\ Ca(OH)_2\ (aq)\ \rightarrow\ Ca(NO_3)_2\ (aq)\ +\ 2\ H_2O\ (l)$$
Ionic:
$$2\ H^+\ (aq) + 2\ NO_3^-\ (aq) + Ca^{2+}\ (aq) + 2\ OH^-\ (aq)\ \rightarrow\ Ca^{2+}\ (aq) + 2\ NO_3^-\ (aq) + 2\ H_2O\ (l)$$
Spectator ions: Ca^{2+} and NO_3^-
Net ionic:
$$2\ H^+\ (aq) + 2\ OH^-\ (aq)\ \rightarrow\ 2\ H_2O\ (l)$$

6⊠   **Oxidation-Reduction (Redox) Reactions**
   A. Oxidation and Reduction
      1. Oxidation - the loss of one or more electrons; increase in oxidation number.
      2. Reduction - the gain of one or more electrons; decrease in oxidation number.
      3. Two processes occur simultaneously
   B. Redox Reaction - electrons are transferred from one substance to another.
      1. The number of electrons lost by the substance being oxidized = the number of electrons gained by the substance being reduced.
   C. Oxidation Number - a means of determining whether the atom is neutral, electron-rich, or electron-poor. Does not necessarily imply ionic charges.
      1. Rules for assigning oxidation number - see pages 127 and 128 of text.

EXAMPLE:
Determine the oxidation number of phosphorus in Na_3PO_4.

SOLUTION:
   1. Assigning an oxidation number of +1 to Na and -2 to the O.
   2. Remembering that the sum of all of the oxidation numbers in the compound = 0.

$Na_3PO_4 = (3\ Na^+)\ (P?)\ (4\ O^{2-})$ $3(+1) + (?) + 4(-2) = 0$ net charge
$$? = 0 - 3(+1) - 4(-2) = +5$$

6⊠   **Identifying Redox Reactions**
   A. Oxidation and reduction occur together.
      1. Whenever an atom loses one or more electrons (is oxidized) another atom must gain those electrons (be reduced).
   B. Reducing Agent - the substance that causes reduction to occur.
      1. Loses one or more electrons.
         a. undergoes oxidation
         b. oxidation number of atom increases
      2. Metals act as reducing agents.

    C. Oxidizing Agent - the substance that causes oxidation to occur.
        1. Gains one or more electrons.
            a. undergoes reduction
            b. oxidation number of atom decreases
        2. Reactive nonmetals act as oxidizing agents.

EXAMPLE:

Identify the species oxidized, the species reduced, the oxidizing agent, and reducing agent in the following reaction:

$$16\ H^+ (aq) + 5\ Sn^{2+} (aq) + 2\ MnO_4^- (aq) \rightarrow 2\ Mn^{2+} (aq) + 5\ Sn^{4+} (aq) + 8\ H_2O\ (l)$$

SOLUTION: To solve this problem, we begin by determining the oxidation number of all species present. The oxidation numbers for H and O stay the same (+1 and –2, respectively). The oxidation number for Sn changes from a +2 to a +4. This is an increase in oxidation number; therefore, Sn is oxidized and is the reducing agent. The oxidation of Mn in MnO_4^- is +7. Remember that the oxidation number of the elements in an ion must add up to the charge on the ion. If O is –2 and the charge on the ion is –1, then

oxidation number of Mn + 4 (-2) = -1; oxidation number of Mn = +7

The oxidation number of Mn changes from a +7 to a +2, resulting in a decrease in oxidation number. Mn^{7+} is the species reduced, and MnO_4^- is the oxidizing agent. (MnO_4^- is the oxidizing agent because Mn^{7+} is present in the form of MnO_4^-.)

The Activity Series

    A. Simple redox process - reaction of an aqueous cation with a free element to produce a different ion and a different element.
        1. Process is dependent upon the ease of oxidation for the element and the ease of reduction of the cation.
    B. Activity Series - ranks the elements in order of their reducing ability in aqueous solution.
        1. Any element higher in the activity series will react with the ion of any element lower in the activity series.
        2. Position of Hydrogen - indicates which metals will react with $H^+(aq)$ to produce H_2 (g).
        3. Most reactive metals - top of the activity series.
        4. Least reactive metals - bottom of the activity series.

EXAMPLE:

Using the activity series, write balanced chemical equations for the following reactions:

$Al\ (s)\ +\ Zn^{2+} (aq) \rightarrow$
$Ni\ (s)\ +\ Mn^{2+} (aq) \rightarrow$
$Ni\ (s)\ +\ HCl\ (aq) \rightarrow$
$Pt\ (s)\ +\ HBr\ (aq) \rightarrow$

SOLUTION: To predict the outcome of these reactions, you must remember that any element higher in the activity series will react with the ion of any element lower in the activity series. Also remember, metals above the H^+ ion in the activity series will displace the hydrogen ion from an acid to form H_2 gas.

$$2 \, Al \, (s) \, + \, 3 \, Zn^{2+} \, (aq) \, \rightarrow \, 2 \, Al^{3+} \, (aq) \, + \, 3 \, Zn \, (s)$$
$$Ni \, (s) \, + \, Mn^{2+} \, (aq) \, \rightarrow \, \text{no reaction;} \; Mn^{2+} \, \text{lies above Ni in the activity series}$$
$$Ni \, (s) \, + \, 2 \, HCl \, (aq) \, \rightarrow \, NiCl_2 \, (aq) \, + \, H_2 \, (g)$$
$$Pt \, (s) \, + \, HBr \, (aq) \, \rightarrow \, \text{no reaction}$$

8⊠ Balancing Redox Reactions by the Oxidation-Number Method

  A. Oxidation number method for balancing redox reactions focuses on the chemical changes involved.

  B. Key - Net change in the total of all oxidation numbers must be zero.

   1. Any increase in oxidation number for the oxidized atoms must be matched by a corresponding decrease in oxidation number for the reduced atoms.

  C. Steps - see page 135 of the text.

EXAMPLE:

Using the oxidation-number method, balance the following reaction which takes place in acidic solution:

$$MnO_4^- \, (aq) \, + \, SO_2 \, (aq) \rightarrow Mn^{2+} \, (aq) \, + \, HSO_4^- \, (aq)$$

SOLUTION:

  1. Unbalanced ionic equation:

$$MnO_4^- \, + \, SO_2 \; \rightarrow Mn^{2+} \, + HSO_4^-$$

  2. Balance all atoms other than hydrogen and oxygen. Same as unbalanced ionic equation.

  3. Assign oxidation numbers to all atoms.

$$MnO_4^- \, + \, SO_2 \; \rightarrow Mn^{2+} \, + HSO_4^-$$

   +7 -2 +4 -2 +2 +1 +6 -2

  4. Atoms which have changed oxidation number:
   Mn: +7 → +2 (gained 5 e⁻); S: +4 → +6 (lost 2 e⁻)

  5. Net increase in oxidation number of oxidized atoms = 2; Net decrease in oxidation number of reduced atoms = 5. Multiply the net increase by 5 and the net decrease by 2.

$$2 \, MnO_4^- \, + 5 \, SO_2 \; \rightarrow \, 2 \, Mn^{2+} \, + \, 5HSO_4^-$$

  6. Reactant side of the equation has two less oxygens, so add 2 H_2O's.

$$2 \, MnO_4^- \, + \, 5 \, SO_2 \, + \, 2 \, H_2O \; \rightarrow \, 2 \, Mn^{2+} \, + \, 5 \, HSO_4^-$$

  7. Reactant side of the equation has one less hydrogen, so add 1 H^+.

$$H^+ \, (aq) \, + \, 2 \, MnO_4^- \, (aq) \, + 5 \, SO_2 \, (aq) \, + \, 2 \, H_2O \, (l) \rightarrow \, 2 \, Mn^{2+} \, (aq) \, + \, 5 \, HSO_4^- \, (aq)$$

  8. Check your answer to make sure all atoms and charges are balanced.

EXAMPLE:

Using the oxidation-number method, balance the following reaction which takes place in basic solution:

$$SO_3^{2-} (aq) + CrO_4^{2-} (aq) \rightarrow SO_4^{2-} (aq) + Cr(OH)_3 (s)$$

SOLUTION:

1. Unbalanced ionic equation:

$$SO_3^{2-} (aq) + CrO_4^{2-} (aq) \rightarrow SO_4^{2-} (aq) + Cr(OH)_3 (s)$$

2. Balance all atoms other than hydrogen and oxygen. Same as unbalanced equation.

3. Assign oxidation numbers to all atoms.

$$\begin{array}{cccc} SO_3^{2-} & + CrO_4^{2-} & \rightarrow SO_4^{2-} & + Cr(OH)_3 \\ \uparrow\uparrow & \uparrow\uparrow & \uparrow\uparrow & \uparrow\uparrow\uparrow \\ +4\,-2 & +6\,-2 & +6\,-2 & +3\,-2\,+1 \end{array}$$

4. Atoms which have changed oxidation number:
   S: $+4 \rightarrow +6$ (lost 2 e^-); Cr: $+6 \rightarrow +3$ (gained 3 e^-)

5. Net increase in oxidation number of oxidized atoms = 2; Net decrease in oxidation number of reduced atoms = 3. Multiply the net increase by 3 and the net decrease by 2.

$$3\,SO_3^{2-} + 2\,CrO_4^{2-} \rightarrow 3\,SO_4^{2-} + 2\,Cr(OH)_3$$

6. Reactant side of the equation has one less oxygen, so add 1 H_2O.

$$H_2O + 3\,SO_3^{2-} + 2\,CrO_4^{2-} \rightarrow 3\,SO_4^{2-} + 2\,Cr(OH)_3$$

7. Reactant side of the equation has four less hydrogens, so add 4 H^+.

$$4\,H^+ + H_2O + 3\,SO_3^{2-} + 2\,CrO_4^{2-} \rightarrow 3\,SO_4^{2-} + 2\,Cr(OH)_3$$

8. Since the reaction takes place in basic solution, we need to add 1 OH^- for every H^+ to both sides of the equation.

$$4\,OH^- + 4\,H^+ + H_2O + 3\,SO_3^{2-} + 2\,CrO_4^{2-} \rightarrow 3\,SO_4^{2-} + 2\,Cr(OH)_3 + 4\,OH^-$$

9. The OH^- and H^+ on the reactant side of the equation can be combined to form H_2O.

$$4\,H_2O + H_2O + 3\,SO_3^{2-} + 2\,CrO_4^{2-} \rightarrow 3\,SO_4^{2-} + 2\,Cr(OH)_3 + 4\,OH^-$$

Combining the water molecules on the reactant side gives

$$5\,H_2O + 3\,SO_3^{2-} + 2\,CrO_4^{2-} \rightarrow 3\,SO_4^{2-} + 2\,Cr(OH)_3 + 4\,OH^-$$

10. Check your answer making sure atoms and charge are balanced.

8⊠   **Balancing Redox Reactions by the Half-Reaction Method**

   A. Half-reaction method of balancing redox reactions focuses on the transfer of electrons.
   B. Key – overall reaction can be broken into two parts or half reactions.
      1. Oxidation part of the reaction.
      2. Reduction part of the reaction.
   B. Steps - see page 139 of the text.

   EXAMPLE:
   Using the half-reaction method balance the following reaction which takes place in acidic medium:

   $$Cr_2O_7^{2-}(aq) + HC_2O_4^-(aq) \rightarrow Cr^{3+}(aq) + CO_2(g)$$

   SOLUTION:
   1. Unbalanced ionic equation: $Cr_2O_7^{2-} + HC_2O_4^- \longrightarrow Cr^{3+} + CO_2$

   2. Two unbalanced half reactions: Chromium is reduced from +6 to +3; C is oxidized from +3 to +4

      $$Cr_2O_7^{2-} \rightarrow Cr^{3+}$$

      $$HC_2O_4^- \rightarrow CO_2$$

   3. Balance each half reaction for atoms other than H and O.

      $$Cr_2O_7^{2-} \rightarrow 2\,Cr^{3+}$$

      $$HC_2O_4^- \rightarrow 2\,CO_2$$

   4. Add H_2O for any oxygens that are needed and H^+ for any hydrogens that are needed.

      $$14\,H^+ + Cr_2O_7^{2-} \rightarrow 2\,Cr^{3+} + 7\,H_2O$$

      $$HC_2O_4^- \rightarrow 2\,CO_2 + H^+$$

   5. Balance each reaction for charge:

      $$6\,e^- + 14\,H^+ + Cr_2O_7^{2-} \rightarrow 2\,Cr^{3+} + 7\,H_2O$$

      $$HC_2O_4^- \rightarrow 2\,CO_2 + H^+ + 2\,e^-$$

   6. Make the electron count the same in both reactions.

      $$6e^- + 14\,H^+ + Cr_2O_7^{2-} \rightarrow 2\,Cr^{3+} + 7\,H_2O$$

      $$3\,HC_2O_4^- \rightarrow 6\,CO_2 + 3\,H^+ + 6\,e^-$$

   7. Add the two half reactions together, canceling anything that appears on both sides of the equation.

$$11 \text{ H}^+ (aq) + \text{Cr}_2\text{O}_7^{2-} (aq) + 3 \text{ HC}_2\text{O}_4^- (aq) \rightarrow 2 \text{ Cr}^{3+} (aq) + 7 \text{ H}_2\text{O} (l) + 6 \text{ CO}_2 (g)$$

8. Check to make sure all atoms and charges are balanced.

EXAMPLE:

Using the half-reaction method, balance the following reaction which takes place in basic solution:

$$\text{CN}^- (aq) + \text{AsO}_4^{3-} (aq) \rightarrow \text{AsO}_2^- (aq) + \text{CNO}^- (aq)$$

SOLUTION:

1. Unbalanced ionic equation: $\text{CN}^- + \text{AsO}_4^{3-} \rightarrow \text{AsO}_2^- + \text{CNO}^-$

2. Two unbalanced half reactions: arsenate is being reduced, while cyanide is being oxidized.

$$\text{CN}^- \rightarrow \text{CNO}^-$$

$$\text{AsO}_4^{3-} \rightarrow \text{AsO}_2^-$$

3. Balance each half reaction for atoms other than H and O.

same as above

4. Add H_2O for any oxygens that are needed and H^+ for any hydrogens that are needed.

$$\text{H}_2\text{O} + \text{CN}^- \rightarrow \text{CNO}^- + 2 \text{ H}^+$$

$$4 \text{ H}^+ + \text{AsO}_4^{3-} \rightarrow \text{AsO}_2^- + 2\text{H}_2\text{O}$$

5. Balance each reaction for charge:

$$\text{H}_2\text{O} + \text{CN}^- \rightarrow \text{CNO}^- + 2 \text{ H}^+ + 2 \text{ e}^-$$

$$2 \text{ e}^- + 4 \text{ H}^+ + \text{AsO}_4^{3-} \rightarrow \text{AsO}_2^- + 2 \text{ H}_2\text{O}$$

6. Make the electron count the same in both reactions.

The oxidation process involves loss of two electrons, and the reduction process involves the gain of two electrons. Therefore, the electron count is the same in both reactions.

7. Add the two half reactions together, canceling anything that appears on both sides of the equation.

$$2 \text{ H}^+ + \text{AsO}_4^{3-} + \text{CN}^- \rightarrow \text{CNO}^- + \text{AsO}_2^- + \text{H}_2\text{O}$$

8. Since the reaction takes place in basic solution, we need to add 1 OH$^-$ for every H$^+$ to both sides of the equation.

$$2 \text{ OH}^- + 2 \text{ H}^+ + \text{AsO}_4^{3-} + \text{CN}^- \rightarrow \text{CNO}^- + \text{AsO}_2^- + \text{H}_2\text{O} + 2 \text{ OH}^-$$

9. The OH$^-$ and H$^+$ on the reactant side of the equation can be combined to form H_2O.

$$2 H_2O + AsO_4^{3-} + CN^- \rightarrow CNO^- + AsO_2^- + H_2O + 2 OH^-$$

Canceling the water on the product side gives
$$H_2O + AsO_4^{3-} + CN^- \rightarrow CNO^- + AsO_2^- + 2 OH^-$$

10. Check your answer to make sure both atoms and charge are balanced.

9⊠   **Redox Titrations**
A. Redox titration - a method for determining the concentration of an oxidizing or reducing agent in solution.
   1. Unknown must react in a 100% yield.
   2. Color change should signal the end of the reaction.
B. Strategy for redox titrations (see figure 4.2, page 141 in your text for the titration of $KMnO_4$ with $H_2C_2O_4$).
   1. Measure a known amount of one substance
   2. Using the mole to mole ratio of the balanced equation, determine the number of moles of the second substrate.
   3. Determine the volume of solution containing the molar amount of the second substance (from the titration).
   4. Determine the molar concentration of the second substance.

EXAMPLE:
When the NaOCl found in a dilute solution of bleach is reacted with I^-, the soluble I_3^- ion is produced. The net ionic equation for this reaction is:

$$3 I^- + OCl^- + 2 H^+ \rightarrow I_3^- + Cl^- + H_2O$$

A bleach sample with a mass of 1.500 g was reacted with excess I^-. The resulting solution was titrated with 0.0500 M $S_2O_3^{2-}$ (thiosulfate ion) using starch as an indicator. The titration required 42.32 mL of the $S_2O_3^{2-}$ solution. The reaction for the titration is:

$$I_3^- + 2 S_2O_3^{2-} \rightarrow 3 I^- + S_4O_6^{2-}$$

How many moles of I_3^- reacted? How many moles of OCl^- reacted? What is the weight percent of NaOCl in the bleach?

SOLUTION: To determine the number of moles of I_3^- that reacted with the thiosulfate ion, we must first calculate the number of moles of thiosulfate used in the titration.

$$\text{mol } S_2O_3^{2-} = 0.0500 \frac{\text{mol } S_2O_3^{2-}}{\text{L soln}} \times 0.04232 \text{ L soln} = 2.12 \times 10^{-3} \text{ mol } S_2O_3^{2-}$$

We know from the balanced equation that 2 moles of $S_2O_3^{2-}$ react with 1 mole of I_3^-. Therefore:

$$\text{mol } I_3^- = 2.12 \times 10^{-3} \text{ mol } S_2O_3^{2-} \times \frac{1 \text{ mol } I_3^-}{2 \text{ mol } S_2O_3^{2-}} = 1.06 \times 10^{-3} \text{ mol } I_3^-$$

From the first chemical reaction given, we know that the mole to mole ratio between I_3^- and OCl^- is 1:1. Therefore, 1.06×10^{-3} moles of OCl^- reacted. To determine the weight percent of NaOCl in the sample, we need to determine the mass of NaOCl present.

$$\text{mass NaOCl} = 1.06 \times 10^{-3} \text{ mol OCl}^- \times \frac{1 \text{ mol NaOCl}}{1 \text{ mol OCl}^-} \times \frac{106.5 \text{ g NaOCl}}{1 \text{ mol NaOCl}} = 0.113 \text{ g NaOCl}$$

We can now determine weight percent:

$$\% \text{ weight} = \frac{0.113 \text{ g NaOCl}}{1.500 \text{ g sample}} \times 100 = 7.53\% \text{ NaOCl}$$

Self Test

This section is intended to test your knowledge of the material covered in this chapter. Think through these problems and make certain you understand what is going on. Ask yourself if your answer makes sense. Many of these questions are linked to the chapter learning goals. Therefore, successful completion of these problems indicates you have mastered the learning goals for this chapter. You will receive the greatest benefit from this section if you use it as a mock exam. You will then discover which topics you have mastered and which topics you need to study in more detail.

True/False

1. $CaCl_2$ is a strong electrolyte.

2. H_3PO_4 is a strong acid.

3. The net ionic equation for the reaction between NaCl and $AgNO_3$ is $Ag^+(aq) + Cl^-(aq) \rightarrow AgCl\ (s)$.

4. In the reaction between NaCl and $Ba(OH)_2$, $BaCl_2$ will precipitate.

5. The driving force for an acid-base neutralization reaction is the formation of the salt.

6. Reduction is the gain of one or more electrons.

7. The reducing agent is the species being reduced.

8. The oxidation number of Br in $NaBrO_3$ is -1.

9. Ca can react with Zn^{2+} to produce Ca^{2+} and Zn.

10. Hg will react with hydrochloric acid to produce H_2 gas.

Multiple Choice

1. Strong bases are:
   a. nonelectrolytes
   b. weak electrolytes
   c. precipitates
   d. strong electrolytes

2. The insoluble chlorides contain:
   a. an alkali metal
   b. NH_4^+
   c. Ba^{2+}
   d. Ag^+, Hg_2^{2+}, or Pb^{2+}

3. Insoluble nitrates contain:
   a. alkali metals
   b. Ba^{2+}
   c. ammonium
   d. none of the above

4. Soluble hydroxide salts contain:
   a. alkali metals
   b. Fe^{3+}
   c. Ag^+
   d. none of the above

5. The weak acid is:
   a. HF(aq)
   b. HCl(aq)
   c. HBr(aq)
   d. HI(aq)

6. The oxidation number of S in K_2SO_3 is:
   a. -2
   b. +4
   c. +6
   d. 0

7. An atom in an uncombined element has an oxidation number of:
   a. the charge of its ion
   b. the group number
   c. 0
   d. none of the above

8. For the reaction $Cr_2O_7^{2-}(aq) + Cl^-(aq) \longrightarrow Cr^{3+}(aq) + Cl_2(g)$, the species being reduced is:
   a. $Cr_2O_7^{2-}$
   b. Cl^-
   c. Cr^{3+}
   d. Cl_2

9. The reducing agent in the reaction $Fe^{2+} + MnO_4^- \rightarrow Fe^{3+} + Mn^{2+}$ is:
   a. Fe^{2+}
   b. MnO_4^-
   c. Fe^{3+}
   d. Mn^{2+}

10. Nickel metal will:
    a. react with K^+ to produce $K(s)$.
    b. react with cold water to produce H_2 gas.
    c. react with Cr^{3+} to produce chromium.
    d. react with $HCl(aq)$ to produce H_2 gas.

Matching

Electrolyte	a. how much of each compound will dissolve in a given amount of solvent at a given temperature.
Strong electrolyte	b. produces H^+ ions in water.
Spectator ions	c. causes reduction to occur.
Solubility	d. dissociates in water to produce an ionic solution.
Oxidation	e. produces OH^- ions in water.
Reduction	f. completely dissociates in water.
Oxidizing Agent	g. loss of electrons.
Reducing Agent	h. ions that do not undergo a change during a reaction.
Acid	i. causes oxidation to occur.
Base	j. gain of electrons.

Fill-in-the-Blank

1. The _____ ranks the element in order of their reducing ability in aqueous solution.

2. When an acid is mixed with a base the reaction is referred to as a

   _____ and the products are

   _____.

3. The concentration of an oxidizing agent or reducing agent in solution can be determined by a

   _____.

4. _____ are processes in which one or more electrons are

   transferred between reaction partners.

5. The _____ provides an indication of whether an atom is neutral, electron-rich or electron-poor.

6. Substances that incompletely dissociate are _____.

7. The _____ method of balancing half-reactions focuses on the

chemical change that occurs.

8. In a redox reaction, the oxidation process and the reduction process can be written as two

_____.

9. The _____ method of balancing redox reactions focuses on the

transfer of electrons.

10. Acids dissociate in aqueous solutions to yield an anion and a _____.

Problems

Write balanced molecular, ionic, and net ionic equations for the reactions between:
1. sodium chloride and lead(II) nitrate
2. perchloric acid and potassium hydroxide
3. potassium carbonate and calcium chloride

Determine the oxidation number for each atom in:
4. hydrogen carbonate ion
5. magnesium sulfate
6. sulfur hexafluoride

For each unbalanced equation given below, identify the species oxidized and the species reduced; identify the oxidizing agent and the reducing agent.

7. $Mg\ (s) + O_2\ (g) \rightarrow MgO\ (s)$

8. $Cr_2O_7^{2-}\ (aq) + Sn^{2+}\ (aq) \rightarrow Cr^{3+}\ (aq) + Sn^{4+}\ (aq)$

9. $FeS\ (s) + NO_3^-\ (aq) \rightarrow NO\ (g) + SO_4^{2-}\ (aq) + Fe^{2+}\ (aq)$

10. $C_2H_4\ (g) + O_2\ (g) \rightarrow CO_2\ (g) + H_2O\ (g)$

Using the activity series, predict the outcome of the following reactions.
11. sodium and cold water
12. iron and steam
13. magnesium and cold water
14. copper and zinc sulfate

Balance the following equations using the oxidation number method.
15. $Sn^{2+} + MnO_4^- \rightarrow Sn^{4+} + Mn^{2+}$ (acidic solution)

16. $S_2O_3^{2-} + Cl_2 \rightarrow SO_4^{2-} + Cl^-$ (acidic solution)

17. $MnO_4^- + C_2O_4^{2-} \rightarrow MnO_2 + CO_3^{2-}$ (basic solution)

Balance the following equations using the method of half reactions.
18. $HSO_3^- + Cr_2O_7^{2-} \rightarrow SO_4^{2-} + Cr^{3+}$ (acidic solution)

19. $Pb(OH)_3^- + OCl^- \rightarrow PbO_2 + Cl^-$ (basic solution)

20. $Br_2 \rightarrow Br^- + BrO_3^-$ (basic solution)

21. Using the balanced equation obtained in number 18 above, calculate the molarity of a $K_2Cr_2O_7$ solution if 28.42 mL of the solution reacts completely with a 25.00 mL solution of 0.3143 M Na_2SO_3.

22. A drunk driver is defined as one who drives with a blood alcohol level of 0.1% by mass or higher. The level of alcohol can be determined by titrating blood plasma with potassium dichromate according to the unbalanced equation:

$$Cr_2O_7^{2-} + C_2H_5OH \rightarrow Cr^{3+} + CO_2$$

Balance the above reaction (which takes place in acidic solution). Assuming that the only substance that reacts with dichromate in blood plasma is alcohol, is a person legally drunk if 6.522 mL of 5.000×10^{-3} M potassium dichromate is required to titrate a 0.50 gram sample of blood plasma?

23. A general chemistry student combines 7.5 g of sodium sulfide with 6.8 g of zinc nitrate.
    a. Write the molecular, ionic, and net ionic equations for this reaction.
    b. How much of the solid will be produced?
    c. How much of the excess reactant will be leftover?
    d. If the experimental yield is 2.53 g, what will the percent yield be?

24. The next week, the same general chemistry student reacted 4.8 g of Mg with 250 mL of 0.75 M copper (II) sulfate.
    a. Write the molecular, ionic, and net ionic equations for this reaction.
    b. How much of the solid will be produced?
    c. How much of the excess reactant will be leftover?
    d. If the experimental yield is 9.2 g, what will the percent yield be?

25. The copper and zinc found in brass can be dissolved by treating a sample of the alloy with nitric acid. The resulting Cu^{2+} ion will react with a solution of potassium iodide by the following equation:

$$2 \, Cu^{2+} \, (aq) + 5 \, I^- \, (aq) \rightarrow 2 \, CuI \, (s) + I_3^- \, (aq)$$

Determine the weight percent of copper in a 0.500 g brass sample, if 20.80 mL of 0.2500 M $Na_2S_2O_3$ is required to titrate the I_3^- produced in the above reaction. (The reaction between I_3^- and $S_2O_3^{2-}$ is found in the example on p. 69)

Solutions - Self Test

True/False
1. T
2. F. H_3PO_4 incompletely dissociates in water and is therefore a weak acid.
3. T
4. F. $BaCl_2$ is a soluble salt: there is no reaction.
5. F. The driving force for an acid-base neutralization reaction is the formation of water.
6. T
7. F. The reducing agent is the species which causes reduction to occur and is therefore oxidized.
8. F. The oxidation number of Br in $NaBrO_3$ is +5. (Na = +1, 3O = -6: +1 + Br + (-6) = 0; Br = +5).
9. T

10. F. Hg is below H_2 in the activity series.

Multiple Choice
1. d
2. d
3. d
4. a
5. a
6. b
7. c
8. a
9. a
10. d

Matching
Electrolyte - d
Strong electrolyte - f
Spectator ions - h
Solubility - a
Oxidation - g
Reduction - j
Oxidizing agent - i
Reducing agent - c
Acid - b
Base - e

Fill-in-the-Blank
1. Activity series
2. Neutralization reaction; salt and water
3. Redox titration
4. Redox reactions
5. Oxidation number
6. Weak electrolytes
7. Oxidation number
8. Half reactions
9. Half-reaction
10. H^+ or H_3O^+ ion

Problems

1. $2\,NaCl\,(aq)\,+\,Pb(NO_3)_2\,(aq) \rightarrow PbCl_2\,(s)\,+\,2\,NaNO_3\,(aq)$ Molecular Eqn.

$2\,Na^+\,(aq)\,+\,2\,Cl^-\,(aq)\,+\,Pb^{2+}\,(aq)\,+\,2\,NO_3^-\,(aq) \rightarrow PbCl_2\,(s)\,+\,2\,Na^+\,(aq)\,+\,2\,NO_3^-\,(aq)$ Ionic Eqn.

$2\,Cl^-\,(aq)\,+\,Pb^{2+}\,(aq) \rightarrow PbCl_2\,(s)$ Net Ionic Eqn.

2. $HClO_4\,(aq)\,+\,KOH\,(aq) \rightarrow KClO_4(aq)\,+\,H_2O\,(l)$ Molecular Eqn.

$H^+\,(aq)\,+\,ClO_4^-\,(aq)\,+\,K^+\,(aq)\,+\,OH^-(aq) \rightarrow K^+\,(aq)\,+\,ClO_4^-\,(aq)\,+\,H_2O\,(l)$ Ionic Eqn.

$H^+\,(aq)\,+\,OH^-\,(aq) \rightarrow H_2O\,(l)$ Net Ionic Eqn.

3. $K_2CO_3(aq) + CaCl_2(aq) \rightarrow 2\,KCl(aq) + CaCO_3(s)$ Molecular Eqn.

   $2\,K^+(aq) + CO_3^{2-}(aq) + Ca^{2+}(aq) + 2\,Cl^-(aq) \rightarrow 2\,K^+(aq) + 2\,Cl^-(aq) + CaCO_3(s)$ Ionic Eqn.

   $CO_3^{2-}(aq) + Ca^{2+}(aq) \rightarrow CaCO_3(s)$ Net Ionic Eqn.

4. HCO_3^-: $(H^+)(C^?)(O^{2-})$ $1(+1) + (?) + 3(-2) = -1$ net charge
                                                   $? = -1 - (+1) - 3(-2) = +4$

5. $MgSO_4$: $(Mg^{2+})(S^?)(O^{2-})$ $(+2) + (?) + 4(-2) = 0$ net charge
                                                   $? = 0 - (+2) - 4(-2) = +6$

6. SF_6: $(S^?)(F^-)$ $(?) + 6(-1) = 0$ net charge
                                        $? = 0 - 6(-1) = +6$

7. $Mg \rightarrow Mg^{2+}$ oxidized, reducing agent; $O_2 \rightarrow O^{2-}$ reduced, oxidizing agent

8. $Cr^{6+}(Cr_2O_7^{2-}) \rightarrow Cr^{3+}$ reduced, oxidizing agent; $Sn^{2+} \rightarrow Sn^{4+}$ oxidized, reducing agent

9. $S^{2-}(FeS) \rightarrow S^{6+}(SO_4^{2-})$ oxidized, reducing agent; $N^{5+}(NO_3^-) \rightarrow N^{2+}(NO)$ reduced, oxidizing agent

10. $C^{2-}(C_2H_4) \rightarrow C^{4+}(CO_2)$ oxidized, reducing agent; $O_2 \rightarrow O^{2-}(CO_2, H_2O)$ reduced, oxidizing agent

11. $Na(s) + H_2O \rightarrow NaOH(aq) + H_2(g)$

12. $Fe(s) + H_2O(g) \rightarrow Fe(OH)_2(s) + H_2(g)$

13. $Mg(s) + H_2O \rightarrow$ no reaction

14. $Cu(s) + Zn(SO_4)(aq) \rightarrow$ no reaction

15. $\underset{\substack{\uparrow \\ +2}}{Sn^{2+}} + \underset{\substack{\uparrow\ \uparrow \\ +7\ -2}}{MnO_4^-} \longrightarrow \underset{\substack{\uparrow \\ +4}}{Sn^{4+}} + \underset{\substack{\uparrow \\ +2}}{Mn^{2+}}$

Sn: $+2 \rightarrow +4$; Mn: $+7 \rightarrow +2$ net increase in oxidation number of oxidized atoms $= 2$; net decrease in oxidation number of reduced atoms $= 5$. Multiply net increase by 5 and the net decrease by 2.

$5\,Sn^{2+} + 2\,MnO_4^- \rightarrow 5\,Sn^{4+} + 2\,Mn^{2+}$

Product side of the equation has eight less oxygens, so add 8 H_2O; Reactant side has 16 less hydrogens, so add 16 H^+.

$5\,Sn^{2+} + 2\,MnO_4^- + 16\,H^+ \rightarrow 5\,Sn^{4+} + 2\,Mn^{2+} + 8\,H_2O$

76

16. $S_2O_3^{2-} + Cl_2 \rightarrow SO_4^{2-} + Cl^-$

$$S_2O_3^{2-} + Cl_2 \rightarrow 2\,SO_4^{2-} + 2\,Cl^-$$
$$\uparrow\uparrow \qquad \uparrow \qquad\quad \uparrow\uparrow \qquad\quad \uparrow$$
$$+2\,-2 \qquad 0 \qquad\quad +6\,-2 \qquad\quad -1$$

S: $+2 \rightarrow +6$; Cl: $0 \rightarrow -1$; net increase in oxidation number of oxidized atoms $= 8$; net decrease in oxidation number of reduced atoms $= 2$. Multiply net increase by 1 and net decrease by 4.

$$S_2O_3^{2-} + 4\,Cl_2 \rightarrow 2\,SO_4^{2-} + 8\,Cl^-$$

Reactant side of the equation has five less oxygens, so add 5 H_2O. Product side of the equation has 10 less hydrogens, so add 10 H^+.

$$5\,H_2O + S_2O_3^{2-} + 4\,Cl_2 \rightarrow 2\,SO_4^{2-} + 8\,Cl^- + 10\,H^+$$

17. $MnO_4^- + C_2O_4^{2-} \rightarrow MnO_2 + CO_3^{2-}$

Remember to first balance all atoms other than hydrogen and oxygen and assign oxidation numbers.

$$MnO_4^- + C_2O_4^{2-} \rightarrow MnO_2 + 2\,CO_3^{2-}$$
$$\uparrow\uparrow \qquad \uparrow\uparrow \qquad\quad \uparrow\ \uparrow \qquad\quad \uparrow\uparrow$$
$$+7\,-2 \qquad +3\,-2 \qquad\ +4\,-2 \qquad\ +4\,-2$$

Mn: $+7 \rightarrow +4$; C: $+3 \rightarrow +4$; net increase in oxidation number of oxidized atoms $= 2$ (each carbon atom has a change of $+1$ in oxidation number; however, there are two carbon atoms so the net increase $= +2$); net decrease in oxidation number of reduced atoms $= 3$. Multiply net increase by 3 and net decrease by 2.

$$2\,MnO_4^- + 3\,C_2O_4^{2-} \rightarrow 2\,MnO_2 + 6\,CO_3^{2-}$$

The reactant side has two less oxygens, so add 2 H_2O. Product side of the equation has four less H^+ so add 4 H^+.

$$2\,H_2O + 2\,MnO_4^- + 3\,C_2O_4^{2-} \rightarrow 2\,MnO_2 + 6\,CO_3^{2-} + 4\,H^+$$

Add 4 OH^- to both sides of the equation to neutralize the H^+.

$$4\,OH^- + 2\,H_2O + 2\,MnO_4^- + 3\,C_2O_4^{2-} \rightarrow 2\,MnO_2 + 6\,CO_3^{2-} + 4\,H^+ + 4\,OH^-$$

Combine the hydrogens and oxygens on the product side to make 4 H_2O and cancel the two water molecules from the reactant side.

$$4\,OH^- + 2\,MnO_4^- + 3\,C_2O_4^{2-} \rightarrow 2\,MnO_2 + 6\,CO_3^{2-} + 2\,H_2O$$

18. Sulfur is oxidized and chromium is reduced, so the two half reactions are:

$$HSO_3^- \rightarrow SO_4^{2-}$$

$$Cr_2O_7^{2-} \rightarrow Cr^{3+}$$

Balance for elements other than O and H.

$$HSO_3^- \rightarrow SO_4^{2-}$$

$$Cr_2O_7^{2-} \rightarrow 2\,Cr^{3+}$$

Add H_2O for any oxygens that are needed and H^+ for any hydrogens that are needed.

$$HSO_3^- + H_2O \rightarrow SO_4^{2-} + 3\,H^+$$

$$Cr_2O_7^{2-} + 14\,H^+ \rightarrow 2\,Cr^{3+} + 7\,H_2O$$

Balance each reaction for charge.

$$HSO_3^- + H_2O \rightarrow SO_4^{2-} + 3\,H^+ + 2\,e^-$$

$$6\,e^- + Cr_2O_7^{2-} + 14\,H^+ \rightarrow 2\,Cr^{3+} + 7\,H_2O$$

Make the electron count the same in both reactions.

$$3(HSO_3^- + H_2O \rightarrow SO_4^{2-} + 3\,H^+ + 2\,e^-)$$

$$6\,e^- + Cr_2O_7^{2-} + 14\,H^+ \rightarrow 2\,Cr^{3+} + 7\,H_2O$$

Add the two half reactions together, canceling anything that appears on both sides of the equation.

$$3\,HSO_3^- + Cr_2O_7^{2-} + 5\,H^+ \rightarrow 3\,SO_4^{2-} + 2\,Cr^{3+} + 4\,H_2O$$

19. Lead is oxidized and Cl is reduced, so the two half reactions are:

$$Pb(OH)_3^- \rightarrow PbO_2$$

$$OCl^- \rightarrow Cl^-$$

Atoms other than O and H are already balanced. Add H_2O for any oxygens that are needed and H^+ for any hydrogens that are needed.

$$Pb(OH)_3^- \rightarrow PbO_2 + H_2O + H^+$$

$$2\,H^+ + OCl^- \rightarrow Cl^- + H_2O$$

Balance each reaction for charge.

$$Pb(OH)_3^- \rightarrow PbO_2 + H_2O + H^+ + 2\,e^-$$

$$2\,e^- + 2\,H^+ + OCl^- \rightarrow Cl^- + H_2O$$

Electron count is the same in both reactions. Add the two half reactions, canceling anything that appears on both sides of the equation.

$$Pb(OH)_3^- + H^+ + OCl^- \rightarrow PbO_2 + 2\,H_2O + Cl^-$$

Add one OH^- to both sides of the equation to neutralize the H^+ on the reactant side.

$$Pb(OH)_3^- + H_2O + OCl^- \rightarrow PbO_2 + 2\,H_2O + Cl^- + OH^-$$

Notice that water is on both the reactant and product side of the equation. The final net reaction is:

$$Pb(OH)_3^- + OCl^- \rightarrow PbO_2 + H_2O + Cl^- + OH^-$$

20. $Br_2 \rightarrow Br^- + BrO_3^-$

The bromine is both oxidized and reduced. The two half reactions are:

$$Br_2 \rightarrow Br^-$$

$$Br_2 \rightarrow BrO_3^-$$

Balance the atoms other hydrogen and oxygen.

$$Br_2 \rightarrow 2\,Br^-$$

$$Br_2 \rightarrow 2\,BrO_3^-$$

Add H_2O for oxygens that are needed and H^+ for hydrogens that are needed.

$$Br_2 \rightarrow 2\,Br^-$$

$$6\,H_2O + Br_2 \rightarrow 2\,BrO_3^- + 12\,H^+$$

Balance each reaction for charge.

$$2\,e^- + Br_2 \rightarrow 2\,Br^-$$

$$6\,H_2O + Br_2 \rightarrow 2\,BrO_3^- + 12\,H^+ + 10\,e^-$$

Make the electron count the same in both reactions.
$$5\,(2\,e^- + Br_2 \rightarrow 2\,Br^-)$$

$$6\,H_2O + Br_2 \rightarrow 2\,BrO_3^- + 12\,H^+ + 10\,e^-$$

Add the two half reactions, canceling anything that is common on both sides of the equation.

$$6\,H_2O + 6\,Br_2 \rightarrow 10\,Br^- + 2\,BrO_3^- + 12\,H^+$$

Add 12 OH^- to both sides of the equation to neutralize the H^+.

$$12\,OH^- + 6\,H_2O + 6\,Br_2 \rightarrow 10\,Br^- + 2\,BrO_3^- + 12\,H^+ + 12\,OH^-$$

Combine the H^+ and OH^- on the product side of the equation to form H_2O and cancel any water molecules that are common to both sides of the equation.

$$12\,OH^- + 6\,Br_2 \rightarrow 10\,Br^- + 2\,BrO_3^- + 6\,H_2O$$

Reduce the coefficients to their smallest whole number ratio.

$$6 \, OH^- + 3 \, Br_2 \rightarrow 5 \, Br^- + BrO_3^- + 3 \, H_2O$$

21. $3 \, SO_3^{2-} + Cr_2O_7^{2-} + 8 \, H^+ \rightarrow 3 \, SO_4^{2-} + 2 \, Cr^{3+} + 4 \, H_2O$

$$\frac{0.3143 \text{ mol Na}_2SO_3}{1 \text{ L}} \times \frac{1 \text{ mol SO}_3^{2-}}{1 \text{ mol Na}_2SO_3} \times 0.02500 \text{ L} = 7.858 \times 10^{-3} \text{ mol SO}_3^{2-}$$

$$7.858 \times 10^{-3} \text{ mol SO}_3^{2-} \times \frac{1 \text{ mol Cr}_2O_7^{2-}}{3 \text{ mol SO}_3^{2-}} = 2.619 \times 10^{-3} \text{ mol Cr}_2O_7^{2-}$$

$$2.619 \times 10^{-3} \text{ mol Cr}_2O_7^{2-} \times \frac{1 \text{ mol K}_2Cr_2O_7}{1 \text{ mol Cr}_2O_7^{2-}} \times \frac{1}{0.02842 \text{ L}} = 0.09216 \frac{\text{mol K}_2Cr_2O_7}{\text{L}}$$

22. Using either the oxidation number or half-reaction method will produce the following balanced equation:

$$16 \, H^+ + 2 \, Cr_2O_7^{2-} + C_2H_5OH \rightarrow 4 \, Cr^{3+} + 2 \, CO_2 + 11 \, H_2O$$

$$\text{mol K}_2Cr_2O_7 = \frac{5.000 \times 10^{-3} \text{ mol K}_2Cr_2O_7}{1 \text{ L soln}} \times 0.006522 \text{ L soln} = 3.26 \times 10^{-5} \text{ mol K}_2Cr_2O_7$$

$$\text{mol C}_2H_5OH = 3.26 \times 10^{-5} \text{ mol K}_2Cr_2O_7 \times \frac{1 \text{ mol C}_2H_5OH}{2 \text{ mol K}_2Cr_2O_7} = 1.63 \times 10^{-5} \text{ mol C}_2H_5OH$$

Knowing moles, we can now calculate grams of C_2H_5OH and determine the % weight of alcohol in the blood.

$$g \, C_2H_5OH = 1.63 \times 10^{-5} \text{ mol C}_2H_5OH \times \frac{46.0 \text{ g C}_2H_5OH}{1 \text{ mol C}_2H_5OH} = 7.50 \times 10^{-4} \text{ g C}_2H_5OH$$

$$\% \text{ weight} = \frac{7.50 \times 10^{-4} \text{ g C}_2H_5OH}{0.50 \text{ g sample}} \times 100 = 0.15\% \, ; \text{ Yes, this person is legally drunk.}$$

23. $Na_2S \, (aq) + Zn(NO_3)_2 \, (aq) \rightarrow 2 \, NaNO_3 \, (aq) + ZnS \, (s)$ molecular eqn.

$2 \, Na^+ \, (aq) + S^{2-} \, (aq) + Zn^{2+} \, (aq) + 2 \, NO_3^- \, (aq) \rightarrow 2 \, Na^+ \, (aq) + 2 \, NO_3^- \, (aq) + ZnS \, (s)$ ionic eqn.

Spectator ions: Na^+ and NO_3^-

$Zn^{2+} \, (aq) + S^{2-} \, (aq) \rightarrow ZnS$

We were given information about both reactants. Therefore, we must determine which reactant is the limiting reactant. Begin by determining the number of moles of each reactant present.

$$\text{mol Na}_2S = 7.5 \text{ g Na}_2S \times \frac{1 \text{ mol Na}_2S}{78.0 \text{ g Na}_2S} = 9.6 \times 10^{-2}$$

$$\text{mol Zn(NO}_3)_2 = 6.8 \text{ g Zn(NO}_3)_2 \times \frac{1 \text{ mol Zn(NO}_3)_2}{189.4 \text{ g Zn(NO}_3)_2} = 3.6 \times 10^{-2} \text{ mol Zn(NO}_3)_2$$

The required mole ratio is 1 mol of Na_2S for every 1 mol $Zn(NO_3)_2$. Therefore, the limiting reactant is $Zn(NO_3)_2$. The mass of ZnS produced is:

$$\text{g ZnS} = 3.6 \times 10^{-2} \text{ mol Zn(NO}_3)_2 \times \frac{1 \text{ mol ZnS}}{1 \text{ mol Zn(NO}_3)_2} \times \frac{97.4 \text{ g ZnS}}{1 \text{ mol ZnS}} = 3.5 \text{ g ZnS}$$

The amount of leftover reactant is:

$$\text{xs Na}_2\text{S} = 0.096 \text{ mol Na}_2\text{S available} - 0.036 \text{ mol Na}_2\text{S reacted} = 0.060 \text{ mol Na}_2\text{S leftover}$$

$$\text{g Na}_2\text{S} = 6.0 \times 10^{-2} \text{ mol Na}_2\text{S} \times \frac{78.0 \text{ g Na}_2\text{S}}{1 \text{ mol Na}_2\text{S}} = 4.7 \text{ g Na}_2\text{S}$$

$$\% \text{ Yield} = \frac{2.53 \text{ g ZnS}}{3.5 \text{ g ZnS}} \times 100 = 72\%$$

24. $Mg\ (s) + CuSO_4\ (aq) \rightarrow MgSO_4\ (aq) + Cu\ (s)$ Molecular eqn.

$Mg\ (s) + Cu^{2+}\ (aq) + SO_4^{2-}\ (aq) \rightarrow Mg^{2+}\ (aq) + SO_4^{2-}\ (aq) + Cu\ (s)$ ionic eqn.

$Mg\ (s) + Cu^{2+}\ (aq) \rightarrow Mg^{2+}\ (aq) + Cu\ (s)$ net ionic eqn.

We were given information about both reactants. Therefore, we must determine which reactant is the limiting reactant. Begin by determining the number of moles of each reactant present.

$$\text{mol Mg} = 4.8 \text{ g Mg} \times \frac{1 \text{ mol Mg}}{24.3 \text{ g Mg}} = 0.20 \text{ mol Mg}$$

$$\text{mol CuSO}_4 = \frac{0.75 \text{ mol CuSO}_4}{1 \text{ L soln}} \times 0.250 \text{ L soln} = 0.19 \text{ mol CuSO}_4$$

The required mole ratio is 1 mol of Mg for every 1 mol $CuSO_4$. Therefore, the limiting reactant is $CuSO_4$. The mass of copper produced is:

$$\text{g CuSO}_4 = 0.19 \text{ mol CuSO}_4 \times \frac{1 \text{ mol Cu}}{1 \text{ mol CuSO}_4} \times \frac{63.5 \text{ g Cu}}{1 \text{ mol Cu}} = 12 \text{ g Cu}$$

The amount of leftover reactant is:

$$\text{xs Mg} = 0.20 \text{ mol Mg available} - 0.19 \text{ mol Mg reacted} = 0.01 \text{ mol Mg leftover}$$

$$\text{g Mg leftover} = 0.01 \text{ mol Mg} \times \frac{24.3 \text{ g Mg}}{1 \text{ mol Mg}} = 0.24 \text{ g Mg}$$

$$\% \text{ Yield} = \frac{9.2 \text{ g Cu}}{12 \text{ g Cu}} \times 100 = 77\%$$

25. We begin this problem by determining the number of moles of I_3^- produced in the reaction between Cu^{2+} and I^-. We can determine the number of moles of I_3^- from the number of moles of $Na_2S_2O_3$ used in the titration.

$$\text{mol } S_2O_3^{2-} = 0.2500 \frac{\text{mol } S_2O_3^{2-}}{\text{L soln}} \times 0.020\,80\,1 \text{ soln} = 0.005\,200 \text{ mol } S_2O_3^{2-}$$

From the balanced equation given in the example on p. 69, we know that 2 moles of $S_2O_3^{2-}$ react with 1 mole of I_3^- Therefore, the number of moles of I_3^- produced in the reaction between Cu^{2+} and I^- is:

$$\text{mol } I_3^- = 0.005\,200 \text{ mol } S_2O_3^{2-} \times \frac{1 \text{ mol } I_3^-}{2 \text{ mol } S_2O_3^{2-}} = 0.002\,600 \text{ mol } I_3^-$$

Knowing the moles of I_3^- produced, we can now calculate the number of moles and mass of Cu^{2+} that reacted.

$$\text{mol } Cu^{2+} = 0.002\,600 \text{ mol } I_3^- \times \frac{2 \text{ mol } Cu^{2+}}{5 \text{ mol } I_3^-} = 0.001\,040 \text{ mol } Cu^{2+}$$

$$\text{g } Cu^{2+} = 0.001\,040 \text{ mol } Cu^{2+} \times \frac{63.5 \text{ g } Cu^{2+}}{1 \text{ mol } Cu^{2+}} = 0.066\,04 \text{ g } Cu^{2+}$$

$$\%Cu = \frac{0.066\,04 \text{ g Cu}}{0.500 \text{ g sample}} \times 100 = 13.2\%$$

CHAPTER 5

PERIODICITY AND ATOMIC STRUCTURE

Chapter Learning Goals

1⊠ Interconvert wavelength, frequency, and energy of electromagnetic radiation.
2⊠ Using the Balmer-Rydberg equation, calculate the wavelength and energy of a photon absorbed or released when an electron changes orbitals.
3⊠ Relate a set of quantum numbers to a particular orbital.
4⊠ Sketch and name each of the *s, p,* and *d* orbitals.
5⊠ Predict ground-state electron configurations for elements and ions; use orbital-filling diagrams to determine the number of unpaired electrons in these species.
6⊠ Write the general valence-shell electron configuration for each group of the periodic table, and identify the blocks in which the elements are located.
7⊠ Given a set of atoms, determine which atom is expected to have the largest radius.
8⊠ Given a set of ions, determine which ion is expected to have the largest radius.

Chapter in Brief

The periodic table is the most important organizing principle in chemistry. This chapter explains why the elements, when placed in order of increasing atomic weight, have a periodic occurrence of chemical and physical properties. This periodicity can be understood by examining the theory used to describe the electronic structure of atoms. Chapter 5 begins the examination of this theory by introducing electromagnetic radiation and the properties of waves. You will then discover how both light and matter can have dual (both wave and particle) properties. With this information, you are introduced to quantum mechanics and quantum numbers, a mathematical theory used to describe the probability of finding an electron in an atom. You are then shown how to use quantum mechanics to determine the electronic configuration of the elements. Finally, you will apply this knowledge to learn how atoms in the same group in the periodic table have similar electronic configurations and how these electronic configurations affects periodic properties such as atomic and ionic radii.

Development of the Periodic Table

A. Mendeleev's hypothesis about organizing known chemical information.
   1. Listed the known elements by atomic weight.
   2. Grouped them together according to their chemical reactivity.
   3. Was able to predict the properties of unknown elements - *eka*-aluminum, *eka*-silicon.

Light and the Electromagnetic Spectrum

A. Electromagnetic radiation - forms of radiant energy (light in all its varied forms).
   1. Electromagnetic spectrum - a continuous range of wavelengths and frequencies of all forms of electromagnetic radiation.
B. Radiant energy - has wavelike properties.
   1. Frequency (ν) - the number of peaks (maxima) that pass by a fixed point per unit time (s^{-1} or Hz).
   2. Wavelength (λ) - the length from one wave maximum to the next.
   3. Amplitude - the height measured from the middle point between peak and trough (maximum and minimum).
C. Speed of light (c) - rate of travel of all electromagnetic radiation in a vacuum.
   1. $c = 3.00 \times 10^8$ m/s.
1⊠   2. Wavelength x frequency = speed.
      λ (m) $\times \nu$ (s^{-1}) $= c$ (m/s)

3. Frequency and wavelength are inversely related.
   a. long λ; low ν
   b. short λ; high ν

EXAMPLE:

Calculate the wavelength, in meters, of radiation with a frequency of 1.18×10^{14} s^{-1}. What region of the electromagnetic spectrum is this?

SOLUTION: We can solve this problem by using the equation $\lambda \times \nu = c$, where $c = 3.00 \times 10^8$ m/s.

$$\lambda = \frac{3.00 \times 10^8 \text{ m/s}}{1.18 \times 10^{14} \text{ s}^{-1}} = 2.54 \times 10^{-6} \text{ m}$$

This particular wavelength is found in the infrared region of the electromagnetic spectrum (see Fig. 5.3, page 156 in your text).

Electromagnetic Radiation and Atomic Spectra

A. Individual atoms give off light when heated or otherwise excited energetically.
   1. Consists of only a few λ.
   2. Line spectrum - series of discrete lines (or wavelengths) separated by blank areas.
   3. Each element has its own unique line spectrum.
B. Balmer – discovered a pattern in atomic line spectra for the hydrogen atom.

   1. All 4 lines in the hydrogen spectrum expressed by $\frac{1}{\lambda} = R\left(\frac{1}{4} - \frac{1}{n^2}\right)$.
      a. R (*Rydberg constant*) = 1.097×10^{-2} nm^{-1}

2⊠
   2. Rydberg - every line in the entire hydrogen spectrum fits the generalized Balmer-Rydberg equation:
      $$\frac{1}{\lambda} = R\left(\frac{1}{m^2} - \frac{1}{n^2}\right); \quad (n > m)$$

EXAMPLE:

Calculate the wavelength of light emitted when an electron falls from the n = 6 to n = 4 levels in the hydrogen atom.

SOLUTION: We solve this problem by using the Rydberg equation.

$$\frac{1}{\lambda} = 1.097 \times 10^{-2} \text{ nm}^{-1}\left(\frac{1}{4^2} - \frac{1}{6^2}\right) = 3.809 \times 10^{-4} \text{ nm}^{-1}$$

$$\lambda = 2.625 \times 10^3 \text{ nm}$$

Particlelike Properties of Electromagnetic Radiation: The Planck Equation

A. Blackbody radiation - the visible glow that solid objects give off when heated.
   1. Intensity does not continue to rise indefinitely as λ decreases.

1⊠
   2. Planck - energy radiated by a heated object is quantized.
      a. radiant energy emitted in discrete units or quanta
      b. $E = h\nu$; $E = \dfrac{hc}{\lambda}$; $h = 6.626 \times 10^{-34}$ J·s (Planck's constant)
      c. high energy radiation - higher ν; shorter λ
      d. low energy radiation - lower ν; longer λ

EXAMPLE:
What energy is emitted in the above example?

SOLUTION: We need to use the equation $E = \dfrac{hc}{\lambda}$ to calculate the energy. However, we first need to convert the wavelength we calculated from nanometers to meters, since the units for the speed of light is in m/s.

$$2.625 \times 10^3 \ nm \times \frac{1 \ m}{1 \times 10^9 \ nm} = 2.625 \times 10^{-6} \ m$$

$$E = \frac{\left(6.626 \times 10^{-34} \ J \cdot s\right)\left(3.00 \times 10^8 \ m/s\right)}{2.625 \times 10^{-6} \ m} = 7.573 \times 10^{-20} \ J$$

B. Photoelectric effect - irradiating a clean metal surface with light causes electrons to be ejected from the metal.
   1. Einstein - beam of light behaves as if it were composed of photons (stream of small particles).
   2. Energy of photons: $E = h\nu$.
      a. energy depends only on frequency of photon
      b. intensity of light beam - measure of number of photons, not energy
C. Light energy can behave as both waves and small particles.
D. Atoms - emit light quanta (photons) of a few specific energies.
   1. Give rise to a line spectrum.

Wavelike Properties of Matter: The De Broglie Equation
A. Einstein - relationship between mass and λ.
$$m = \frac{E}{c^2} = \frac{hc/\lambda}{c^2} = \frac{h}{\lambda c}$$
B. de Broglie - matter can behave in some respects like light.
   1. Both light and matter are wavelike as well as particlelike.
   2. Relationship between λ of an electron or of any other particle or object of mass *m* moving at velocity *v*:
$$m = \frac{h}{\lambda v}; \qquad \lambda = \frac{h}{mv}$$
C. Dual wave/particle description of light and matter is a mathematical model that accounts for atomic properties and behavior.

The Quantum Mechanical Description of the Hydrogen Atom
A. Heisenberg Uncertainty Principle - both the position (Δx) and the velocity (Δmv) of an electron cannot be known beyond a certain level of precision.
   1. $(\Delta x)(\Delta mv) \geq \dfrac{h}{4\pi}$
B. Schrödinger - quantum mechanical model of the atom.
   1. Abandon idea of an electron as a small particle moving around the nucleus in a defined path.
   2. Concentrate on the electron's wavelike properties.
C. Quantum mechanical model of atomic structure.
   1. Mathematical form is a wave equation.
   2. Wave function (ψ) or orbital - solutions to wave equation.
      a. has a specific energy
      b. contains information about an electron's position in 3D space

    c. ψ^2 - gives the probability of finding an electron within a given region in space

    d. defines a volume of space around the nucleus where there is a high probability of finding an electron

    e. says nothing about the electron's path or movement

3⊠ **Wave Functions and Quantum Numbers**

A. Wave function - contains a set of three variables, quantum numbers.

    1. Describes the energy level of an orbital.

    2. Defines the shape and orientation of the region in space where the electron is most likely to be found.

B. Principal quantum number (n) - describes the size and energy level of the orbital.

    1. A positive integer ($n = 1,2,3,4,..$).

    2. As value of n increases.

        a. increase in the number of allowed orbitals

        b. increase in size of the orbitals

        c. increase of the energy of the electron in the orbital

    3. Shell - grouping of orbitals according to the principal quantum number.

C. Angular-momentum quantum number (ℓ) - defines the 3D shape of the orbital.

    1. Integral value from 0 to $n - 1$.

    2. Within each shell, there are n different shapes for orbitals.

    3. Subshells - grouping of orbitals according to the angular-momentum quantum number.

    4. Referred to by letter rather than by number.

    quantum number ℓ: 0   1   2   3   4 ...

    subshell notation: s   p   d   f   g ...

D. Magnetic quantum number (m_ℓ) defines the spatial orientation of the orbital along a standard set of coordinate axes.

    1. Integral value from $-\ell$ to $+\ell$.

    2. Within each subshell (same n and same ℓ) there are $2\ell + 1$ different spatial orientations.

E. Energy level of various orbitals.

    1. Hydrogen - energy levels depend only on n.

    2. Multielectron atoms - energy levels depend on both n and ℓ.

EXAMPLE:

Give the values of all possible quantum numbers of a 3d subshell.

SOLUTION:

For a 3d subshell, $n = 3$, $\ell = 2$; and $m_\ell = -2, -1, 0, 1, 2$.

Atomic Spectra Revisited

A. Electron in an atom:

    1. Occupies an orbital.

    2. Each orbital has a specific energy.

    3. The energies available to electrons are quantized.

        a. have only the specific energy values associated with the orbital

B. Addition of energy to an atom causes:

    1. An electron to jump from a lower-energy orbital to a higher-energy orbital.

        a. the atom becomes excited

    2. Excited atom is unstable.

          a. electron returns to a lower-energy level by emission of amount of energy equal to the energy difference between the higher and lower-energy orbitals

          b. since the energies of the orbitals are quantized, the amount of energy emitted is quantized

      3. Calculate energy differences between orbitals by measuring the frequencies emitted.

2⊠   C. Balmer-Rydberg equation - variables m and n corresponds to the principal quantum numbers of the two orbitals involved in the transition.

      1. n - principal quantum number of the outer-shell (orbital the transition is from).

      2. m - principal quantum number of the inner-shell (orbital the transition is to).

4⊠ **Orbital Shapes**

    A. s orbitals (see Fig. 5.13, page 175 in text).

      1. Spherical.

      2. Probability of finding an electron depends only on the distance of the electron from the nucleus.

      3. Node - a surface of zero probability.

         a. intrinsic property of a wave - zero amplitude at node

    B. p orbitals (see Fig. 5.14, page 176 in text).

      1. Dumbbell shaped.

      2. Electron distribution concentrated in identical lobes on either side of the nucleus.

      3. Nodal plane cuts through the nucleus.

      4. Three p orbitals oriented along the x, y, and z axes. (p_x, p_y, p_z).

    C. d and f orbitals

      1. d orbitals (see Fig. 5.17, page 177 in text).

         a. $4 - d_{xy}, d_{xz}, d_{yz}, d_{x^2-y^2}$ - are cloverleaf shaped

            i. four lobes of maximum electron probability separated by two nodal planes through the nucleus

         b. d_{z^2} - similar in shape to a p_z orbital with an additional donut-shaped region of electron probability in the xy plane

      2. f orbitals - eight lobes of maximum electron probability.

         a. three nodal planes through the nucleus

Orbital Energy Levels in Multielectron Atoms

    A. Energy level of an orbital in multielectron atoms depends on both n and ℓ.

      1. Energy difference due to electron-electron repulsions.

         a. outer-shell electrons are pushed farther away from the nucleus and are held less tightly

         b. partially cancels the electron-nucleus attractions

         c. electrons are shielded from the nucleus by the other electrons

    B. Effective nuclear charge, Z_{eff} - net nuclear charge actually felt by an electron.

      1. $Z_{eff} = Z_{actual}$ - electron shielding.

      2. Lower than the actual nuclear charge.

      3. For the same shell, the lower value of ℓ corresponds to a higher value of Z_{eff}.

         a. corresponds to lower energy for the orbital

      4. Useful for explaining various chemical phenomena.

Electron Spin and the Pauli Exclusion Principle

    A. Fourth quantum number – m_s.

      1. Related to electron spin.

      2. Has two values: $+1/2$ (↑) or $-1/2$ (↓).

B. Pauli exclusion principle: no two electrons in an atom can have the same four quantum numbers.
　　1. Only 2 electrons with opposite spins per orbital.

5⊠　**Electron Configurations of Multielectron Atoms**
　　A. Electron configuration - describes the orbitals that are occupied by the electrons in an atom.
　　B. Aufbau principle.
　　　　1. Fill the lowest-energy orbitals first.
　　　　2. Only two electrons with opposite spin per orbital.
　　　　3. Follow Hund's rule.
　　　　　　a. If two or more orbitals with the same energy are available, put one electron with parallel spin in each until all are half full.
　　C. Ground-state configuration - lowest-energy electron configuration.
　　D. Degenerate orbitals - orbitals with the same energy level.
　　E. Orbital filling diagrams - electrons are represented by arrows.

EXAMPLE:
　Give the electron configuration for phosphorus.

SOLUTION: Using the orbital filling diagram given in Fig. 5.11 on page 171 in the text, we find that the electron configuration for phosphorus is $1s^2 \, 2s^2 2p^6 3s^2 3p^3$.

EXAMPLE:
　Give the orbital filling diagram for phosphorus.

　SOLUTION: The orbital filling diagram for phosphorus would be:

$$\begin{array}{ccccccccc}
\uparrow\downarrow & \uparrow\downarrow & \uparrow\downarrow & \uparrow\downarrow & \uparrow\downarrow & \uparrow\downarrow & \uparrow & \uparrow & \uparrow \\
\hline
1s & 2s & 2p_x & 2p_y & 2p_z & 3s & 3p_x & 3p_y & 3p_z
\end{array}$$

　　F. Shorthand version - give the symbol of the noble gas in the previous row to indicate electrons in filled shells and then specify only those electrons in unfilled shells.

EXAMPLE:
　Give the shorthand electron configuration for phosphorus.

SOLUTION: The shorthand electron configuration for phosphorus would be [Ne] $3s^2 \, 3p^3$.

　　G. Valence shell electrons - outermost shell of electrons.
　　　　1. Elements in each group of the periodic table have similar valence-shell electron configurations.
　　　　2. Most loosely held.
　　　　3. Determine an element's properties.
　　H. Similar electron configurations explain why the elements in a given group have similar chemical behavior.
6⊠　　I. Blocks of elements in the periodic table - depends upon the valence orbitals being filled.
　　　　1. *s*-block elements - Groups 1A and 2A (filling of an *s* orbital).
　　　　2. *p*-block elements - Groups 3A through 8A (filling of *p* orbitals).
　　　　3. *d*-block elements - transition metals (filling of *d* orbitals).
　　　　4. *f*-block elements - lanthanide and actinide elements (filling of *f* orbitals).

EXAMPLE:

Give the general electron configuration for the elements in Group 5A.

SOLUTION:

The elements in group 5A are p block elements. The p orbitals begin filling with the 3A elements; therefore, we know that the group 5A elements contain 3 p electrons. The electron configuration for these elements is ns^2np^3.

Some Anomalous Electron Configurations
A. Half-filled and filled subshells have an unusual stability.
  1. Leads to anomalies in electron configurations.
B. Anomalies occur where the energy differences between subshells are small.
  1. Transfer of an electron from one subshell to another lowers the total energy of the atom.
    a. due to decrease in electron-electron repulsions

EXAMPLE:

Give the electron configuration of Ag.

SOLUTION: Silver is one of the elements that will transfer an electron from one subshell to another in order to lower the total energy of the atom. Therefore, the electron configuration of Ag is $[Kr]5s^14d^{10}$.

Electron Configuration of Ions
A. Main-group elements.
  1. Electrons lost by a metal come from the highest-energy occupied orbital.
  2. Electrons gained by a nonmetal go into lowest-energy unoccupied orbital.
B. Transition metals.
  1. Lose valence-shell s electrons first, then d electrons.

EXAMPLE:

Give the electron configuration for Al^{3+}, Se^{2-}, and Fe^{2+}.

SOLUTION: When writing the electron configuration of an ion, remember that the electrons are lost from the highest-energy occupied orbital and gained by the lowest-energy unoccupied orbital. Transition metals lose electrons from the s orbital first.

Al^{3+}: $1s^22s^22p^6$
Se^{2-}: $1s^22s^22p^63s^23p^64s^23d^{10}4p^6$
Fe^{2+}: $1s^22s^22p^63s^23p^63d^6$

7⊠   Electron Configurations and Periodic Properties: Atomic Radii
A. Radius of an atom - half the distance between the nuclei of two identical atoms when they are covalently bonded together.
B. Atomic radius increases down a group.
  1. Successively larger valence-shell orbitals are occupied.
C. Atomic radius decreases across a period.
  1. Due to increase in Z_{eff} for valence-shell electrons across a period.
  2. The value of Z_{eff} is dependent upon the amount of shielding felt by an electron.
    a. amount of shielding depends on both the shell and subshell of the other electrons.

8⊠ **Electron Configurations and Periodic Properties: Ionic Radii**
    A. Cations - radii are smaller than for neutral atoms.
        1. Electrons are removed from larger, valence-shell orbitals.
        2. Increase in Z_{eff} when electrons are removed.
    B. Anions - radii are larger than for neutral atoms.
        1. Decrease in Z_{eff} when electrons are added.
        2. Increase in electron-electron repulsions.

Self-Test

This section is intended to test your knowledge of the material covered in this chapter. Think through these problems and make certain you understand what is going on. Ask yourself if your answer makes sense. Many of these questions are linked to the chapter learning goals. Therefore, successful completion of these problems indicates you have mastered the learning goals for this chapter. You will receive the greatest benefit from this section if you use it as a mock exam. You will then discover which topics you have mastered and which topics you need to study in more detail.

True/False

1. Frequency and wavelength are directly related.

2. High energy radiation consists of higher frequency and shorter wavelength.

3. The energy of photons depends both on the frequency and number of photons.

4. The line spectra of atoms are due to the emission of light quanta of a few specific energies.

5. The quantum mechanical description of the hydrogen atom is based on the idea of an electron as a small particle moving around the nucleus in a defined path.

6. Wave functions define a volume of space around the nucleus where there is a high probability of finding an electron.

7. The principal quantum number defines the three-dimensional shape of the orbital.

8. The energy level of various orbitals depends only on the value of n.

9. The increase in atomic radius down a group is due to an increase in Z_{eff}.

10. The radius of a cation is larger than the radius of the neutral atom.

Multiple Choice

1. The quantum number ℓ is referred to as
    a. the principle quantum number
    b. the angular momentum quantum number
    c. the magnetic quantum number
    d. the spin quantum number

2. The Heisenberg uncertainty principle states that
    a. no two electrons can have the same four quantum numbers.
    b. if two or more orbitals are equal in energy, each is half-filled before any one of them is completely filled.
    c. the lowest energy orbital is filled first.
    d. you can never know both the position and the velocity of an electron beyond a certain level of precision.

3. For the lanthanide elements, the valence electrons are found in a
    a. s subshell
    b. p subshell
    c. d subshell
    d. f subshell

4. The valence shell of the atom is
    a. the lowest energy shell
    b. the most probable excited state
    c. the outermost shell
    d. represented by a noble gas when writing the electron configuration in shorthand.

5. High energy radiation is associated with
    a. higher v and longer λ
    b. higher v and shorter λ
    c. lower v and longer λ
    d. the speed of light

6. The quantum numbers associated with a $2p$ electron are
    a. $n = 2$, $\ell = 0$, $m_\ell = 0$
    b. $n = 2$, $\ell = 1$, $m_\ell = 0$
    c. $n = 2$, $\ell = 1$, $m_\ell = -1, 0, 1$
    d. $n = 2$, $\ell = 1$, $m_\ell = -2, -1, 0, 1, 2$

7. The element Au is a
    a. s block element
    b. p block element
    c. d block element
    d. f block element

8. The electron configuration for Cl is
    a. $1s^2 2s^2 2p^6 2d^7$
    b. $1s^2 2s^2 2p^6 3s^2 3p^5$
    c. $[Ar]3s^2 3p^5$
    d. $1s^2 2s^2 2p^6 3s^2 3p^7$

9. The electron configuration for Cr^{2+} is
    a. $[Ar]4s^2 3d^2$
    b. $[Kr]3d^4$
    c. $[Ar]4s^1 3d^3$
    d. $[Ar]3d^4$

10. The atom or ion with the larger radius is
    a. $Na > Cs$
    b. $Na > Na^+$
    c. $Na > Al$
    d. $Na < Li$

Matching

Planck	a. can never know both the position and the velocity of an electron beyond a certain level of precision.
Photoelectric effect	b. quantum mechanical model of the atom which concentrates on the electron's wavelike properties.
Einstein	c. organized the known elements by atomic weight, and grouped them together according to their chemical reactivity.
de Broglie	d. radiant energy is emitted in discrete units or quanta.
Schrödinger	e. no two electrons in an atom can have the same four quantum numbers.
Heisenberg	f. beam of light behaves as if it were composed of photons.
Pauli	g. set of rules that guides the filling of atomic orbitals.
Hund	h. dual wave/particle description of light and matter.
Aufbau principle	i. irradiating a clean metal surface with light causes electrons to be ejected.
Mendeleev	j. if two or more orbitals with the same energy are available, put one electron in each with parallel spin until all are half full.

Fill-in the Blank

1. A continuous range of wavelengths and frequencies of all forms of electromagnetic radiation is referred to as the _____.

2. Each element has its _____ line spectrum.

3. The visible glow that solid objects give off when heated is called _____.

4. Atomic properties and behavior can be accounted for by the _____ description of light and matter.

5. Wave functions contain information about an electron's _____.

6. The grouping of orbitals according to the principal quantum number is referred to as a _____.

7. When an electron returns to a lower-energy level from a higher energy level energy equal to _____ is emitted.

8. The probability of finding an electron in an *s* orbital depends only upon the _____ _____.

9. The _____ is the net nuclear charge actually felt by an electron.

10. The _____ describes the orbitals which are occupied by the electrons in an atom.

11. The lowest-energy electron configuration is the _____.

12. Elements in each group of the periodic table have similar _____ _____.

13. The similar chemical behavior of elements in a given group is due to _____ _____.

14. The transfer of an electron from one subshell to another can lower the _____.

15. When transition metals form cations, the electrons are lost first from the _____ and then from the _____.

16. Across a period, the atomic radius _____.

17. The radius of an anion is _____ than the radius of the neutral atom.

Problems

1. What is the wavelength of infrared light (in nanometers) with a frequency of 1.58×10^{13} Hz? What is the energy of this light?

2. When lithium ions are excited, they emit radiation at $\lambda = 670.8$ nm. What is the frequency of this radiation? What is the color of the radiation emitted?

3. What energy in kJ/mol is associated with a wavelength of 130 nm?

4. How much energy will be released when an electron falls from the $n = 6$ shell to the $n = 2$ shell in the hydrogen atom?

5. What are the allowed quantum numbers for a 4*p* subshell?

6. What is the orbital notation for an electron with $n = 3$, and $\ell = 1$?

7. What quantum numbers are associated with the 5*g* subshell?

8. Lines in the Pfund series of the hydrogen spectrum are caused by emission energy when the electron falls from outer shells to the third shell. Using the Rydberg equation, calculate the wavelength (in nanometers) and the energy (in kilojoules/mol) of the first line in the Pfund series.

9. If a hydrogen atom in the ground state is excited by light with a wavelength of 97.3 nm, how much energy has been absorbed? Which shell is the electron excited to?

10. Give the ground state electron configuration for sulfur, strontium, lead, and nickel atoms. Draw the orbital-filling diagrams for the valence-shell electrons of atoms of these elements. How many unpaired electrons does each atom have?

11. What is the general valence-shell electron configuration for the group 5A elements? Which block are these elements located in?

12. Using the shorthand version, give the ground-state electron configuration for Al^{3+}, Fe^{3+}, and Rb^+.

13. Identify the atoms with the following electron configurations:
$[Ar]4s^2 3d^7$
$[Xe]6s^2 4f^{14} 5d^{10} 6p^2$
$[Kr]5s^2$

14. Group the following atoms by increasing atomic radius: Ir, Ba, Pb, and Er.

15. Practice writing the electron configuration for as many elements as possible.

Solutions

True/False

1. F. Frequency and wavelength are inversely related.
2. T
3. F. The energy of the photons depends only on the frequency.
4. T
5. F. Quantum mechanics concentrates on the wave properties of an electron.
6. T
7. F. The principal quantum number defines the size and energy level of the electron.
8. F. The energy levels of various orbitals in the hydrogen atom depend only on the value of n; in multielectron atoms, the energy levels of various orbitals depends on both the value of n and ℓ.
9. F. The increase in atomic radius down a group is due to the fact that successively larger valence-shell orbitals are occupied.
10. F. The radius of a cation is smaller than the radius of a neutral atom.

Multiple Choice
1. b
2. d
3. d
4. c
5. b
6. c
7. c
8. b
9. d
10. b and c

Matching

Planck – d
Photoelectric effect – i
Einstein – f
de Broglie – h
Schrödinger – b
Heisenberg – a
Pauli – e
Hund – j
Aufbau principle – g
Mendeleev – c

Fill-in the Blank

1. electromagnetic spectrum
2. own unique
3. blackbody radiation
4. dual wave/particle
5. position in three dimensional space
6. shell
7. the energy difference between the higher and lower-energy orbitals
8. distance of the electron from the nucleus
9. effective nuclear charge
10. electron configuration
11. ground state configuration
12. chemical properties
13. similar valence-shell electron configurations
14. total energy of the atom.
15. valence-shell *s* orbital; *d* orbitals
16. decreases
17. larger

Problems

1. $\lambda \times \nu = c; \quad \lambda = \dfrac{c}{\nu}; \quad \lambda = \dfrac{3.00 \times 10^8 \, m/s}{1.58 \times 10^{13} \, s^{-1}} = 1.90 \times 10^{-5} \, m = 1.90 \times 10^4 \, nm$

$E = h\nu; \quad E = (6.626 \times 10^{-34} \, J \cdot s) \times (1.58 \times 10^{13} \, s^{-1}) = 1.05 \times 10^{-20} \, J$

2. $\lambda \times \nu = c; \quad \nu = \dfrac{6.708 \times 10^{-7} \, m}{3.00 \times 10^8 \, m/s} = 2.236 \times 10^{-15} \, s^{-1}$

The color of this radiation is red.

3. $E = \dfrac{hc}{\lambda}; \quad E = \dfrac{(6.626 \times 10^{-34} \, J \cdot s) \times (3.00 \times 10^8 \, m/s)}{1.30 \times 10^{-7} \, m} = 1.53 \times 10^{-18} \, J$

This represents the energy of one photon at a wavelength of 130 nm. The energy of 1 mol of photons is calculated in the following manner.

$\left(1.53 \times 10^{-18} \, \dfrac{J}{photon}\right) \times \left(6.022 \times 10^{23} \, \dfrac{photon}{mol}\right) = 9.21 \times 10^5 \, \dfrac{J}{mol}$

4. $\dfrac{1}{\lambda} = 1.097 \times 10^{-2} \text{ nm}^{-1}\left(\dfrac{1}{2^2} - \dfrac{1}{6^2}\right) = 2.438 \times 10^{-3}\text{ nm}^{-1}$; $\lambda = 410.2$ nm

$E = \dfrac{(6.626 \times 10^{-34}\text{ J} \cdot \text{s}) \times (3.00 \times 10^8\text{ m/s})}{4.102 \times 10^{-7}\text{ m}} = 4.846 \times 10^{-19}\text{ J}$

5. For a $4p$ orbital the allowed quantum numbers are $n = 4$; $\ell = 1$; and m_ℓ can be +1, 0 or –1.

6. $3p$

7. $n = 5$, $\ell = 4$, m_ℓ can be +4, +3, +2, +1, 0, -1, -2, -3, or –4.

8. The first line in the Pfund series results from the electron falling from the $n = 4$ to the $n = 3$ shell.

$\dfrac{1}{\lambda} = 1.097 \times 10^{-2}\text{ nm}^{-1}\left(\dfrac{1}{3^2} - \dfrac{1}{4^2}\right) = 5.332 \times 10^{-4}\text{ nm}^{-1}$; $\lambda = 1.875 \times 10^3$ nm

$E = \dfrac{\left(6.626 \times 10^{-34}\text{ J} \cdot \text{s}\right)\left(3.00 \times 10^8\text{ m/s}\right)}{1.875 \times 10^{-6}\text{ m}} = 1.060 \times 10^{-19}\text{ J}$

This represents the energy of one photon at a wavelength of 1.875 x 10³ nm. We must now calculate the energy of one mol of photons.

$1.060 \times 10^{-19}\ \dfrac{\text{J}}{\text{photon}} \times \dfrac{6.022 \times 10^{23}\text{ photons}}{1\text{ mol}} \times \dfrac{1\text{ kJ}}{1000\text{ J}} = 63.84$ kJ / mol

9. $E = \dfrac{(6.626 \times 10^{-34}\text{ J} \cdot \text{s}) \times (3.00 \times 10^8\text{ m/s})}{9.73 \times 10^{-8}\text{ m}} = 2.043 \times 10^{-18}\text{ J}$

We know that the hydrogen atom is in the ground state, therefore, m = 1. Use the Balmer-Rydberg equation to solve for n.

$\dfrac{1}{97.3\text{ nm}} = 1.097 \times 10^{-2}\text{ nm}^{-1}\left(\dfrac{1}{1} - \dfrac{1}{n^2}\right)$; $n = 4$

10. S: $1s^2\,2s^2\,2p^6\,3s^2\,3p^4$ [Ne] $\underset{3s}{\uparrow\downarrow}$ $\underset{}{\uparrow\downarrow}$ $\underset{3p}{\uparrow}$ $\uparrow$; 2 unpaired e⁻

Sr: $1s^2\,2s^2\,2p^6\,3s^2\,3p^6\,4s^2\,3d^{10}\,4p^6\,5s^2$ [Kr] $\underset{5s}{\uparrow\downarrow}$; 0 unpaired e⁻

Pb: $1s^2\,2s^2\,2p^6\,3s^2\,3p^6\,4s^2\,3d^{10}\,4p^6\,5s^2 4d^{10}\,5p^6\,6s^2\,4f^{14}\,5d^{10}\,6p^2$

[Xe] $\underset{4f}{\uparrow\downarrow\ \uparrow\downarrow\ \uparrow\downarrow\ \uparrow\downarrow\ \uparrow\downarrow\ \uparrow\downarrow\ \uparrow\downarrow}$ $\underset{5d}{\uparrow\downarrow\ \uparrow\downarrow\ \uparrow\downarrow\ \uparrow\downarrow\ \uparrow\downarrow}$ $\underset{6s}{\uparrow\downarrow}$ $\underset{6p}{\uparrow\ \uparrow}$ 2 unpaired e⁻

Ni: $1s^2\,2s^2\,2p^6\,3s^2\,3p^6\,4s^2\,3d^8$ [Ar] $\underset{4s}{\uparrow\downarrow}$ $\underset{3d}{\uparrow\downarrow\ \uparrow\downarrow\ \uparrow\downarrow\ \uparrow}$ $\uparrow$; 2 unpaired e⁻

11. The general valence-shell electron configuration for group VA is $ns^2\,np^3$. These elements are located in the p block.

12. Al^{3+}: [Ne] Fe^{3+}: [Ar] $3d^5$ Rb^+: [Kr]

13. $[Ar]4s^2 3d^7$: Co
    $[Xe]6s^2 4f^{14} 5d^{10} 6p^2$: Pb
    $[Kr]5s^2$: Sr

14. Ba > Er > Ir > Pb

15. Check your answer with Table 5.20, page 182 in your text.

CHAPTER 6

IONIC BONDS AND SOME MAIN-GROUP CHEMISTRY

Chapter Learning Goals

1⊠ For any two elements predict which has the higher first ionization energy.
2⊠ For any two elements, predict which has the higher second, third, fourth, etc. ionization energy.
3⊠ For any two elements predict which has the more negative first electron affinity.
4⊠ Identify the energies involved in a Born-Haber calculation of lattice energy. Know whether these energies are positive or negative, large or small, and use the Born-Haber cycle to calculate the lattice energy of an ionic compound.
5⊠ On the basis of ionic charges and ionic radii, predict which of two ionic compounds should have the greater lattice energy.
6⊠ Know which alkali and alkaline earth metals form a) oxides, b) peroxides, and c) superoxides, upon reaction with oxygen. Assign oxidation numbers to the oxygen atoms in these compounds.
7⊠ Give the formulas of products formed when alkali metals react with halogens, hydrogen, nitrogen, oxygen, water, and ammonia. Balance the equations.
8⊠ Give the formulas of products formed when alkaline earth metals react with halogens, hydrogen, oxygen, and water. Balance the equations.
9⊠ Give the formulas of products formed when aluminum reacts with halogens, nitrogen, oxygen, acids, and bases. Balance the equations.
10⊠ Balance redox equations representing reactions studied in this chapter. Identify which species are oxidized and which are reduced and which elements have undergone a change in oxidation state.
11⊠ Give the formulas of products formed when halogens react with metals, hydrogen, and other halogens.
12⊠ Use the octet rule to generalize the chemistry of each family of elements studied in this chapter. Know when to expect the octet rule to be valid and when it can fail.
13⊠ Give the noble gas configuration of cations and anions in ionic compounds.

Chapter in Brief

Chemical bonds are the forces that hold atoms together. There are two types: ionic bonds and covalent bonds. This chapter concentrates on the reasons ionic bonds are formed and the energies involved in the formation of these bonds. You will learn that the underlying reason for the formation of ions and ionic compounds is the valence-shell configuration of the elements, which determines the amount of energy needed for losing or gaining electrons. Knowing the periodic trends in the energies for ion formation, you will learn how to predict which elements form cations or anions and if the formation of ionic compounds is energetically favorable. You will also learn about the Born-Haber cycle, a series of hypothetical steps used to calculate the overall energy involved in the formation of ionic compounds. Finally, you will learn about the chemistry of the group 1A - 3A, 7A, and 8A elements and how the chemistry of these elements is governed by the octet rule.

Ionization Energy

    A. Ionization energy (E_i) - the amount of energy required to remove the outermost electron from an isolated neutral atom in the gaseous state.
    B. Periodic trends in ionization energy.
        1. Minimum E_i - group IA alkali metals.
        2. Maximum E_i - group VIIIA noble gases.
        3. E_i increases across a period.

4. E_i decreases down a group.
C. Periodicity due to electron configurations.
  1. Single s electron in valence-shell of alkali metals feels a low Z_{eff}.
    a. electron is loosely held
    b. energy needed to remove electron is low
  2. Electrons in filled valence-subshell of noble gas elements feel a high Z_{eff}.
    a. electrons are tightly held
    b. radius of atom shrinks
    c. energy needed to remove electron is high
  3. Increase in atomic number down a group.
    a. value of n increases along with the average distance of the electron from the nucleus
    b. valence-shell electrons are less tightly held
    c. E_i decreases
D. Minor irregularities occur from left to right across a row of the periodic table.
  1. Group 2A elements - E_i (Be) > E_i (B) (due to electron configuration).
    a. $2p$ electron of boron is shielded somewhat by the $2s$ electrons
      i. feels a smaller Z_{eff}
      ii. more easily removed
  2. Group 6 A elements - E_i (N) >E_i (O) (due to electron configuration).
    a. $2p$ electron removed from nitrogen
      i. removed from a half-filled orbital (stable configuration)
    b. $2p$ electron removed from oxygen
      i. removed from a filled orbital
      ii. electrons in filled orbitals are forced together and have a slightly higher energy
      iii. easier to remove electrons

1⊠ **EXAMPLE:**
Which has the higher ionization energy: Na or Al, Al or In?

SOLUTION: The ionization energy increases across the periodic table and decreases down the table. Therefore, E_i (Al) > E_i(Na) and E_i (Al) > E_i (In)

Higher Ionization Energies
A. Remove two, three, or even more electrons sequentially from an atom.
B. Each ionization step requires successively larger amounts of energy.
  1. Harder to remove a negatively charged electron from a positively charged ion.
C. Large jumps in successive ionization energies of the elements (see Table 6.1, page 204 in text).
  1. Due to electron configuration.
    a. high degree of stability associated with filled s and p sublevels
    b. valence-shell electrons are easily lost during ionization
    c. core electrons - relatively difficult to remove
D. Valence-shell electron configuration of an atom controls its chemistry.

2⊠ **EXAMPLE:**
Which ion has the higher third ionization energy, Mg^{2+} or Al^{2+}?

SOLUTION: For Mg^{2+} to lose another electron, the electron must come from the $2p$ subshell which is filled. This filled subshell has a high degree of stability and removing an electron will require a large amount of energy. Al^{2+}, on the other hand, can obtain a noble gas configuration by losing another electron. Therefore, the third ionization energy of Mg^{2+} is greater than the third ionization energy of Al^{2+}.

Electron Affinity

    A. Electron affinity (E_{ea}) - the energy change that occurs when an electron is added to an isolated atom in the gaseous state.

    B. E_{ea} - has a negative value.
        1. Energy is usually released when an atom adds an electron.
        2. The more negative the value, the greater the tendency of the atom to accept an electron.
            a. the more stable the anion
        3. $E_{ea} > 0$ for atoms that form an unstable anion on addition of an electron.
            a. no experimental measurements can be made.

3⊠   C. Periodic trends of E_{ea} - related to the electron configurations.
        1. Sign and magnitude of E_{ea} due to offsetting factors.
            a. negative E_{ea} due to attraction between the additional electron and the atomic nucleus
            b. positive E_{ea} due to repulsions between the additional electron and the electrons in the atom
        2. Halogens - large, negative E_{ea}.
            a. high Z_{eff}
            b. room in valence-shell for an additional electron
            c. high attraction between additional electron and atomic nucleus
        3. Noble-gas elements - positive E_{ea}.
            a. filled s and p sublevels
            b. additional electron goes into the next higher shell
                i. feels a low Z_{eff}
            c. small attraction between the additional electron and the atomic nucleus
            d. high electron-electron repulsions
        4. Alkaline earth metals – $E_{ea} \approx 0$.
            a. filled s subshell
            b. added electron goes into a p subshell of higher energy
            c. relatively low Z_{eff}

3⊠   *EXAMPLE:*
        Determine which has the larger E_{ea}: phosphorus or arsenic; phosphorus or sulfur.

        SOLUTION: To determine which element has the larger E_{ea}, we need to determine which element has the larger Z_{eff}. We learned in Chapter 5 that Z_{eff} increases across the periodic table and decreases down the periodic table. Therefore, phosphorus E_{ea} (P) > E_{ea} (As) and E_{ea} (S) > E_{ea} (P).

Ionic Bonds and the Formation of Ionic Solids

    A. An element with a low E_i can transfer an electron to an element with a negative E_{ea}.
        1. Produces a cation and an anion.
        2. Cation and anion are attracted together by electrostatic forces.
        3. Creates an ionic bond.
        4. Form a three-dimensional network of ions.
            a. each cation is surrounded by and attracted to many anions
            b. each anion is surrounded by and attracted to many cations
    B. Formation of ionic bonds leads to a large gain in stability.
        1. Overcomes unfavorable energy change of electron transfer.
4⊠   C. Born-Haber cycle - a series of hypothetical steps each of which contributes to the overall energy change during the formation of an ionic compound.
        1. Net process is the sum of the individual steps.
        2. Individual steps may involve:
            a. sublimation of a metal atom to a gaseous atom
            b. dissociation of nonmetal molecules
            c. ionization of isolated metal atoms

        d. formation of nonmetal anions

        e. formation of solid ionic compound.

5⊠   D. Lattice energy (U) - the sum of the electrostatic interaction energies between ions in a solid.

        1. Refers to the breakup of a crystal into individual ions.

           a. has a positive value

           b. when using in Born-Haber cycle, change the sign since the last step is the reverse of the breakup of the crystal

        2. Compounds with same anion but different cations.

           a. trend for increasing lattice energies parallels trend of decreasing cation size

        3. Compounds with same cation but different anion.

           a. trend for increasing lattice energies parallels trend of decreasing anion size

        4. Compounds of ions with higher charges have greater lattice energies than compounds of ions with lower charges.

EXAMPLE:

For CsI:

a. Calculate the energy change in kJ/mol if cesium atoms react with iodine atoms to yield isolated Cs^+ and I^- ions. Is the reaction favorable or unfavorable? (E_{ea} (I) = -295.4 kJ/mol; E_i (Cs) = 375.3 kJ/mol) Is this a favorable energy change?

b. Calculate the net energy change in kJ/mol that takes place on formation of CsI(s) from the elements: Cs (s) + 1/2 I_2 (g) → CsI (s) using the following information: Heat of sublimation for Cs = 77.6 kJ/mol, Bond dissociation energy for I_2 = 106.8 kJ/mol, E_{ea} for I = -295.4 kJ/mol, E_i for Cs = 375.3 kJ/mol; Lattice energy for CsI = 602 kJ/mol. Is this a favorable energy change?

SOLUTION:

a. E = -295.4 kJ/mol + 375.3 kJ/mol = 79.9 kJ/mol. The reaction is unfavorable.

b. sublimation of metal atom

    Cs (s) → Cs (g)   77.6 kJ/mol

    dissociation of nonmetal atom

    1/2 I_2 (g) → I (g)   53.4 kJ/mol

    ionization of the metal

    Cs (g) → Cs^+ + e^- 375.3 kJ/mol

    formation of nonmetal anion

    I (g) + e^- → I^- (g)   -295.4 kJ/mol

    formation of the solid compound

    Cs^+ (g) + I^- (g) → CsI (s)   -602.0 kJ/mol

E = 77.6 kJ/mol + 53.4 kJ/mol + 375.3 kJ/mol + (-295.4 kJ/mol) + (-602.0 kJ/mol) = -391.1 kJ/mol. This reaction is favorable.

EXAMPLE:

Which has the larger lattice energy: $MgCl_2$ or $CaCl_2$, $AlCl_3$ or $AlBr_3$, $FeCl_2$ or $FeCl_3$.

SOLUTION: For compounds which have the same anion but different cations, lattice energies increase with decreasing cation size. Since the ionic radius of Ca^{2+} is greater than the ionic radius of Mg^{2+}, then U ($MgCl_2$) > U ($CaCl_2$). For compounds which have the same cation but different anions, lattice energies increase with decreasing anion size. The ionic radius of Cl^- is less than the ionic radius of Br^-; therefore, U ($AlCl_3$) > U ($AlBr_3$). Compounds of ions with higher charge have greater lattice energies; therefore, U ($FeCl_3$) > U ($FeCl_2$).

The Alkali Metals

A. General Properties:
   1. Valence-shell electron configuration - ns^1.
      a. common ion - M^+ (dominates chemistry)
      b. lowest ionization energies of all the elements
      c. most powerful reducing agents in the periodic table
   2. Low densities and melting points.
   3. Metals.
      a. bright, silvery solids
      b. malleable
      c. good conductors of electricity
      d. soft enough to cut with a knife
   4. Very reactive.
      a. must be stored under oil to prevent their instantaneous reaction with oxygen and water
   5. Occur in nature as salts.

B. Produced by reaction of their chloride salts.
   1. Lithium and sodium – produced by electrolysis.
      a. an electric current is passed through the molten salt.
   2. Potassium, rubidium, and cesium – produced by chemical reduction.

C. Reactions of Alkali Metals.
   1. Halogens.
      a. general reaction: $2 M (s) + X_2 (g) \rightarrow 2 MX (s)$
      b. products - halides, crystalline ionic salts
      c. a decrease in ionization energy leads to an increase in reactivity
   2. Hydrogen.
      a. general reaction: $2 M (s) + H_2 (g) \rightarrow 2 MH (s)$
      b. products - hydrides (ox. # of H = -1)
      c. sluggish reaction at room temperature
   3. Nitrogen.
      a. only lithium reacts: $6 Li (s) + N_2 (g) \rightarrow 2 Li_3N (s)$
   4. Oxygen.
      a. all react rapidly
      b. form different kinds of products
         i. Li → oxide (ox. # of O = -2)
         ii. Na → peroxide (ox. # of O = -1
         iii. K, Rb, Cs → superoxide (ox. # O = -1/2)
   5. Water.
      a. general reaction: $2 M (s) + 2 H_2O (l) \rightarrow 2 M^+ (aq) + 2 OH^- (aq) + H_2 (g)$
      b. reactivity increases down the group
      c. produces an alkaline solution
      d. redox process
   6. Ammonia.
      a. general reaction: $2 M (s) + 2 NH_3 (l) \rightarrow 2 M^+ (soln) + 2 NH_2^- (soln) + H_2 (g)$
      b. analogous to reaction between metal and water
      c. solutions have powerful reducing properties
         i. MNH_2 - metal amide

The Alkaline-Earth Metals

A. General Properties.
   1. Similar to alkali metals.
   2. Valence-shell electron configuration - ns^2.
      a. common ion - M^{2+}

        b. powerful reducing agents
- 3. Soft, silvery metals.
- 4. Higher melting points and densities than alkali metals.
- 5. Less reactive toward oxygen and water than alkali metals.
- 6. Occur in nature as salts.

B. Produced by reduction of their salts.

8⊠   C. Reactions of Alkaline-Earth Metals.
- 1. Same kinds of redox reactions as alkali metals.
- 2. Less reactive than alkali metals.
  - a. E_i (alkaline-earth) $> E_i$ (alkali)
- 3. Halogens $\rightarrow$ ionic halide salts, MX_2.

6⊠   4. Oxygen $\rightarrow$ oxides, MO.
  - a. Sr, Ba $\rightarrow$ peroxides, MO_2
- 5. Water $\rightarrow$ metal hydroxides, $M(OH)_2$.

The Group 3A Elements: Aluminum

A. General properties.
- 1. Valence-shell electron configuration - ns^2np^1.
- 2. B - semimetal.
- 3. Others - silvery metals.
  - a. soft
  - b. good conductors of electricity
- 4. Al - common ion - Al^{3+}.
  - a. reducing agent

B. Most common element – aluminum.
- 1. Name comes from alum, $KAl(SO_4)_2 \cdot 12\ H_2O$.
  - a. medicinal salt
- 2. Occurs in many common minerals and gemstones.
  - a. sapphire and ruby impure forms of $Al_2\ O_3$
  - b. impurity creates color
    - i. Cr – impurity in ruby
    - ii. Fe and Ti impurity in sapphire
- 3. Commercially obtained from bauxite, $Al_2O_3 \cdot xH_2O$.
- 4. Reducing agent.
  - a. loses all three valence electrons

9⊠   C. Reactions of aluminum.
- 1. Halogens $\rightarrow$ AlX_3.
  - a. vigorous at room temperature
  - b. releases large amount of heat
- 2. Oxygen $\rightarrow$ Al_2O_3.
  - a. vigorous at room temperature only on surface
- 3. Nitrogen $\rightarrow$ AlN.
- 4. Acids and bases $\rightarrow$ Al^{3+} + H_2.

The Halogens

A. General Properties.
- 1. Exist as diatomic molecules.
- 2. Valence-shell electron configuration - ns^2np^5.
  - a. gain electrons
  - b. powerful oxidizing agents
  - c. large negative E_{ea}
- 3. Occur in nature as salts and minerals.

B. Produced by oxidation of their anions.

11⊠   C. Reactions.
    1. Most reactive elements in the periodic table.
    2. Less reactive going down the periodic table.
    3. Metals → metal halides, MX_n.
    4. Hydrogen → hydrogen halides, HX.
      a. behave as acids when dissolved in water
    5. Other halogens → XY.
      a. interhalogen compounds
      b. redox process – lighter, more reactive element is oxidizing agent and the heavier, less reactive element is the reducing agent
      c. properties intermediate between parent elements
      d. act as strong oxidizing agents in redox reactions

The Noble Gases

A. General Properties.
    1. Colorless, odorless, unreactive gases.
    2. Valence-shell electron configuration - ns^2np^6.
      a. stable configuration
      b. do not form cations or anions
      c. normally do not undergo redox reactions
      d. commercial use - applications that require inert atmospheres
B. Reactions of Noble Gases.
    1. Kr and Xe react only with fluorine.
      a. xenon fluorides - powerful oxidizing agents
    2. Lack of reactivity due to valence-shell configuration.
      a. large E_i
      b. small E_{ea}

12⊠   ## The Octet Rule

A. Octet rule - main-group elements tend to undergo reactions that leave them with eight valence electrons.
    1. 8 - "magic number".
      a. filled octet - electrons are tightly held by a high Z_{eff}
    2. Failure of rule - elements on right side of periodic table in third or lower period.
      a. due to availability of d orbitals
13⊠   B. Factors which determine the formation of a cation or anion.
    1. Cation - electrons are shielded from nucleus by core electrons.
      a. electrons feel low Z_{eff}
      b. electrons easily lost
      c. reach next lower noble-gas configuration - more difficult to lose another electron
    2. Anion - electrons are poorly shielded.
      a. electrons feel high Z_{eff}
      b. can gain electrons because of Z_{eff}
      c. reach noble-gas configuration - low-energy orbitals no longer available

Self-Test

This section is intended to test your knowledge of the material covered in this chapter. Think through these problems and make certain you understand what is going on. Ask yourself if your answer makes sense. Many of these questions are linked to the chapter learning goals. Therefore, successful completion of these problems indicates you have mastered the learning goals for this chapter. You will receive the greatest benefit from this section if you use it as a mock exam. You will then discover which topics you have mastered and which topics you need to study in more detail.

True/False

1. The elements that have the maximum E_i in the periodic table are the noble gases.

2. E_i increases across a period because the value of n increases along with the average distance of the electron from the nucleus.

3. The valence-shell electron configuration of an atom controls its chemistry.

4. The tendency of the atom to accept an electron decreases as the value of E_{ea} becomes more negative.

5. An element with a high E_i can transfer an electron to an atom with a negative E_{ea}.

6. The formation of an ionic bond leads to a large gain in stability.

7. The lattice energy refers to the breakup of a crystal into individual ions and has a positive value.

8. The alkaline earth metals are more reactive than the alkali metals because the E_i (alkaline earth metal) > E_i (alkali metal).

9. The halogens have the valence-shell configuration ns^2np^5 and are powerful reducing agents.

10. In a filled octet, the electrons are tightly held by a high Z_{eff}.

Multiple Choice

1. Ionization energies
   a. increase across a period and down a group.
   b. decrease across a period and down a group.
   c. increase across a period and decrease down a group.
   d. decrease across a period and increase down a group.

2. The element with the higher ionization energy is:
   a. Rb > Li
   b. Rb > Sr
   c. Rb > Ca
   d. Rb > Cs

3. The element with the more negative electron affinity is:
   a. Se > S
   b. Se > Br
   c. Se > Cl
   d. Se > Te

4. The ionic compound with the higher lattice energy is:
   a. NaCl > LiCl
   b. CaS > BaS
   c. NaI > NaF
   d. MgS > MgO

5. The alkali metals
   a. are the most powerful oxidizing agents in the periodic table.
   b. have high densities and melting points.
   c. occur in nature as the free elements.
   d. are good conductors of electricity.

6. When potassium reacts with oxygen, a(n):
   a. oxide, K_2O, is formed.
   b. no reaction occurs.
   c. peroxide, KO, is formed.
   d. superoxide, KO_2, is formed.

7. The oxidation number of oxygen in a peroxide is:
   a. -1
   b. -1/2
   c. -2
   d. none of the above

8. The alkaline earth metals:
   a. are more reactive toward oxygen and water than the alkali metals.
   b. occur in nature as salts.
   c. are dull, hard metals.
   d. are powerful oxidizing agents.

9. The group 3A elements:
   a. are all semimetals.
   b. are all metals.
   c. have the valence-shell configuration ns^2np^1.
   d. show an increase in ionization energy going down the group.

10. The noble gases:
    a. are odorless, colorless, reactive gases.
    b. have no commercial use.
    c. normally do not undergo redox reactions.
    d. have an unstable electron configuration.

Matching

Ionization energy

a. the energy change that occurs when an electron is added to an isolated atom in the gaseous state.

Electron affinity

b. Main-group elements tend to undergo reactions that leave them with eight valence electrons.

Born-Haber cycle

c. the amount of energy required to remove the outermost electron from an isolated atom in the gaseous state.

Lattice energy

d. the sum of the electrostatic interaction energies between ions in a solid.

Octet rule

e. a series of hypothetical steps, each of which contributes to the overall energy change during the formation of an ionic compound.

Fill-in-the-Blank

1. Alkali metals have the lowest _____ of all the elements.

2. When the alkali metals react with halogens, a decrease in ionization energy leads to an increase in

   _____ .

3. Alkaline earth metals are _____ reactive than the alkali metals because _____

   _____ .

4. For compounds with the same anion but different cation, the trend for increasing lattice energies

   parallels the trend of _____ .

5. The oxidation number of each oxygen atom in the superoxide ion is _____ .

6. When the alkaline earth elements react with water, the product is _____ .

7. The halogens act as _____ in redox reactions.

8. When halogens react with each other to form interhalogen compounds, the _____

   _____ is the oxidizing agent.

9. Xenon fluorides act as powerful _____ .

10. Elements on the right side of the periodic table in the third or lower period can have an expanded octet

    due to _____ .

Problems

1. Arrange the elements Ge, Si, and P in order of increasing ionization energies.

2. Which has the larger fourth ionization energy, In or Sn?

3. Explain why fluorine has a higher E_{ea} than its neighbors on either side.

4. For $MgCl_2$:
   a. Calculate the energy change in kJ/mol if magnesion atoms react with a chlorine molecule to yield isolated Mg^{2+} and Cl^- ions. Is the reaction favorable or unfavorable? E_{ea} (Cl) = -348.6 kJ/mol; E_{i1} (Mg) = 738 kJ/mol; E_{i2} = 1451 kJ/mol)

   b. Calculate the net energy change in kJ/mol that takes place on formation of $MgCl_2$ (s) from the elements, Mg (s) + Cl_2 (g) → $MgCl_2$ (s) using the E_{ea} and E_i given above along with the additional following information: heat of sublimation for Mg = 149.6 kJ/mol, bond dissociation energy for Cl_2 = 243 kJ/mol, lattice energy for $MgCl_2$ = 2526 kJ/mol.

9⊠

5. Balance the following redox reactions. Identify which species are oxidized and which are reduced and which elements have undergone a change in oxidation state.
   a. lithium reacts with water to produce lithium hydroxide
   b. in the presence of water, chlorine oxidizes iodine and produces hydrochloric acid and iodic acid.

6. Predict the outcome of the following reactions:
   a. $Ca\ (s)\ +\ Br_2\ (l)\ \rightarrow$

   b. $Al\ (s)\ +\ N_2\ (g)\ \rightarrow$

   c. $K\ (s)\ +\ O_2\ (g)\ \rightarrow$

   d. $Na\ (s)\ +\ H_2\ (g)\ \rightarrow$

   e. $H_2\ (g)\ +\ I_2\ (g)\ \rightarrow$

7. Which has the higher lattice energy?
   a. $MgCl_2$ or $AlCl_3$
   b. $AlCl_3$ or $GaCl_3$
   c. $SnCl_2$ or $PbBr_2$

8. Determine the oxidation number of the halogens in the following compounds: ClF_3, IF_7, ICl_5, IBr.

9. The reaction used to manufacture berrylium is given below:

   $BeF_2\ (l)\ +\ Mg\ (l)\ \rightarrow\ Be\ (l)\ +\ MgF_2\ (l)$

   How much berrylium is produced when 4.50 g of BeF_2 reacts with 8.23 g of Mg?

10. Write and balance the chemical equations for the formation of ClF_3 and BrF_5 from the halogens.

Solutions

True/False

1. T
2. F. The E_i increases across a period because Z_{eff} increases and the radius of the atoms shrink.
3. T
4. F. The tendency of the atom to accept an electron increases as the value of E_{ea} becomes more negative.
5. F. An element with a low E_i can transfer an electron to an atom with a negative E_{ea}.
6. T
7. T
8. F. The alkaline earth metals are less reactive than the alkali metals because the E_i (alkaline earth metals) $> E_i$ (alkali metals).
9. F. The halogens are powerful oxidizing agents.
10. T

Multiple Choice
1. c
2. d
3. d
4. b
5. d
6. d
7. a
8. b
9. c
10. c

Matching

Ionization energy - c
Electron affinity - a
Born-Haber cycle - e
Lattice energy - d
Octet rule - b

Fill-in-the-Blank
1. ionization energy
2. reactivity
3. less; E_i (alkaline earth metals) > E_i (alkali metals)
4. decreasing cation size
5. $-\frac{1}{2}$
6. a metal hydroxide, $M(OH)_2$
7. oxidizing agents
8. lighter, more reactive halogen
9. oxidizing agents
10. the availability of a *d* subshell

Problems

1. P > Si > Ge; ionization energies increase across a period (P > Si) and decrease down a group (Si > Ge).

2. Sn: The fourth electron is a valence electron in Sn but a core electron in In.

3. Fluorine has a higher Z_{eff} than oxygen; neon is a noble gas and has a filled valence shell, making it very difficult to gain an electron.

4. a. $E = 738$ kJ/mol + 1451 kJ/mol + 2(-348.6 kJ/mol) = 1493 kJ/mol
   (Note, the E_{ea} is multiplied by 2 since there are two Cl⁻ ions in $MgCl_2$.) This reaction is unfavorable.

   b. $Mg\ (s) \rightarrow Mg\ (g)$ 149.6 kJ/mol
      $Cl_2\ (g) \rightarrow 2\ Cl\ (g)$ 243.0 kJ/mol
      $Mg\ g) \rightarrow Mg^+(g) + e^-$ 738.0 kJ/mol
      $Mg^+(g) \rightarrow Mg^{2+}\ (g) + e^-$ 1451.0 kJ/mol
      $2\ Cl\ (g) + 2\ e^- \rightarrow 2Cl^-\ (g)$ -697.2 kJ/mol
      $Mg^{2+}\ (g) + 2Cl^-(g) \rightarrow MgCl_2\ (s)$ -2526.0 kJ/mol

   $E = 149.6$ kJ/mol + 243.0 kJ/mol + 738.0 kJ/mol + 1451.0 kJ/mol + (-697.2 kJ/mol) + (-2526 kJ/mol) = -641.6 kJ/mol

5. a. $Li(s) + H_2O \rightarrow LiOH(aq) + H_2(g)$
   Half reactions: $2(Li(s) \rightarrow Li^+(aq) + e^-)$
   $2 H_2O + 2e^- \rightarrow 2 OH^-(aq) + H_2(g)$
   Balanced reaction: $2 Li(s) + 2 H_2O \rightarrow 2 Li^+(aq) + 2 OH^-(aq) + H_2(g)$

   b. $Cl_2(g) + I_2(s) + H_2O \rightarrow HCl(aq) + HIO_3(aq)$
   Half reactions: $(Cl_2(g) + 2 H^+ + 2 e^- \rightarrow 2 HCl(aq))5$
   $I_2(s) + 6 H_2O \rightarrow 2 HIO_3(aq) + 10 H^+ + 10 e^-$
   Balanced reaction: $5 Cl_2(g) + I_2(s) + 6 H_2O \rightarrow 10 HCl(aq) + 2 HIO_3(aq)$

6. a. $2 Cs(s) + Br_2(l) \rightarrow 2 CsBr(s)$
   b. $2 Al(s) + N_2(g) \rightarrow 2 AlN(s)$
   c. $K(s) + O_2(g) \rightarrow KO_2(s)$
   d. $2 Na(s) + H_2(g) \rightarrow 2 NaH(s)$
   e. $H_2(g) + I_2(g) \rightarrow 2 HI(g)$

7. a. $AlCl_3$; Al^{3+} has the higher charge
   b. $AlCl_3$; Al^{3+} is the smaller cation
   c. $SnCl_2$; both Sn^{2+} and Cl^- have the smaller size

8. ClF_3: ox. # $Cl = +3$; ox. # $F = -1$
   IF_7: ox. # $I = +7$; ox. # $F = -1$
   ICl_5: ox. # $I = +5$; ox. # $Cl = -1$
   IBr: ox. # $I = +1$; ox. # $Br = -1$

9. We know we need to work this problem as a limiting reagent problem because we were given data for both reactants. First, we need to calculate the number of moles of each reactant present and determine which reactant is the limiting reactant.

$$4.50 \text{ g BeF}_2 \times \frac{1 \text{ mol BeF}_2}{47.0 \text{ g BeF}_2} = 0.0957 \text{ mol BeF}_2$$

$$8.23 \text{ g Mg} \times \frac{1 \text{ mol Mg}}{24.3 \text{ g Mg}} = 0.339 \text{ mol Mg}$$

The BeF_2:Mg mole ratio is 1:1. Therefore, BeF_2 is the limiting reagent since we need 0.339 mol BeF_2 to react with 0.339 mol Mg. We can now calculate the amount of Be produced in this reaction.

$$0.0957 \text{ mol BeF}_2 \times \frac{1 \text{ mol Be}}{1 \text{ mol BeF}_2} \times \frac{9.0 \text{ g Be}}{1 \text{ mol Be}} = 0.861 \text{ g Be}$$

10. $Cl_2(g) + 3 F_2(g) \rightarrow 2 ClF_3(g)$; $Br_2(l) + 5 F_2(g) \rightarrow 2 BrF_5(g)$

CHAPTER 7

COVALENT BONDS AND MOLECULAR STRUCTURE

Chapter Learning Goals

1⊠ From a list of compounds, predict which are ionic and which are molecular.

2⊠ Using only the periodic table, predict which of two elements is more electronegative.

3⊠ Using only the periodic table, predict whether a given bond is ionic, polar covalent, or nonpolar covalent.

4⊠ Using a table of electronegativities, predict which of two bonds is expected to be more polar.

5⊠ Write Lewis symbols for atoms and tell how many electrons must be shared to enable the atom to achieve a completed valence shell. Give the symbol of the noble gas with the same number of valence electrons.

6⊠ For each atom in a Electron-dot structure give the number of bonded electron pairs and the number of nonbonded electron pairs.

7⊠ For a given Electron-dot structure, give the number of single bonds, double bonds, and triple bonds. Give the bond order of each bond.

8⊠ Draw Electron-dot structures of molecules and polyatomic ions, recognizing when multiple bonding and resonance structures are needed.

9⊠ Determine the formal charge on each atom in a resonance structure and use the formal charges to select the best resonance structure.

10⊠ Use the VSEPR model to predict the geometries of molecules and polyatomic ions, including those with more than one central atom.

11⊠ For molecules and polyatomic ions, sketch and identify the orbitals used by each atom to form bonds. Show which orbital overlaps result in σ bonds and which result in π bonds.

12⊠ Sketch a molecular orbital diagram for a diatomic molecule. Use the molecular orbital diagram to determine the number of unpaired electrons and to calculate the bond order of the molecule described.

Chapter in Brief

The most important kind of bond in all of chemistry is the covalent bond, which involves the sharing of electrons between atoms. To understand this bonding concept, you need to first learn about the type of interactions that occur between atoms and how to draw Electron-dot structures, a useful tool for visualizing the interactions of electrons in a molecule. You will also learn how to determine the type of bonding that occurs in a molecule based on bond polarities, to predict the relative importance of resonance structures based on formal charge calculations, and to determine molecular shapes using VSEPR theory. Once you have mastered molecular shape determinations, you will be able to apply this knowledge to valence bond theory and to ascertain the type of hybridization predicted by valence bond theory. Finally, you will be introduced to molecular orbital theory, a more sophisticated and, at times, more accurate bonding description.

The Covalent Bond

   A. Formation of a covalent bond.
   1. Two atoms come close together and electrostatic interactions begin to develop.
      a. two nuclei repel each other; electrons repel each other
      b. each nucleus attracts the electrons; electrons attract both nuclei
   2. Attractive forces > repulsive forces, then covalent bond is formed.
   B. Distance between the two atoms affects the magnitude of the various attractive and repulsive forces.

       1. Bond length - the optimum point where net attractive forces are maximized.
         a. each covalent bond has a characteristic length that leads to maximum stability
         b. can predict from atomic radii

Strengths of Covalent Bonds

    A. Formation of covalent bonds leads to lower energy.
    B. Bond dissociation energy - the amount of energy necessary to break a chemical bond in an isolated molecule in the gaseous state (positive value).
       1. Equals the amount of energy released when the bond forms (negative value).
    C. Bonds between the same pairs of atoms usually have similar bond dissociation energies.

1⊠   **A Comparison of Ionic and Covalent Compounds**

    A. Ionic Compounds.
       1. High-melting solids.
         a. must overcome every ionic attraction in the entire crystal
    B. Covalent Compounds.
       1. Low-melting solids, liquids, or even gases.
         a. attractive forces between different molecules (intermolecular forces) are relatively weak

Polar Covalent Bonds: Electronegativity

    A. Polar covalent bonds - the bonding electrons are attracted somewhat more strongly by one atom in a bond.
       1. Electrons are not completely transferred.
       2. More electronegative atom - δ-.
       3. Less electronegative atom - δ+.
2⊠   B. Electronegativity (EN) - the ability of an atom in a molecule to attract the shared electrons in a bond.
       1. Metallic elements - low electronegativities.
       2. Halogens and other elements in upper-right-hand corner of periodic table - high electronegativities.
3,4⊠   C. Predicting bond polarity.
       1. Atoms with similar electronegativities ($\Delta EN \leq 0.4$) - form nonpolar bonds.
       2. Atoms whose electronegativities differ by more than 2 - form ionic bonds.
       3. Atoms whose electronegativities differ by less than 2 - form polar covalent bonds.

    EXAMPLE:
    Predict whether the following bonds are polar, nonpolar or ionic: Si — Cl, Si — H, and Ca — F.

    SOLUTION: Si — Cl - polar ($3.0 - 1.8 = 1.2$); Si — H – nonpolar ($2.1 - 1.8 = 0.3$);
    Ca — F – ionic ($4.0 - 1.0 = 3.0$)

5⊠   **Electron-Dot Structures**

    A. Electron-dot structure - represents how an atom's valence electrons are distributed in a molecule.
    B. Filled *s* and *p* subshells for each atom in a molecule leads to maximum stability.
    C. Use the octet rule to determine the number of covalent bonds an element will form.
       1. Only unpaired electrons can form bonds.
       2. An atom will share electrons until it is surrounded by four pairs of electrons (an octet) or has no more electrons to share.
    D. Electron-dot structures for elements.
       1. Electrons are represented by dots.

        a. first, place the dots, one to each side, until all four sides are occupied
        b. next, pair the dots until all the valence-shell electrons are used up
           i. the pairing of dots does not correspond to the pairing of electrons in the electron configuration

6⊠   c. lone pairs (nonbonded pairs) – electron pairs not used in bonding
        d. bonding pairs – share electrons

EXAMPLE:
Draw the electron-dot structure for C and N.

SOLUTION:

$$\cdot \overset{\displaystyle \cdot}{\underset{\displaystyle \cdot}{C}} \cdot \qquad \cdot \overset{\displaystyle \cdot\cdot}{\underset{\displaystyle \cdot}{N}} \cdot$$

7⊠   E. Multiple covalent bonds - atoms share more than one pair of electrons.
      1. Double bond - atoms share two pairs of electrons.
      2. Triple bond - atoms share three pairs of electrons.
      3. Bond order - refers to the number of electron pairs between atoms.
      4. Shorter and stronger than single bonds.
    F. Coordinate covalent bonds - one atom donates both electrons (a lone pair) to another atom that has a vacant valence orbital.
      1. N, O, P, and S frequently form coordinate covalent bonds.

Electron-Dot Structures of Polyatomic Molecules
    A. Compounds of second-row elements: C, H, N, O.
      1. Octet rule almost always applies.
      2. Easy to predict the number of bonds formed by each element.
        a. number of bonds formed = number of unpaired dots in the Lewis formula
      3. Indicate a covalent bond by a line.
    B. Compounds with elements beyond the second row.
      1. Third period and lower have unfilled d orbitals.
        a. expand their valence-shell beyond an octet

6,7⊠   C. Steps for drawing Electron-dot structures (see pages 251 and 252 in text).
      1. Atoms with less than four unpaired dots will not form an octet.
      2. When determining the connection keep in mind that usually the central atom is written first.
        a. when unsure, always assign the most symmetrical structure to the atom.

EXAMPLE:
Draw the electron-dot stucture of IF_5.

SOLUTION:
To draw the Electron-dot structure of IF_5:

1. Total number of valence-shell electrons = 7 (from I) + 35 (from 5 F) = 42

2. Determine the connections. In this case, I is the central atom. (This leads to the most symmetrical structure.)

$$\begin{array}{c}
F \\
F \diagdown \mid \diagup F \\
\mid \\
F \diagup I \diagdown F
\end{array}$$

3. Ten of the 42 valence electrons are used in forming the 5 I-F bonds, leaving 32. Thirty of these electrons are used in forming the octet around each fluorine.

4. There are two remaining electrons which are placed on the iodine.

8⊠   **Electron-Dot Structures and Resonance**
        A. More than one Electron-dot structure for a molecule.
            1. Resonance hybrid - average of the various possible Electron-dot structures for a molecule.
                a. resonance indicated by a straight double-headed arrow (↔)

EXAMPLE:
Draw the resonance structures for CO_3^{2-}.

SOLUTION:
    1. Total number of valence-shell electrons = 4 (from C) + 18 (from 3 O) + 2 (from the 2- charge) = 24

    2. Determine the connections. In this anion, C is the central atom.

3. Six of the 24 electrons are used for forming the three C — O bonds, leaving 18. All 18 of these electrons are used to form the octet around the three oxygen atoms.

4. There are no more electrons to distribute; however, carbon does not have a completed octet. To form a completed octet, we can borrow a pair of lone electrons from one of the oxygen atoms. However, since all three carbon – oxygen bonds are equal, we must draw three different resonance structures, each of which has a C=O bond between carbon and a different oxygen.

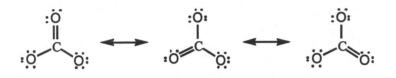

9⊠   **Formal Charges**
   A. Result from electron bookkeeping.
      1. Formal charge = (number of valence electrons in free atom) - 1/2(number of bonding electrons) - (number of nonbonding electrons).
      2. For ions, the sum of the formal charges on all the atoms is equal to the overall charge on the ion.
   B. Use to evaluate the relative importance of different resonance structures.
      1. Structures with the smallest formal charges are more stable.
      2. Structures which have the more negative formal charge on the more electronegative atom are more stable.

10⊠   **Molecular Shapes: The VSEPR Model**
   A. Molecular shape - determined by the numbers of valence electrons around the atoms.
   B. Valence-shell electron-pair repulsion model - predicts approximate shape of a molecule.
      1. Bonded and lone electrons occupy charge clouds.
      2. Count the number of charge clouds surrounding the atom.
      3. Predict the geometry.
         a. charge clouds will orient themselves so that they stay as far away as possible from each other
         b. see Table 7.4, pages 264 and 265 in text

EXAMPLE:

Predict the shape of AsF_5.

1. First, draw the Electron-dot structure of the molecule.

$$\begin{array}{c} :\ddot{F}: \\ :\ddot{F} \diagdown \; | \; \diagup \ddot{F}: \\ As \\ :\ddot{F} \diagup \quad \diagdown \ddot{F}: \end{array}$$

2. There are five charge clouds around arsenic, all of which are attached to atoms. Therefore, the molecular shape is trigonal bipyramidal.

11⊠   C. Shapes of larger molecules
      1. Use rules summarized in Table 7.4 to describe the geometry around each atom.

EXAMPLE:

Butyric acid has the formula $CH_3CH_2CH_2COOH$. It is the organic constituent that gives rancid butter its smell. The structural formula for this compound is:

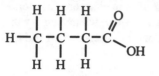

Describe the geometry around each carbon atom.

SOLUTION:

The first three carbons from left to right all have four charge clouds around them. Therefore, these three carbon atoms have a tetrahedral geometry around them. The carbon atom that has a double bonded oxygen and a hydroxide group bonded to it only has three charge clouds around it. Therefore, the geometry around this carbon atom is trigonal planar.

Valence Bond Theory

A. Easily visualized orbital picture of how electron pairs are shared in a covalent bond.
   1. Singly occupied valence orbital on one atom overlaps a singly occupied valence orbital on another atom.
   2. The greater the overlap, the stronger the bond.
      a. for orbitals other than *s* orbitals, gives a direction to the bond
   3. Each of the bonded atoms maintain its own orbital, but shares the electron.

Hybridization and sp^3 Hybrid Orbitals

A. Hybrid orbitals - the combination of wave functions for atomic orbitals which form a new set of equivalent wave functions.
B. Combine one *s* with three *p* - form four equivalent new orbitals - sp^3.
   1. Each hybrid has two lobes; one larger than the other.
   2. Four large lobes point towards the corners of a tetrahedron.
   3. Bonds formed with sp^3 orbitals are very strong.
C. A tetrahedral arrangement of charge clouds always implies sp^3 hybridization.

11⬚ Other Kinds of Hybrid Orbitals

A. Geometries in Table 7.4 - accounted for by a specific kind of hybridization.
B. σ bond - a bond that is formed by head-on overlap and that has its shared electrons centered about the axis between the two nuclei.
C. π bond - a bond that is formed by side-by-side overlap and that has its shared electrons occupy a region above and below a line connecting the two nuclei.
D. Given geometry implies the necessary hybridization (see Table 7.5, page 274 in text).

Molecular Orbital: The Hydrogen Molecule

A. Molecular orbital model.
   1. More complex bonding model than valence bond.
   2. Sometimes bonding description offers better agreement with experimental observations.
B. Atomic orbitals - a wave function whose square gives the probability of finding the electron within a given region of space in an atom.
   1. Same atom - can combine to form hybrids.
   2. Different atoms - overlap to form covalent bonds.

      a. orbitals and the electrons in them remain localized on specific atoms

C. Molecular orbital - a wave function whose square gives the probability of finding the electron within a given region of space in a molecule.

    1. Specific energy levels and specific shapes.

    2. Occupied by a maximum of two electrons with opposite spins.

D. Orbital interactions in molecular orbital theory.

    1. Additive interaction.

        a. for H_2 - formation of a molecular orbital that is egg-shaped

        b. denoted σ

        c. lower in energy than the two isolated $1s$ orbitals

        d. bonding molecular orbital

            i. electrons in bonding MO spend most of their time in the region between the two nuclei

            ii. bond the atoms together

    2. Subtractive interaction.

        a. for H_2 - formation of a molecular orbital that has a node between the two atoms

        b. denoted σ^*

        c. higher in energy than the two isolated $1s$ orbitals

        d. antibonding molecular orbital

            i. electrons in antibonding MO do not occupy the central region between the nuclei

            ii. don't contribute to bonding

E. Bond order = 1/2(number of bonding electrons - number of antibonding electrons).

F. Number of molecular orbitals formed = the number of atomic orbitals combined.

12⊠ Molecular Orbital Theory of Other Diatomic Molecules

A. Paramagnetic - substances with unpaired electrons are attracted by magnetic fields.

B. Diamagnetic - substances whose electrons are all spin-paired.

C. Molecular orbital of O_2.

    1. $2s$ orbitals interact giving σ_{2s} and σ^*_{2s} molecular orbitals.

    2. $2p$ orbitals on the internuclear axis.

        a. interact head-on

        b. form σ_{2p} and σ^*_{2p} molecular orbitals

    3. Remaining $2p$ orbitals perpendicular to the internuclear axis.

        a. interact in sideways manner

        b. form π_{2p} and π^*_{2p} molecular orbitals

    4. See Figure 7.20, page 281 in text for energy levels of the molecular orbitals.

D. Additive interaction of atomic orbitals leads to a bonding molecular orbital.

E. Subtractive interaction of atomic orbitals leads to the antibonding molecular orbital with a node between the nuclei.

F. Molecular orbital diagrams require some experience to generate.

Combining Valence Bond Theory and Molecular Orbital Theory

A. Molecules with resonance structures.

    1. σ-bond framework - localized; describe with valence bond theory.

    2. π-bond framework - delocalized; describe with molecular orbital theory.

Self-Test

This section is intended to test your knowledge of the material covered in this chapter. Think through these problems and make certain you understand what is going on. Ask yourself if your answer makes sense. Many of these questions are linked to the chapter learning goals. Therefore, successful completion of these problems indicates you have mastered the learning goals for this chapter. You will receive the greatest benefit from this section if you use it as a mock exam. You will then discover which topics you have mastered and which topics you need to study in more detail.

True/False

1. A covalent bond is formed when the attractive forces between the two atoms is greater than the repulsive forces between the two atoms.

2. Energy is needed for a covalent bond to form.

3. The bond order refers to the number of electrons between atoms.

4. The pairing of dots in the Electron-dot structures corresponds to the pairing of electrons in the electron configuration.

5. N, O, and P frequently form coordinate covalent bonds.

6. All elements in the periodic table obey the octet rule when they undergo bonding.

7. Electrons are transferred in polar covalent bonds.

8. Resonance structures that have the more negative formal charge on the more electronegative atom are more stable.

9. The shape of a molecule can be determined from the number of charge clouds surrounding the central atom.

10. The type of hybrid orbitals used by the central atom in a molecule can be determined from the geometry of the molecule.

Fill-in-the-Blank

1. A representation of the distribution of an atom's valence electrons in a molecule is referred to as the

   _____.

2. An atom will share electrons until _____.

3. In a multiple covalent bond, the atoms share _____

   _____.

4. The _____ can expand their valence-shell beyond an

   octet.

5. The _____ represents the average of the various possible Electron-

   dot structures of a molecule.

6. Nonpolar bonds are formed by atoms with _____.

7. In valence bond theory, a singly occupied valence orbital on one atom _____

   _____.

8. A trigonal bipyramidal arrangement of charge clouds always implies _____ hybridization.

9. Orbital interactions in molecular orbital theory are both _____.

10. The number of molecular orbitals formed is equal to _____.

Matching

Bond length	a. the ability of an atom in a molecule to attract the shared electrons in a bond.
Bond dissociation energy	b. predicts approximate shape of a molecule.
Coordinate covalent bond	c. the amount of energy necessary to break a chemical bond in an isolated molecule in the gaseous state.
Polar covalent bond	d. a bond in which one atom donates both electrons to another atom that has a vacant valence orbital.
Electronegativity	e. a bond that is formed by head-on overlap and that has its shared electrons centered about the axis between the two nuclei.
VSEPR Model	f. a wave function whose square gives the probability of finding the electrons within a given region of space in a molecule.
Hybrid orbitals	g. substance whose electrons are all spin paired.
σ bond	h. substance with unpaired electrons.
π bond	i. the optimum point where net attractive forces are maximized.
Molecular orbital	j. bonds in which the bonding electrons are attracted somewhat more strongly by one atom in a bond.
Paramagnetic	k. the combinations of wave functions for atomic orbitals which form a new set of equivalent wave functions.
Diamagnetic	l. a bond that is formed by side-by-side overlap and that has its shared electrons occupying a

region above and below a line connecting the two nuclei.

Problems

1. Draw the Electron-dot structures for a) the chromate ion, b) IF_6^+, c) ClF_3, d) H_2F^+, e) PF_4^-, f) XeF_4 and g) BF_3.

2. Draw all resonance structures for SO_3.

3. Predict whether the following bonds are nonpolar, polar covalent, or ionic: C-H, Na-Cl, C-N, O-H, F-F.

4. Calculate the formal charges on S and O in SO_3.

5. Predict the molecular geometries of the molecules in #1.

6. Predict the shape around the carbon atoms in formaldehyde, CH_3CHO. The structural formula for formaldehyde is:

7. Predict the hybridization of the central atom in the molecules in #1.

8. Calculate the bond order for N_2, O_2, and F_2 using the molecular orbital diagram found in Fig. 7.20 in your text.

Solutions

True/False
1. T
2. F. The formation of a covalent bond leads to lower energy; therefore, energy is released when a covalent bond is formed.
3. F. The bond order refers to the number of electron pairs between atoms.
4. F. The pairing of dots in a Electron-dot structure does not correspond to the pairing of electrons in the electron configuration.
5. T
6. F. Some elements are able to expand their octets because they have unfilled *d* orbitals; other elements will have less than an octet.
7. F. Polar covalent bonds do not involve the complete transfer of electrons.
8. T
9. T
10. T

Fill-in-the-Blank
1. Electron-dot structure
2. it is surrounded by four pairs of electrons or has no more electrons to share.
3. more than one pair of electrons
4. elements beyond the second row
5. resonance hybrid
6. similar electronegativities
7. overlaps a singly occupied valence orbital on another atom.
8. sp^3d
9. additive and subtractive
10. the number of atomic orbitals combined.

Matching
Bond length - i
Bond dissociation energy - c
Coordinate covalent bond - d
Polar covalent bond - j
Electronegativity - a
VSEPR model - b
hybrid orbitals - k
σ bond - e
π bond - l
Molecular orbital - f
Paramagnetic - h
Diamagnetic - g

Problems

1. a) CrO_4^{2-}; # valence electrons = 32; # bonding electrons = 8; # electrons used for O's octet = 24

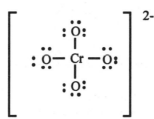

b) IF_6^+; # valence electrons = 48; # bonding electrons = 12; # electrons used for F's octet = 36

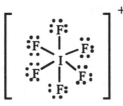

c) ClF_3; # valence electrons = 28; # bonding electrons = 6; # electrons used for F's octet = 18; # of electrons left over = 4

d) H_2F^+; # valence electrons = 8; # bonding electrons = 4; # electrons left over = 4

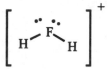

e) PF_4^-; # valence electrons = 34; # bonding electrons = 8; # electrons used for F's octet = 24; # electrons left over = 2

f) XeF_4; # valence electrons = 36; # bonding electrons = 8; # electrons used for F's octet = 24; # electrons leftover = 4

g) BF_3; # valence electrons = 24. # gonding electrons = 6; # electrons used for F's octet = 18. (Remember that B is one of the few elements that will have less than an octet around it.)

2. SO_3; # valence electrons = 24; # bonding electrons = 6; # electrons used for O's octet = 18

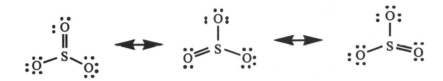

3. C-H; EN (C) = 2.5, EN (H) = 2.1; ΔEN = 0.4, nonpolar
   Na-Cl; EN (Na) = 0.9; EN (Cl) = 3.0; ΔEN = 2.1, ionic
   C-N; EN (C) = 2.5; EN (N) = 3.0; ΔEN = 0.5; slightly polar
   O-H; EN (O) = 3.5; EN (H) = 2.1; ΔEN = 1.4; polar
   F-F; EN (F) = 4.0; ΔEN = 0; nonpolar

4. S: 6 valence electrons; 8 bonding electrons; Formal charge = 6 - 4 = 2
   double bonded O: 6 valence electrons; 4 bonding electrons; 4 nonbonding electrons; Formal charge = 6 - 2 - 4 = 0
   single bonded O: 6 valence electrons; 2 bonding electrons; 6 nonbonding electrons; Formal charge = 6 - 1 - 6 = -1.

5. CrO_4^{2-}; number of charge clouds = 4; number of bonds = 4; number of lone pairs = 0; shape = tetrahedral

   IF_6^+; number of charge clouds = 6, number of bonds = 6; number of lone pairs = 0; shape = octahedral

   ClF_3; number of charge clouds = 5; number of bonds = 3; number of lone pairs = 2; shape = T shaped

   H_2F^+; number of charge clouds = 4; number of bonds = 2; number of lone pairs = 2; shape = bent

   PF_4^-; number of charge clouds = 5; number of bonds = 4; number of lone pairs = 1; shape = seesaw

   XeF_4; number of charge clouds = 6; number of bonds = 4; number of lone pairs = 2; shape = square planar

   BF_3; number of charge clouds = 3; number of bonds = 3; shape = trigonal planar

6. The carbon which is bonded to three H atoms and one C atom is surrounded by four charge clouds; therefore the geometry around this carbon is tetrahedral. The other carbon atom is surrounded by three charge clouds; therefore the geometry around this carbon is trigonal planar.

7. CrO_4^{2-}; number of charge clouds = 4; hybridization = sp^3

   IF_6^+; number of charge clouds = 6; hybridization = sp^3d^2

   ClF_3; number of charge clouds = 5; hybridization = sp^3d

   H_2F^+; number of charge clouds = 4; hybridization = sp^3

123

PF_4^-; number of charge clouds = 5; hybridization = sp^3d

XeF_4; number of charge clouds = 6; hybridization = sp^3d^2

BF_3; number of charge clouds = 3; hybridization = sp^2

8. N: bond order = 1/2(8 - 2) = 3; O: bond order = 1/2(8 - 4) = 2; F: bond order = 1/2(8 - 6) = 1

CHAPTER 8

THERMOCHEMISTRY: CHEMICAL ENERGY

Chapter Learning Goals

1⊠ Differentiate between the concepts of heat and temperature.

2⊠ Identify a state function.

3⊠ Define and calculate PV work. Know whether work is being done by the system or on the system.

4⊠ Differentiate between energy and enthalpy and perform calculations interconverting the two. From ΔH or ΔE tell whether energy is being lost from or gained by the system.

5⊠ Given a balanced chemical equation and enthalpy change for a chemical reaction, calculate the enthalpy change per mole or per gram of each reactant and product.

6⊠ Perform calculations involving specific heat (or molar heat capacity), heat flow, and temperature change.

7⊠ Perform calculations involving Hess's law.

8⊠ Use standard heats of formation to calculate a standard heat of reaction.

9⊠ Use bond dissociation energies to approximate a standard heat of reaction.

10⊠ Predict whether entropy increases or decreases for a chemical reaction or physical change.

11⊠ Use the equation $\Delta G = \Delta H - T\Delta S$ to determine whether the forward reaction or the reverse reaction is favored.

12⊠ Use ΔH and ΔS to determine the temperature at which a reversible system is at equilibrium.

Chapter in Brief

This chapter introduces you to the concept of thermochemistry: the heat changes that take place during reactions. You begin the study of this topic by learning the difference between heat and energy, and the types of energy changes that can take place. You are then introduced to the Law of Conservation of Energy, the First Law of Thermodynamics, and the concept of state functions. With this background, you will then learn how to calculate the internal energy of the system using $P\Delta V$ work and how the internal energy of the system is related to the enthalpy (ΔH) of the system. You will spend much of the rest of the chapter exploring how to use specific heat calculations in the laboratory and how to use Hess's law, standard heats of formation and bond dissociation energies to calculate heats of reaction. Finally, you are introduced to the topics of entropy and free energy, topics that will be explored in more detail in later chapters.

Heat and Energy
    A. Energy - the capacity to do work or supply heat.
       1. Energy = work + heat
    B. Kinetic energy - the energy of motion.
       1. $E_K = 1/2mv^2$
    C. Potential energy - stored energy.
    D. Joule - SI unit for energy.
       1. $1\text{ J} = 1(\text{kg·m}^2)/\text{s}^2$
    E. Calorie - the amount of energy necessary to raise the temperature of 1 g of water by $1°$ C.
       1. 1 cal = 4.184 J.
       2. Nutritional calorie (Calorie); 1 Cal = 1000 cal = 1 kcal = 4.184 kJ.

Energy Changes and Energy Conservation
    A. Law of Conservation of Energy: Energy can be neither created nor destroyed. It can only be converted from one form into another.

1⊠   B. Many forms of energy.
            1. Thermal energy.
            2. Heat - energy transferred from one object to another as the result of a temperature difference between them.
            3. Temperature - a measure of the kinetic energy of molecular motion.
            4. Chemical energy - a type of potential energy in which the chemical bonds of molecules act as the storage medium.
         C. First Law of Thermodynamics: The energy of the universe is constant.

Internal Energy and the First Law of Thermodynamics
         A. System - everything we focus on in an experiment.
         B. Surroundings - everything other than the system.
         C. Internal energy - energy of the system.
            1. System isolated from the surroundings - no energy transfer to the surroundings; $\Delta E = 0$.
            2. System is not isolated from the surrounding - energy flow to or from the surroundings:
                $\Delta E = E_{final} - E_{initial}.$
         D. Energy changes are measured from the point of view of the system.
            1. Energy flows out of the system to surroundings - negative value.
            2. Energy flows into the system from the surroundings - positive value.

2⊠   E. State Functions - a function or property whose value depends only on the present state (condition) of the system, not on the path used to arrive at that condition.
            1. For any state function, the overall change is zero if the system returns to its original condition.

3⊠ **Expansion Work**
         A. Work - the distance (d) moved times the force (F) that opposes the motion
         B. Expansion work (PV work) - work done as the result of a volume change in the system.
            1. $w = P \times \Delta V$
            2. Expansion of the system.
                a. system does work on the surroundings
                b. ΔE is negative
                c. $w = -P\Delta V$
                   i. work is negative
            3. Contraction of the system.
                a. surroundings do work on the system
                b. ΔE is positive
                c. $w = -P\Delta V$
                   i. work is positive

Energy and Enthalpy
         A. Total energy change of a system.
            1. $\Delta E = q + w$ (q = heat).
            2. $\Delta E = q + (-P\Delta V)$.
         B. Amount of heat transferred.
            1. $q = \Delta E + P\Delta V$.
         C. Reactions carried out with constant volume.
            1. $\Delta V = 0$; no PV work is done.
            2. $q_v = \Delta E$.
         D. Reactions carried out at constant pressure.
            1. $\Delta V \neq 0$; energy change due to both heat transfer and PV work.
            2. $q_p = \Delta E + P\Delta V$.
(L. Goal 5)   E. Enthalpy of a system - name given to the quantity $E + PV$.

1. $\Delta H = \Delta E + P\Delta V$.
2. State function.
3. $\Delta H = H_{products} - H_{reactants}$.
4. Amount of heat released in a specific reaction depends on the actual amounts of reactants.
5. Physical states of reactants and products must be specified.
6. Temperature and pressure must be specified.

The Thermodynamic Standard State

A. The value of the enthalpy change ΔH reported for a reaction represents the amount of heat released when reactants are converted to products in the molar amounts represented by coefficients of the balanced equation.
   1. The actual amount of heat involved in a reaction depends on the actual amounts of reactants.
   2. The physical states of reactants and products must be specified.
   3. Temperature and pressure also must be reported.
B. Thermodynamic Standard State – 298.15 K (25° C) 1 atm pressure of each gas, 1 M concentration (for solutions)
   1. Allows different reactions to be compared.
   2. Indicated by addition of a superscript o.
C. Standard enthalpy of reaction – an enthalpy change measured under standard conditions.

Enthalpies of Physical and Chemical Change

A. Enthalpies of physical change.
   1. Heat of fusion - the amount of heat required for melting.
   2. Heat of vaporization - amount of heat required for evaporation.
   3. Sublimation - the direct conversion of a solid to a vapor without going through a liquid state.
      a. heat of sublimation = heat of fusion + heat of vaporization
B. Enthalpies of chemical change.
   1. Heats of reaction - enthalpies of chemical change.
   2. Endothermic reactions.
      a. $H_{products} > H_{reactants}$
      b. heat flow into the system from the surroundings
      c. ΔH is positive
   3. Exothermic reactions.
      a. $H_{products} < H_{reactants}$
      b. heat flow to the surroundings from the system
      c. ΔH is negative
   4. ΔH^o values for a given equation.
      a. the equation is balanced for the number of moles of reactants and products
      b. all substances are in their standard states
      c. physical state of each substance is specified
      d. refer to the reaction going in the direction written
         i. reverse the direction of the reaction, change the sign of ΔH^o

EXAMPLE:
Sulfuric acid is produced by reacting sulfur trioxide with water according to the equation:

$$SO_3\ (g) + H_2O\ (l) \rightarrow H_2SO_4\ (l) \qquad \Delta H^o = -131.8\ \text{kJ/mol}$$

How much heat is evolved when 75.0 g of SO_3 reacts with a stoichiometric amount of H_2O?

SOLUTION:

The ΔH° reported is for the reaction of 1 mol of SO_3. To answer this question, we need to calculate how many moles of SO_3 are found in 75.0 g of SO_3.

$$75.0 \text{ g } SO_3 \times \frac{1 \text{ mol } SO_3}{80.0 \text{ g } SO_3} = 0.938 \text{ mol } SO_3$$

We can now calculate the amount of heat evolved when 75.0 g of SO_3 reacts.

$$-131.8 \frac{\text{kJ}}{1 \text{ mol } SO_3} \times 0.938 \text{ mol } SO_3 = -124 \text{ kJ}$$

6⊠   Calorimetry and Heat Capacity

A. Calorimetry – an experimental technique which allows the energy change associated with a chemical or physical process to be determined.
1. A temperature change is observed when a system gains or loses energy in the form of heat.
2. Carried out in a calorimeter
3. For an exothermic reaction:
    a. amount of heat released by the reaction = amount of heat gained by calorimeter + amount of heat gained by solution.
B. Heat capacity (C) - the amount of heat required to raise the temperature of an object or substance a given amount.
1. $C = \dfrac{q}{\Delta T}$
2. extensive property
C. Specific heat - the amount of heat necessary to raise the temperature of exactly 1 g of a substance by exactly $1^\circ C$.
1. $q = (\text{specific heat}) \times (\text{mass of substance}) \times (\Delta T)$
D. Molar heat capacity (C_m) - the amount of heat necessary to raise the temperature of 1 mole of a substance by $1^\circ C$.
1. $q_m = (C_m) \times (\text{moles of substance}) \times (\Delta T)$

EXAMPLE:

A student mixes 50.0 mL of 0.400 M $CuSO_4$ at 23.35° C with 50.0 mL of 0.600 M NaOH also at 23.35° C in a coffee cup calorimeter with a heat capacity of 25.0 J/$^\circ$C. The final temperature of the reaction is 26.65° C. Calculate the amount of heat evolved in this reaction. (The density of the solution is 1.02 g/mL. The specific heat of water may be used for the specific heat of the solution.)

SOLUTION:

For an exothermic reaction the amount of heat released by the reaction = the amount of heat gained by calorimeter plus the amount of heat gained by solution. We will begin by determining the amount of heat gained by the solution. (We first need to know the mass of the solution.)

$$100 \text{ mL soln} \times \frac{1.02 \text{ g soln}}{1 \text{ mL soln}} = 102 \text{ g soln}$$

$$102 \text{ g soln} \times 4.184 \frac{J}{g \cdot ^\circ} \times \left(26.65^\circ \text{ C} - 23.35^\circ \text{ C}\right) = 1410 \text{ J}$$

The amount of heat gained by the calorimeter is:

$$25.0 \frac{J}{^\circ C} \times \left(26.65^\circ\, C - 23.35^\circ\, C\right) = 82.5\ J$$

The amount of heat released by the solution is:

1410 J + 82.5 J = 1492.5 J

This value should be reported as −1492.5 J since the reaction is exothermic.

7⊠   **Hess's Law**
   A. Hess's law - the overall enthalpy change for a reaction is equal to the sum of the enthalpy
       changes for the individual steps in the reaction.
       1. Reactants and products in the individual steps can be added and subtracted like algebraic
           quantities in determining the overall equation.

EXAMPLE:
   Calculate ΔH for the reaction: NO (g) + O (g) → NO_2 (g) given the following information:

$$NO\ (g)\ +\ O_3\ (g)\ \rightarrow\ NO_2\ (g)\ +\ O_2\ (g) \qquad \Delta H = \text{-200 kJ}$$
$$O_3\ (g)\ \rightarrow\ \tfrac{3}{2} O_2\ (g) \qquad \Delta H = \text{-143 kJ}$$
$$O_2\ (g)\ \rightarrow\ 2\,O\ (g) \qquad \Delta H = \text{498 kJ}$$

SOLUTION:
   We will use the first equation as written since we know that the overall equation has 1 mol of NO
   as a reactant. We need to reverse the third equation and divide by 2 because we also need 1 mol
   of O as a reactant. Finally, we will need to cancel both O_3 and O_2 from the first and third
   equations. This requires that we reverse the second equation. Remember, whatever we do to the
   balanced chemical equation, we have to also do to the value of ΔH. When we reverse the
   direction of a chemical equation, we change the sign of ΔH. If we divide the coefficients in an
   equation by 2, we divide the value of ΔH by 2.

$$NO\ (g)\ +\ O_3\ (g)\ \rightarrow\ NO_2\ (g)\ +\ O_2\ (g) \qquad \Delta H = \text{-200 kJ}$$
$$O\ (g)\ \rightarrow\ \tfrac{1}{2} O_2\ (g) \qquad \Delta H = \text{-249 kJ}$$
$$\tfrac{3}{2} O_2\ (g)\ \rightarrow\ O_3\ (g) \qquad \Delta H = \text{+143 kJ}$$

$$NO\ (g)\ +\ O\ (g)\ \rightarrow\ NO_2\ (g) \qquad \Delta H = \text{-306 kJ}$$

8⊠   **Standard Heats of Formation**
   A. Standard heat of formation - The enthalpy change ΔH_f^o for the hypothetical formation of 1
       mol of a substance in its standard state from the most stable forms of it constituent elements
       in their standard states.
       1. The most stable forms of all elements in their standard state have $\Delta H_f^o = 0$.
   B. The standard enthalpy change for any chemical reaction is found by subtracting the sum of
       the heats of formation of the reactants from the sum of the heats of formation of the products.
       1. $\Delta H^o_{rxn} = \Sigma(\Delta H^o_{products}) - \Sigma(\Delta H^o_{reactants})$

EXAMPLE:
   Calculate ΔH^o for the reaction $2\ Na_2O_2\ (s)\ +\ 2\ H_2O\ (l)\ \rightarrow\ 4\ NaOH\ (s)\ +\ O_2\ (g)$ using ΔH_f^o
   found in Appendix B of your text.

SOLUTION:

Subtract the total heats of formation of the reactants from the total heats of formation from the products.

$$\Delta H^o{}_{rxn} = [(4 \times \Delta H_f^o \text{ NaOH } (s)] - [(2 \times \Delta H_f^o \text{ Na}_2\text{O}_2 (s)) + (2 \times \Delta H_f^o \text{ H}_2\text{O } (l))]$$
$$= [(4 \times -425.6 \text{ kJ})] - [(2 \times -510.9 \text{ kJ}) + (2 \times -285.8 \text{ kJ})]$$
$$= -109 \text{ kJ}$$

Bond Dissociation Enthalpies

A. Bond dissociation enthalpies - enthalpy changes, ΔH^o for the corresponding bond-breaking reactions.
   1. $\Delta H^o = D =$ bond dissociation energy.
   2. Always positive; always need energy to break a bond.
9⊠   B. $\Delta H^o{}_{rxn} = D(\text{bonds broken}) - D(\text{bonds formed})$.

Fuel Efficiency and Heats of Combustion

A. Heat of combustion (ΔH_c) - amount of energy released on burning a substance.
B. Fuel efficiency - calculate ΔH_c in kJ/g or kJ/mL.
   1. can compare efficiency for different fuels.

An Introduction to Entropy

A. Spontaneous process - a process that proceeds on its own without any continuous external influence.
   1. Need either a release of energy or an increase in disorder of the system.
B. Entropy - (S) - the amount of molecular disorder or randomness in a system.
   1. Increase in entropy - ΔS has a positive value.
   2. $\Delta S = S_{final} - S_{initial}$.
10⊠   C. Spontaneous process.
   1. Favored by decrease in H (negative ΔH).
   2. Favored by increase in S (positive ΔS).

An Introduction to Free Energy

11⊠   A. Gibbs free-energy change (ΔG); $\Delta G = \Delta H - T\Delta S$.
   1. Sign of ΔG used as a criterion for determining spontaneity of a process.
      a. ΔG negative - spontaneous
      b. ΔG positive - nonspontaneous
12⊠   B. Temperature dependence ($T\Delta S$) term for ΔG.
   1. Spontaneity of some processes depends on temperature.
   2. Low temperatures - ΔH dominates and controls spontaneity.
   3. High temperatures - $T\Delta S$ dominates and controls spontaneity.
C. $\Delta G = 0$; process is at equilibrium.
   1. Balanced between spontaneous and nonspontaneous.
   2. $T = \dfrac{\Delta H}{\Delta S}$.

Self-Test

This section is intended to test your knowledge of the material covered in this chapter. Think through these problems and make certain you understand what is going on. Ask yourself if your answer makes sense. Many of these questions are linked to the chapter learning goals. Therefore, successful completion of these problems indicates you have mastered the learning goals for this chapter. You will receive the greatest benefit from this section if you use it as a mock exam. You will then discover which topics you have mastered and which topics you need to study in more detail.

True/False

1. When the system is isolated from the surroundings, $\Delta E > 0$.

2. If energy flows out of the system to the surroundings, $\Delta E < 0$.

3. For any state function, the overall change is zero if the system returns to its original condition.

4. If a system expands doing PV work, the system does work on the surroundings and $w = P\Delta V$.

5. For reactions carried out at a constant pressure, the energy change is due only to PV work.

6. In an endothermic reaction, heat flows into the system from the surroundings and $H_{products} > H_{reactants}$.

7. For a reaction to be spontaneous, there must always be a release of energy.

8. Specific heat is an extensive property.

9. The most stable forms of all elements in their standard state have $\Delta H_f^o < 0$.

10. For a spontaneous process, ΔG is positive.

Matching

Energy	a. the direct conversion of a solid to a vapor without going through a liquid state.
Temperature	b. everything we focus on in an experiment.
System	c. name given to the quantity $E + PV$.
State function	d. a process that proceeds on its own without any continuous external influence.
Work	e. the amount of heat required to raise the temperature of an object or substance a given amount.
Enthalpy	f. a function or property whose value depends only on the present state (condition) of the system, not on the path used to arrive at that condition.

Heat of fusion   g. the amount of energy released on burning a substance.

Sublimation   h. the amount of molecular disorder or randomness in a system.

Heat capacity   i. the overall enthalpy change for a reaction is equal to the sum of the enthalpy changes for the individual steps in the reaction.

Hess's Law   j. the capacity to do work or supply heat.

Heat of combustion   k. the distance moved times the force that opposes the motion.

Spontaneous process   l. a measure of the kinetic energy of molecular motion.

Entropy   m. the amount of heat required for melting.

Fill-in-the-Blank

1. Chemical energy is a type of _____ in which the chemical bonds of the molecules act as _____.

2. The energy transferred from one object to another as the result of a temperature difference between them is referred to as _____.

3. The statement "the energy of the universe is constant" is referred to as _____ _____.

4. The change in the internal energy of the system is positive when energy flows _____ _____.

5. When a system contracts, the _____ do work on the _____, ΔE is _____ and $w =$ _____.

6. The thermodynamic standard state conditions are _____.

7. Enthalpies of chemical change are referred to as _____.

8. During an exothermic process, heat flows _____ from the _____ and ΔH is _____.

9. The enthalpy change ΔH_f^o for the hypothetical formation of 1 mol of a substance in its standard state from the most stable forms of its constituent elements in their standard states is referred to as the _____.

10. Bond dissociation enthalpies are the _____ _____.

11. The _____ is used as a criterion for determining the spontaneity of a process.

Problems

1. If 750 mL of a gas is compressed to 350 mL under a constant external pressure of 5.00 atm, and if the gas absorbs 15 kJ, what are the values of q, w, and ΔE for the gas? Is work being done on the system or by the system? What is the value of ΔE for the surroundings?

2. What are the values of ΔH and ΔE when a gas at 2.50 atm expands after the addition of 575 J of heat and does 200 J of work on the surroundings?

3. For the reaction

$$10 \, N_2O \, (g) \; + \; C_3H_8 \, (g) \; \rightarrow \; 10 \, N_2 \, (g) \; + \; 3 \, CO_2 \, (g) \; + \; 4 \, H_2O \, (g)$$

$\Delta H = $ -2862.7 kJ. How much heat is evolved from this reaction when 3.98 g of N_2O reacts with a stoichiometric amount of propane?

4. How much heat is needed to raise the temperature of 78.0 g of iron 15.0°C?

5. When 7.75 g of NH_4NO_3 is dissolved in 110 g of water at 25.00°C in a calorimeter, 25.8 kJ/ mol NH_4NO_3 is absorbed. Assuming that the specific heat of the solution is the same as that of pure water, calculate the final temperature of the solution.

6. Calculate the standard enthalpy change for the reaction

$$2 \, Al \, (s) \; + \; Fe_2O_3 \, (s) \; \rightarrow \; 2 \, Fe \, (s) \; + \; Al_2O_3 \, (s)$$

from the following:

$$2 \, Al \, (s) \; + \; \tfrac{3}{2} \, O_2 \, (g) \; \rightarrow \; Al_2O_3 \, (s) \qquad\qquad \Delta H^\circ = \text{-1676 kJ}$$

$$2 \, Fe \, (s) \; + \; \tfrac{3}{2} \, O_2 \, (g) \; \rightarrow \; Fe_2O_3 \, (s) \qquad\qquad \Delta H^\circ = \text{-824.2 kJ}$$

7. For the following reactions, calculate ΔH°_{rxn} from the standard heats of formation found in Appendix B.

$$SO_3 \, (g) \; + \; H_2O \, (l) \; \rightarrow \; H_2SO_4 \, (l)$$

$$2 \, KClO_3 \, (s) \; \rightarrow \; 2 \, KCl \, (s) \; + \; 3 \, O_2 \, (g)$$

8. Calculate the standard heat of reaction for the reaction

$$CH_3CH{=}CH_2 \; + \; HCl \; \rightarrow \; CH_3CHClCH_3$$

using the bond dissociation energies found in Table 7.1 in your text. The strength of a C=C bond is 635 kJ/mol.

9. Using standard heats of formation, calculate the heat of combustion of propane in kJ/mol and kJ/g.

10. Determine the sign of ΔS for the following reactions:

$$CaO\ (s)\ +\ 2\ NH_4Cl\ (s)\ \rightarrow\ 2\ NH_3\ (g)\ +\ CaCl_2\ (s)$$

$$BaCl_2\ (aq)\ +\ Na_2SO_4\ (aq)\ \rightarrow\ BaSO_4\ (s)\ +\ 2\ NaCl\ (aq)$$

11. Calculate the enthalpy change for the formation of ethane from graphite and hydrogen given the following thermochemical equations:

$$C\ (graphite)\ +\ O_2\ (g)\ \rightarrow\ CO_2\ (g) \qquad \Delta H^\circ = -393.5\ kJ$$

$$H_2\ (g)\ +\ \tfrac{1}{2}\ O_2\ (g)\ \rightarrow\ H_2O\ (l) \qquad \Delta H^\circ = -285.8\ kJ$$

$$2\ C_2H_6\ (g)\ +\ 7\ O_2\ (g)\ \rightarrow\ 4\ CO_2\ (g)\ +\ 6\ H_2O\ (l) \qquad \Delta H^\circ = -3119.6\ kJ$$

12. When hydrazine (N_2H_4) reacts with hydrogen peroxide, nitrogen gas and water are produced. Write a balanced reaction and determine ΔH° for the reaction from the following thermochemical data:

$$N_2H_4\ (l)\ +\ O_2\ (g)\ \rightarrow\ N_2\ (g)\ +\ 2\ H_2O\ (l) \qquad \Delta H^\circ = -621.6\ kJ$$

$$H_2\ (g)\ +\ \tfrac{1}{2}\ O_2\ (g)\ \rightarrow\ H_2O\ (l) \qquad \Delta H^\circ = -285.8\ kJ$$

$$H_2\ (g)\ +\ O_2\ (g)\ \rightarrow\ H_2O_2\ (l) \qquad \Delta H^\circ = -187.8\ kJ$$

13. For the reaction $SiO_2\ (s)\ +\ 2\ C\ (graphite)\ +\ 2\ Cl_2\ (g)\ \rightarrow\ SiCl_4\ (g)\ +2\ CO\ (g)$, $\Delta H^\circ = 32.9$ kJ/mol and $\Delta S^\circ = 226.5$ J/mol K. Calculate ΔG° for this reaction.

14. For the reaction $N_2\ (g)\ +\ O_2\ (g)\ \rightarrow\ 2\ NO\ (g)$, $\Delta G^\circ = 173.1$. Calculate the temperature at which this reaction becomes spontaneous if $\Delta H^\circ = 180.4$ kJ/mol and $\Delta S^\circ = 421.4$ J/mol K.

Solutions

True/False
1. F. $\Delta E = 0$ when the system is isolated from the surroundings.
2. T
3. T
4. F. If a system expands doing PV work, $w = -P\Delta V$.
5. F. For reactions carried out at constant pressure, the energy change is due to both heat transfer and PV work.
6. T
7. F. A reaction can absorb energy and still be spontaneous if the entropy of the reaction is positive and $T\Delta S > \Delta H$.
8. T
9. F. The most stable forms of all elements in their standard state have $\Delta H_f^\circ = 0$.
10. F. ΔG is negative for a spontaneous process.

Matching
Energy - j
Temperature - l
System - b
State function - f

Chapter 8 - Thermochemistry: Chemical Energy

Work - k
Enthalpy - c
Heat of fusion - m
Sublimation - a
Heat capacity - e
Hess's Law - i
Heat of combustion - g
Spontaneous process - d
Entropy - h

Fill-in-the-Blank
1. potential energy; the storage medium
2. heat
3. the First Law of Thermodynamics
4. into the system from the surroundings
5. surroundings; system; positive, $P\Delta V$
6. 1 atm pressure of each gas, 298.15 K, and 1 M concentration for solutions.
7. heats of reaction
8. to the surroundings; system; negative
9. standard heat of formation
10. enthalpy changes for the corresponding bond-breaking reactions
11. Gibbs free energy

Problems

1. Since we know the gas absorbs 15 kJ of heat, we know that $q = +15$ kJ. To calculate work, we use the equation $w = P\Delta V$ where $P = 5.00$ atm and $\Delta V = (750$ mL - 350 mL$) = 400$ mL.

   $w = 5.00$ atm $\times (0.400$ L$) = 2.00$ L·atm

   $$2.00\,\text{L}\cdot\text{atm} \times 101\frac{\text{J}}{\text{L}\cdot\text{atm}} = 202\,\text{J}$$

   Work is being done on the system since the change in volume is due to a contraction of the system. Knowing q and w allows us to calculate ΔE.

   For the system, $\Delta E = q + w$; $\Delta E = 15$ kJ + 0.202 kJ = 15 kJ.

   For the surroundings, $\Delta E = -15$ kJ

2. $\Delta H = 575$ J; $\Delta E = \Delta H - P\Delta V = 575$ J - (-200 J) = 775 J

3. $\Delta H = -2862.7$ kJ for the reaction of 10 mol N_2O. 3.98 g N_2O is equivalent to 0.0905 mol N_2O.

   $$\text{heat evolved} = 0.0905\,\text{mol} \times \frac{-2862.7\,\text{kJ}}{1\,\text{mol}} = -259\,\text{kJ}$$

4. From Table 8.1 in the text, we find that the specific heat of iron is 0.450 J/g·°C.

   $$0.450\frac{\text{J}}{\text{g}\cdot\text{°C}} = \frac{\text{heat}}{78.0\,\text{g} \times 15.0\text{°C}}; \quad \text{heat} = 527\,\text{J}$$

5. In this problem, the heat of the reaction is determined from the temperature change of the known quantity of solution in the calorimeter. The heat absorbed by the reaction is equal to the heat released by the solution or $q_{rxn} = -q_{soln}$. We can solve for the final temperature by using the equation

$q_{rxn} = -q_{soln} = -[(\text{sp. heat}) \times (\text{mass of soln}) \times (\Delta T)]$

7.75 g of NH_4NO_3 is equivalent to 0.0969 mol NH_4NO_3; therefore $q_{rxn} = 2.50$ kJ

$2,500 \, J = -4.18 \dfrac{J}{g \cdot {}^\circ C} \times 117.75 \, g \times \Delta T; \quad \Delta T = -5.08 \, {}^\circ C$

$\Delta T = T_f - T_i; \quad -5.08 \, {}^\circ C = T_f - 25.00 \, {}^\circ C; \quad T_f = 19.92 \, {}^\circ C$

6. From the overall reaction we know that we want to have 2 mol of Al and 1 mol of Fe_2O_3 on the reactant side and 1 mol of Al_2O_3 and 2 mol of Fe on the product side. We can use the first reaction as given since that reaction has 2 mol of Al on the reactant side. We need to reverse the second reaction that is given since we need the 1 mol of Fe_2O_3 on the reactant side. Remember, if you reverse the reaction, you must reverse the sign of ΔH°.

$2 \, Al \, (s) + \tfrac{3}{2} O_2 \, (g) \rightarrow Al_2O_3 \, (s)$ $\Delta H^\circ = -1676$ kJ

$Fe_2O_3 \, (s) \rightarrow 2 \, Fe \, (s) + \tfrac{3}{2} O_2 \, (g)$ $\Delta Ho = +824.2$ kJ

_____

$2 \, Al \, (s) + Fe_2O_3 \, (s) \rightarrow Al_2O_3 \, (s) + 2 \, Fe \, (s)$ $\Delta H^\circ = -852$ kJ

Note that the $\tfrac{3}{2} O_2$ molecules on the reactant side of the first reaction cancel out with the $\tfrac{3}{2} O_2$ molecules on the product side of the second reaction.

7. $SO_3 \, (g) + H_2O \, (l) \rightarrow H_2SO_4 \, (l)$

$\Delta H^\circ_{rxn} = [(-814.0 \text{ kJ})] - [(-395.7 \text{ kJ}) + (-285.8 \text{ kJ})] = -132.5$ kJ

$2 \, KClO_3 \, (s) \rightarrow 2 \, KCl \, (s) + 3 \, O_2 \, (g)$

$\Delta H^\circ_{rxn} = [(2 \times -436.7 \text{ kJ})] - [(2 \times -397.7 \text{ kJ})] = -78.0$ kJ

8. $CH_3CH{=}CH_2 + HCl \rightarrow CH_3CHClCH_3$

Reactant: break 1 H-Cl bond (432 kJ) and 1 C=C bond (635 kJ); Products: form 1 C-H bond (410 kJ) 1 C-Cl bond (330 kJ), and 1 C-C bond (350 kJ)

$\Delta H^\circ_{rxn} = [D(\text{H-Cl}) + D(\text{C=C})] - [D(\text{C-H}) + D(\text{C-Cl}) + D(\text{C-C})] = [432 + 635] - [410 + 330 + 350]$
$= -23$ kJ

9. $\Delta H^\circ_{rxn} = [(3 \times -393.5 \text{ kJ}) + (4 \times -285.8 \text{ kJ})] - [(-105)] = -2218.7$ kJ

-2218.7 kJ represents the heat released when 1 mol of C_3H_8 reacts.

$\dfrac{-2218.7 \, kJ}{1 \, mol \, C_3H_8} \times \dfrac{1 \, mol \, C_3H_8}{44.0 \, g \, C_3H_8} = -50.4 \dfrac{kJ}{g \, C_3H_8}$

10. $CaO\ (s)\ +\ 2\ NH_4Cl\ (s)\ \rightarrow\ 2\ NH_3\ (g)\ +\ CaCl_2\ (s)$ ΔS = positive
     1 mole   2 moles   2 moles   1 mole
     solid   solid   gas   solid

   $BaCl_2\ (aq)\ +\ Na_2SO_4\ (aq)\ \rightarrow\ BaSO_4\ (s)\ +\ 2\ NaCl\ (aq)$ ΔS = negative
   3 moles ions   3 moles ions   1 mole solid   4 moles ions

11. From the overall reaction, we know that we need 2 moles of graphite and 3 moles of hydrogen on the reactant side of the equation and 1 mole of ethane on the product side of the equation. Therefore, we need to multiply the first equation by 2 and the second equation by 3. We need to reverse the third equation and multiply it by $\frac{1}{2}$. Remember, whatever we do to the coefficients we also do to the value of ΔH^o. If we reverse an equation, we change the sign of ΔH^o.

   $2\ C\ (graphite)\ +\ 2O_2\ (g)\ \rightarrow\ 2CO_2\ (g)$ ΔH^o = -787 kJ

   $3\ H_2\ (g)\ +\ \frac{3}{2}O_2\ (g)\ \rightarrow\ 3H_2O\ (l)$ ΔH^o = -857.4 kJ

   $2CO_2\ (g)\ +\ 3H_2O\ (l)\ \rightarrow\ C_2H_6\ (g)\ +\ 7/2O_2\ (g)$ ΔH^o = 1559.8 kJ

   $2\ C\ (graphite)\ +\ H_2\ (g)\ \rightarrow\ C_2H_6\ (g)$ ΔH^o = -84.6 kJ

12. Balanced overall reaction: $N_2H_4\ (l)\ +\ 2\ H_2O_2\ (l)\ \rightarrow\ N_2\ (g)\ +\ 4\ H_2O\ (l)$

   From the above overall reaction, we know that we need 1 mole of hydrazine and 2 moles of hydrogen peroxide on the reactant side of the equation and that we need 1 mole of nitrogen and 4 moles of water on the product side. We can use the first equation as given. The third equation should be reversed and multiplied by 2, and the second equation should be multiplied by 2. Remember, whatever we do to the coefficients we also do to the value of ΔH^o. If we reverse an equation, we change the sign of ΔH^o.

   $N_2H_4\ (l)\ +\ O_2\ (g)\ \rightarrow\ N_2\ (g)\ +\ 2\ H_2O\ (l)$ ΔH^o = -621.6 kJ

   $2\ H_2O_2\ (l)\ \rightarrow\ 2H_2\ (g)\ +\ 2O_2\ (g)$ ΔH^o = 375.6 kJ

   $2H_2\ (g)\ +\ O_2\ (g)\ \rightarrow\ 2\ H_2O\ (l)$ ΔH^o = -571.6 kJ

   $N_2H_4\ (l)\ +\ 2\ H_2O_2\ (l)\ \rightarrow\ N_2\ (g)\ +\ 4\ H_2O\ (l)$ ΔH^o = -817.6 kJ

13. $\Delta G^o = \Delta H^o - T\Delta S^o$; When using this equation, be careful with your units. ΔH^o is reported in kJ/mol while ΔS^o is reported in J/mol K.

   ΔG^o = 32.9 kJ/mol – [298.15 K × 0.2265 J/mol K] = -34.6 kJ/mol

   ΔG^o is negative; therefore the reaction is spontaneous.

14. When ΔG^o = 0 the reaction changes from a nonspontaneous reaction to a spontaneous reaction. Therefore, we can calculate the temperature by using the equation:

   $$T = \frac{\Delta H^o}{\Delta S^o};\quad T = \frac{180.4\ kJ/mol}{0.4214\ kJ/mol \cdot K} = 428.1\ K = 154.9\ ^oC$$

CHAPTER 9

GASES: THEIR PROPERTIES AND BEHAVIOR

Chapter Learning Goals

1⊠ Explain how the height of a liquid in a barometer depends on the density of the liquid.
2⊠ Interconvert units of pressure.
3⊠ Know how to determine the pressure of a gas using a manometer.
4⊠ Use the ideal-gas law to calculate pressure, volume, moles of gas, or temperature, given the other three variables.
5⊠ Use the ideal-gas law to calculate final pressure, volume, moles of gas, or temperature from initial pressure, volume, moles of gas, and temperature.
6⊠ Perform stoichiometric calculations relating the mass of a reactant to the mass, moles, and volume or pressure of a gaseous product.
7⊠ Use the ideal-gas law to calculate the molar mass of a gas.
8⊠ Use the ideal-gas law to calculate the density of a gas.
9⊠ Use Dalton's law to calculate the partial pressure of a gas in a mixture.
10⊠ Use the Kinetic Molecular Theory of gases to explain each of the gas laws.
11⊠ Use Graham's law to calculate the relative rates of effusion of two different gases.
12⊠ State the conditions under which a gas is expected to behave ideally or nonideally.

Chapter in Brief

This chapter looks at the behavior of gases and how that behavior can be explained. You begin with a general description of gases and how pressures are measured. You then will learn how the behavior of gases can be defined by the four variables pressure, temperature, volume, and the number of moles and how these variables are related through the gas laws and the ideal-gas law. You will learn how to apply the ideal-gas law to stoichiometric calculations and calculations involving the density and molar mass of a gas. You will also learn how the ideal-gas law can be used to describe the behavior of a mixture of gases. Once you are able to describe the behavior of gases through the use of the gas laws, you are introduced to the Kinetic Molecular Theory, which explains the reason for that behavior. You will also be introduced to the concepts of effusion and diffusion and the difference between ideal and real gases. Finally, you will take a brief look at some of the chemistry of the atmosphere.

Gases and Gas Pressure

    A. Gases - constituent atoms or molecules have little attraction for one another.
        1. Free to move about in available volume.
    B. Some properties of gases.
        1. Mixtures are always homogeneous.
            a. very weak attraction between gas molecules
            b. identity of neighbor is irrelevant
        2. Compressible - volume contracts when pressure is applied.
            a. 0.10% of volume of gas is occupied by molecules

1⊠   C. Pressure - force exerted per unit area.
        1. SI unit = Pascal (Pa).
            a. $1 \text{ Pa} = 1 \text{ N/m}^2$ ($1 \text{ N} = 1 \text{ (kg·m)/s}^2$)
        2. Alternative units.
            a. millimeters of mercury (mm Hg) or mm Hg
            b. atmosphere (atm)

3. 1 atm = 760 mm Hg = 101,325 Pa.
D. Atmospheric pressure - pressure created from the mass of the atmosphere pressing down on the earth's surface.
   1. Standard atmospheric pressure at sea level - 760 mm Hg.
E. Measuring pressure.
   1. Barometer.
      a. (density of liquid in g/cm^3) x (height of liquid in sealed tube in cm) ×

$$\left(\frac{10^4 \text{ cm}^2}{\text{m}^2} \times \frac{1 \text{ kg}}{10^3 \text{ g}} \times 9.80665 \, \text{m}/_{\text{s}^2} \right) = \text{pressure in pascals}$$

   2. Manometer - U-tube filled with mercury with one end connected to the gas-filled container and the other end open to the atmosphere.
      a. $P_{gas} = P_{atm}$; liquid level in both arms is equal.
      b. $P_{gas} > P_{atm}$; liquid level in the arm connected to the gas-filled cylinder will be lower
         i. $P_{gas} = P_{atm} + P_{Hg}$ (P_{Hg} = the difference in the heights of the two mercury columns)
      c. $P_{gas} < P_{atm}$; liquid level in the arm open to the atmosphere will be lower
         i. $P_{gas} + P_{Hg} = P_{atm}$

EXAMPLE:
What is the pressure in mm Hg inside a container of gas connected to a mercury-filled, open-ended manometer when the level in the arm connected to the container is 22.4 mm Hg higher than the level in the arm open to the atmosphere, and the atmospheric pressure reading outside the apparatus is 672.2 mm Hg?

SOLUTION:
$P_{gas} < P_{atm}$ since the level of Hg is lower in the arm open to the atmosphere; therefore $P_{gas} + P_{Hg} = P_{atm}$ and $P_{gas} = 672.2$ mm Hg - 22.4 mm Hg = 649.8 mm Hg

The Gas Laws
A. Different gases show similar physical behavior.
   1. Defined by four variables - pressure, temperature, volume, and number of moles.
      a. relationships of variables - gas laws
B. Boyle's law - relationship between volume and pressure at constant temperature.
   1. $V \propto 1/P$.
C. Charles' law - relationship between volume and temperature at constant pressure.
   1. $V \propto T$ (temperature is expressed in Kelvin).
D. Avogadro's law - relationship between volume and amount of gas at constant pressure and temperature.
   1. $V \propto n$ (where n = number of moles of gas).
   2. 1 mol of gas at 273.15 K and 1.00 atm = 22.4 L of gas.

The Ideal-Gas Law
A. Describes how the volume of a gas is affected by changes in pressure, temperature, and amount.
   1. $PV = nRT$; $R = gas \; constant = 0.08206 \, \dfrac{\text{L} \cdot \text{atm}}{\text{K} \cdot \text{mol}}$.

B. Standard temperature and pressure (STP) T = 273.15 K; P = 1 atm.

139

4⊠ **EXAMPLE:**

What is the volume of 3.57 g of O_2 at a temperature of $18.5°$ C and a pressure of 0.563 atm?

SOLUTION:

$$V = \frac{nRT}{P}; \quad n = 3.57 \text{ g } O_2 \times \frac{1 \text{ mol } O_2}{32.0 \text{ g } O_2} = 0.112 \text{ mol } O_2; \quad T = 273.15 + 18.5 = 291.6 \text{ K}$$

$$V = \frac{0.112 \text{ mol } O_2 \times 0.08206 \dfrac{L \cdot atm}{mol \cdot K} \times 291.6 \text{ K}}{0.563 \text{ atm}} = 4.76 \text{ L}$$

5⊠ **EXAMPLE:**

What would be the volume of a 125 mL sample of gas at $23.5°$ C and a pressure of 754 mm Hg, if the gas is compressed to 725 mm Hg at a temperature of $18.7°$ C?

SOLUTION:

In this problem the number of moles of gas remains constant so we can use the following equation:

$$nR = \left(\frac{PV}{T}\right)_{initial} = \left(\frac{PV}{T}\right)_{final}; \quad V_{final} = \frac{T_{final} \times (PV)_{initial}}{T_{initial} \times P_{final}};$$

$$V_{final} = \frac{291.8 \text{ K} \times 754 \text{ mm Hg} \times 125 \text{ mL}}{296.6 \text{ K} \times 725 \text{ mm Hg}} = 128 \text{ mL}$$

Note that the units mm Hg cancel, so it isn't necessary to convert to atm.

6,7,8⊠ Stoichiometric Relationships with Gases

A. Stoichiometric calculations involve the application of the ideal-gas law.

EXAMPLE:

How many liters of oxygen are needed to completely react with 15.75 g of propane (C_3H_8) at $400°$ C and 3.75 atm?

SOLUTION:

$$C_3H_8 (g) + 5 O_2 (g) \rightarrow 3 CO_2 (g) + 4 H_2O (g)$$

$$15.75 \text{ g } C_3H_8 \times \frac{1 \text{ mol } C_3H_8}{44.0 \text{ g } C_3H_8} = 0.358 \text{ mol } C_3H_8$$

$$0.358 \text{ mol } C_3H_8 \times \frac{5 \text{ mol } O_2}{1 \text{ mol } C_3H_8} = 1.79 \text{ mol } O_2$$

$$V = \frac{1.79 \text{ mol} \times 0.08206 \dfrac{L \cdot atm}{mol \cdot K} \times 673.2 \text{ K}}{3.75 \text{ atm}} = 26.4 \text{ L}$$

B. Calculations to determine the density of a gas.

EXAMPLE:
What is the density of 0.275 g of NO at 758 mm Hg and 23.5° C?

SOLUTION:
To solve for density, we need to know the volume that the gas occupies. We can use the ideal-gas law to solve for volume if we first convert grams of NO to moles of NO.

$$0.275 \text{ g NO} \times \frac{1 \text{ mol NO}}{30.0 \text{ g NO}} = 0.009\ 17 \text{ mol NO}$$

$$V = \frac{0.009\ 17 \text{ mol NO} \times 0.082\ 06 \dfrac{\text{L} \cdot \text{atm}}{\text{mol} \cdot \text{K}} \times 296.6 \text{ K}}{0.997 \text{ atm}} = 0.224 \text{ L}$$

$$d = \frac{0.275 \text{ g NO}}{0.224 \text{ L NO}} = 1.23 \text{ g} / \text{L}$$

C. Calculations to determine the molar mass of a gas.

EXAMPLE:
The density of a gas was found to be 3.79 g/L at 45.0° C and 2.25 atm. What is the molar mass of the gas?

SOLUTION:
To solve this problem with the information given, we need to do a little thinking. First, molar mass is the number of grams divided by the number of moles, n.

$$\text{molar mass} = \frac{\text{g}}{n}.$$

We can rearrange this equation so that the number of moles of a gas is equal to the grams of gas divided by its molecular weight or $n = \dfrac{\text{g}}{\text{molar mass}}$. We can also rearrange the ideal gas law to solve for the number of moles of gas.

$$n = \frac{PV}{RT}$$

We can now equate the last two equations and rearrange them so that we are solving for the molar mass of the gas. (Keep in mind that density is grams/volume.)

$$\frac{\text{g}}{\text{molar mass}} = \frac{PV}{RT}; \quad \text{molar mass} = \frac{\text{g}}{V} \times \frac{RT}{P} \text{ or molar mass} = d \times \frac{RT}{P}$$

$$\text{molar mass} = 3.79 \frac{\text{g}}{\text{L}} \times \frac{0.082\ 06 \dfrac{\text{L} \cdot \text{atm}}{\text{mol} \cdot \text{K}} \times 318.2 \text{ K}}{2.25 \text{ atm}} = 44.0 \frac{\text{g}}{\text{mol}}$$

141

Partial Pressure and Dalton's Law

A. Gas laws apply to mixtures of gases.

B. Dalton's law of partial pressures - $P_{total} = P_1 + P_2 + P_3 + ...$ at constant V, T, where P_1, P_2, ... refer to the pressures of the individual gases in the mixture.

C. Partial pressures - refer to the pressure each individual gas would exert if it were alone in the container (P_1, P_2, ...).

  1. Total pressure depends on the total molar amount of gas present.

  2. Mole fraction (X) - the number of moles of the component divided by the total number of moles in the mixture.

    a. Mole fraction $(X) = \dfrac{\text{moles of component}}{\text{total moles in mixture}}$

  3. $P_1 = X_1 \cdot P_{total}$.

9⊠   ***EXAMPLE:***

A 5.0 L flask at 25°C contains N_2 at a partial pressure of 0.28 atm, He at a partial pressure of 0.12 atm, and Ne at a partial pressure of 0.56 atm. What is the total pressure of the mixture? What is the mole fraction of each gas?

SOLUTION: $P_{total} = 0.28\ \text{atm} + 0.12\ \text{atm} + 0.56\ \text{atm} = 0.96\ \text{atm}$

$$X_{N_2} = \frac{0.28\ \text{atm}}{0.96\ \text{atm}} = 0.29 \ ; \quad X_{He} = \frac{0.12\ \text{atm}}{0.96\ \text{atm}} = 0.12 \ ; \quad X_{Ne} = \frac{0.56\ \text{atm}}{0.96\ \text{atm}} = 0.58\ \text{atm}$$

The Kinetic Molecular Theory of Gases

A. Model that explains the behavior of gases.

B. Assumptions.

  1. A gas consists of particles in constant random motion.

  2. Most of the volume of a gas is empty space.

  3. The attractive forces between molecules of a gas are negligible.

  4. The total kinetic energy of the gas particles is constant at constant T.

  5. Avg. K.E. $\propto T$.

C. Above assumptions can be used to explain the gas laws (see page 356 in your text).

Graham's Law: Diffusion and Effusion of Gases

A. Consequences of constant motion and high velocities of gas particles.

  1. Gases mix rapidly when they come in contact.

    a. diffusion - mixing of different gases by random molecular motion and with frequent collisions

    b. effusion - a process in which gas molecules escape through a tiny hole in a membrane without collisions

B. Graham's law - the rate of effusion of a gas is inversely proportional to the square root of its molar mass.

  1. Rate $\propto \dfrac{1}{\sqrt{M}}$.

  2. Two gases at the same temperature and pressure - $\dfrac{\text{Rate}_1}{\text{Rate}_2} = \sqrt{\dfrac{M_2}{M_1}}$.

    a. can use different rates of effusion to separate a gas mixture into separate components

11⊠   *EXAMPLE:*
Calculate the ratio of effusion rates of NO_2 and SO_3 from the same container at the same temperature and pressure.

SOLUTION: $\dfrac{\text{Rate of effusion of } NO_2}{\text{Rate of effusion of } SO_3} = \sqrt{\dfrac{80.0 \text{ g } SO_3 / \text{mol}}{46.0 \text{ g } NO_2 / \text{mol}}} = 1.32$

12⊠   **The Behavior of Real Gases**
   A. Ideal Gas.
      1. No attractive forces between molecules - true at low pressures.
      2. Molecular volume - most of the volume of a gas is empty space - true at low pressures.
   B. Real Gas.
      1. Attractive forces are more important at higher pressures
         a. decreases the volume of the gas from that predicted by the ideal gas law
      2. Molecular volume - actual volume of a gas at high pressure is larger than predicted by the ideal-gas law.
   C. Two effects - molecular volume and intermolecular forces cancel out at intermediate pressures.

The Earth's Atmosphere
   A. Troposphere - the region nearest the earth's surface.
      1. Has greatest effect on the earth's surface.
      2. Air pollution.
         a. release of unburned hydrocarbon molecules
         b. production of NO - produces photochemical smog
      3. Acid rain.
         a. results from the production of SO_2
         b. $2 SO_2 (g) + O_2 (g) \rightarrow 2 SO_3 (g)$; $SO_3 (g) + H_2O (l) \rightarrow H_2SO_4 (aq)$
      4. The greenhouse effect and global warming.
         a. CO_2 absorbs radiation from the sun
         b. recent increase in the concentration of atmospheric carbon dioxide
         c. may lead to widespread global warming
   B. Ozone layer - an atmospheric band stretching from about 20 to 40 km above the earth's surface.
      1. Absorbs intense ultraviolet radiation from the sun.
         a. shields high-energy solar radiation from reaching the earth's surface
      2. Ozone depletion - due to the presence in the stratosphere of chlorofluorocarbons (CFCs).
         a. reaction sequence - a chain reaction in which the generation of a few chlorine atoms leads to the destruction of a great many ozone molecules.

Self-Test

This section is intended to test your knowledge of the material covered in this chapter. Think through these problems and make certain you understand what is going on. Ask yourself if your answer makes sense. Many of these questions are linked to the chapter learning goals. Therefore, successful completion of these problems indicates you have mastered the learning goals for this chapter. You will receive the greatest benefit from this section if you use it as a mock exam. You will then discover which topics you have mastered and which topics you need to study in more detail.

Multiple Choice

1. The SI unit for pressure is
   a. Newton
   b. mm Hg
   c. Pascal
   d. atmosphere

2. When the liquid level in the arm connected to the gas-filled cylinder is lower than the liquid level in the arm open to the atmosphere in an open end manometer
   a. $P_{gas} = P_{atm}$
   b. $P_{gas} > P_{atm}$
   c. $P_{gas} < P_{atm}$
   d. $P_{gas} = P_{Hg}$

3. The gas law that states the relationship between volume and pressure at constant temperature is
   a. Boyle's law
   b. Charles' law
   c. Avogadro's law
   d. Ideal-gas law

4. The gas law that states the relationship between volume and amount of gases at constant pressure and temperature is
   a. Boyle's law
   b. Charles' law
   c. Avogadro's law
   d. Ideal-gas law

5. The model used to explain the behavior of gases is
   a. Ideal-Gas law
   b. Dalton's law of partial pressure
   c. Kinetic Molecular Theory
   d. Graham's law of effusion

6. The mixing of different gases by random molecular motion and with frequent collisions is
   a. diffusion
   b. effusion
   c. confusion
   d. stirring

7. Acid rain is a consequence of
   a. the release of unburned hydrocarbon molecules
   b. the production of SO_2
   c. the production of HNO_3
   d. an increased concentration of CO_2 in the atmosphere

8. Different gases
   a. can only be described based upon their chemical properties
   b. show similar physical behavior
   c. produce heterogeneous mixtures when combined
   d. occupy different volumes when one mole is present

9. The region of the atmosphere that has the greatest effect on the earth's surface is
   a. the ozone layer
   b. the mesosphere
   c. the troposphere
   d. the thermosphere

Matching

Gas	a. the rate of effusion of a gas is inversely proportional to the square root of its molar mass.
Pressure	b. describes how the volume of a gas is affected by changes in pressure, temperature, and amount of gas.
Atmospheric pressure	c. the pressure each individual gas would exert if it were alone in the container.
Boyle's law	d. a process in which gas molecules escape through a tiny hole in a membrane without collisions.
Charles' law	e. a model that explains the behavior of gases.
Avogadro's law	f. the total pressure exerted by a mixture of gases in a container at constant V and T is equal to the sum of the pressures exerted by each individual gas in the container.
Ideal-gas law	g. mixing of different gases by random molecular motion and with frequent collisions.
Dalton's law	h. states the relationship between volume and pressure at constant temperature and constant amount of gas.
Partial pressures	i. states the relationship between volume and amount at constant pressure and temperature.
Kinetic Molecular Theory	j. a substance whose constituent atoms or molecules have little attraction for one another.
Graham's law	k. force exerted per unit area.
Diffusion	l. pressure created from the mass of the atmosphere pressing down on the earth's surface.
Effusion	m. states the relationship between volume and temperature at constant pressure.

Fill-in-the-Blank

1. A gas can be compressed because _____ .

2. If $P_{gas} < P_{atm}$, in an open end manometer, the liquid level in the arm connected to the gas-filled cylinder will be _____ the liquid in the arm open to the atmosphere.

3. Standard temperature and pressure are defined to be _____ .

4. The total pressure of a mixture of gases depends on _____

   _____ .

5. Mole fraction is _____

   _____ .

6. The average kinetic energy of a gas is proportional to the _____ .

7. A gas mixture can be separated into its constituent components by using the different

   _____ of the gases in the mixture.

8. At higher pressures, the intermolecular forces of gases _____ causing the

   volume of the gas to _____ .

9. The actual volume of a gas at very high pressure is _____ than that predicted by the ideal-gas law.

10. The depletion of the ozone layer is due to the presence of _____ in the stratosphere.

Problems

1. An open-ended manometer containing mercury is connected to a container of gas. What is the pressure of the gas (in mm Hg) when: a) the level of mercury in the arm connected to the gas is 38 mm lower than in the arm connected to the atmosphere and the atmospheric pressure is 784 mm Hg; and b) the level of mercury in the arm connected to the gas is 23 mm higher than in the arm connected to the atmosphere and the atmospheric pressure is 747 mm Hg?

2. Assume that you are using an open-ended manometer filled with silicon oil rather than mercury. What is the gas pressure in mm Hg if the level of silicon oil in the arm connected to the bulb is: a) 140 mm lower and P_{atm} = 760 mm Hg; and b) 175 mm higher and P_{atm} = 760 mm Hg? The density of the oil is 1.30 g/mL and the density of mercury is 13.6 g/mL.

3. A 500 mL flask contains 0.40 g of O_2 at a temperature of 23.5° C. What is the pressure of the gas?

4. Calculate the mass in grams of 250 mL of NO_2 at STP.

5. A student carried out a reaction in the lab in which one of the products was a gas. She collected the gas for analysis and found that it contained 82.8% carbon and 17.2% hydrogen. She also observed that 350 mL of the gas at 23° C and 757 mm Hg had a mass of 0.835 g. a) What is the empirical formula of the gas? b) What is the molar mass of the gas? c) What is its molecular formula?

6. What is the density of 3.45 g of H_2S gas at 25° C and 770 mm Hg?

7. What is the molar mass of a gas with a density of 1.49 g/L at 745 mm Hg and 18° C?

8. The oxidation of ammonia is an important reaction in the production of fertilizers,

$$4 NH_3 (g) + 5 O_2 (g) \rightarrow 4 NO (g) + 6 H_2O (g)$$

How many liters of NO at 500°C and 735 mm Hg can be produced from 75 L of O_2 at 100°C and 650 mm Hg?

9. 175 mL of O_2 at 30°C and 723 mm Hg were mixed with 275 mL of NH_3 at 50°C and 613 mm Hg and were transferred to a 500 mL vessel where they underwent reaction according to the above chemical equation. What will be the total pressure (in mm Hg) in the reaction vessel at 200°C after the reaction goes to completion?

10. Calculate the volume of carbon dioxide measured at STP that would be produced from the combustion of 8.92 g of propane (C_3H_8).

11. A gas occupying 575 mL at 23°C is compressed to 350 mL at constant pressure. What is the final temperature?

12. A gas mixture consisting of CH_4, C_3H_8, and C_4H_{10} has a total pressure of 2.0 atm. What are the mole fractions of each gas if the partial pressures are 0.68 atm, 1.05 atm, and 0.27 atm, respectively?

13. Three gases consisting of 4.0 g O_2, 3.0 g CO_2, and an unknown amount of N_2, were added to the same 15.0 L container to give a total pressure of 950 mm Hg at 28.0° C. Calculate a) the total number of moles of gas in the container, b) the mole fraction of each gas, c) the partial pressure of each gas, and d) the number of grams of N_2 in the container.

14. Explain Boyle's law using the Kinetic Molecular Theory.

15. Calculate the ratio of effusion rates of He and Ar from the same container at the same temperature and pressure.

Solutions

Multiple Choice
1. c
2. b
3. a
4. c
5. c
6. a
7. b
8. b
9. c

Matching
Gas - j
Pressure - k

Atmospheric pressure - 1
Boyle's law - h
Charles' law - m
Avogadro's law - i
Ideal-gas law - b
Dalton's law - f
Partial pressures - c
Kinetic Molecular Theory - e
Graham's law - a
Diffusion - g
Effusion - d

Fill-in-the-Blank

1. only 0.10% of the volume of a gas is occupied by the molecules
2. higher than
3. 273.15 K and 1 atm
4. the total molar amount of gas present, temperature, volume, and the partial pressure of gases in a mixture.
5. the number of moles of a component of a gas mixture divided by the total number of moles in the mixture.
6. absolute temperature
7. rates of effusion
8. increases; decrease.
9. larger
10. chlorofluorocarbons

Problems

1. a. If the level of mercury in the arm connected to the gas is lower, then $P_{gas} = P_{atm} + P_{Hg}$; $P_{gas} =$ 784 mm Hg + 38 mm Hg = 822 mm Hg.
    b. If the level of mercury in the arm connected to the gas is higher, then $P_{gas} + P_{Hg} = P_{atm}$; $P_{gas} =$ 747 mm Hg - 23 mm Hg = 724 mm Hg

2. The relationship between the heights of columns of fluids in a manometer can be stated as

$$h_b = h_a \times \frac{d_a}{d_b}$$

In the above equation, $h_a = P_{Hg}$ and h_b = the height of the silicon oil column.

a. 140 mm oil $\times \dfrac{1.30 \text{ g} / \text{mL oil}}{13.6 \text{ g} / \text{mL Hg}} = 13.4$ mm Hg ; $P_{gas} = P_{atm} + P_{Hg} = 760$ mm Hg + 13.4 mm Hg = 773 mm Hg

b. 175 mm oil $\times \dfrac{1.30 \text{ g} / \text{mL oil}}{13.6 \text{ g} / \text{mL Hg}} = 16.7$ mm Hg ; $P_{gas} + P_{Hg} = P_{atm}$; 760 mm Hg - 16.7 mm Hg = 743 mm Hg

3. $0.40 \text{ g O}_2 \times \dfrac{1 \text{ mol O}_2}{32.0 \text{ g O}_2} = 0.0125 \text{ mol O}_2$;

$$P = \dfrac{0.0125 \text{ mol O}_2 \times 0.082\ 06\ \dfrac{\text{L} \cdot \text{atm}}{\text{mol} \cdot \text{K}} \times 296.6 \text{ K}}{0.500 \text{ L}} = 0.61 \text{ atm}$$

4. $n = \dfrac{1 \text{ atm} \times 0.250 \text{ L}}{0.082\ 06\ \dfrac{\text{L} \cdot \text{atm}}{\text{mol} \cdot \text{K}} \times 273 \text{ K}} = 0.0112 \text{ mol}$; $0.0112 \text{ mol NO}_2 \times \dfrac{46.0 \text{ g NO}_2}{1 \text{ mol NO}_2} = 0.515 \text{ g NO}_2$

5. a. Assume a 100 g sample and convert 82.8 g C and 17.2 g H to moles.

$$82.8 \text{ g C} \times \dfrac{1 \text{ mol C}}{12.0 \text{ g C}} = 6.90 \text{ mol C} ; \quad 17.2 \text{ g H} \times \dfrac{1 \text{ mol H}}{1.01 \text{ g H}} = 17.0 \text{ mol H} ;$$

This gives rise to a 6.9:17.0 C to H mole ratio.

$$\dfrac{17.0}{6.9} = 2.5 ;$$

For every mole of carbon there are 2.5 moles of hydrogen or a 1:2.5 mole ratio. However, formulas must consist of whole numbers, so we need to multiply both numbers in the ratio by 2 which gives a 2:5 mole ratio and the empirical formula C_2H_5.

b. To solve for molar mass, we need to find the number of moles of gas since we already know the mass of the gas.

$$n = \dfrac{0.996 \text{ atm} \times 0.350 \text{ L}}{0.082\ 06\ \dfrac{\text{L} \cdot \text{atm}}{\text{mol} \cdot \text{K}} \times 296 \text{ K}} = 0.0144 \text{ mol} ; \quad \dfrac{0.835 \text{ g}}{0.0144 \text{ mol}} = 58.0 \text{ g / mol}$$

c. We find the molecular formula by finding the multiplier (dividing the molar mass of the compound by the molar mass of C_2H_5).

$$\dfrac{58.1 \text{ g / mL}}{29.0 \text{ g / mL}} = 2 ; \text{ The molecular formula is } C_4H_{10}.$$

6. $3.45 \text{ g H}_2\text{S} \times \dfrac{1 \text{ mol H}_2\text{S}}{34.0 \text{ g H}_2\text{S}} = 0.101 \text{ mol H}_2\text{S}$;

$$V = \dfrac{0.101 \text{ mol H}_2\text{S} \times 0.082\ 06\ \dfrac{\text{L} \cdot \text{atm}}{\text{mol} \cdot \text{K}} \times 298 \text{ K}}{1.01 \text{ atm}} = 2.45 \text{ L}$$

$$d = \dfrac{3.45 \text{ g}}{2.45 \text{ L}} = 1.41 \text{ g / mL}$$

7. $\text{molar mass} = 1.49\ \dfrac{\text{g}}{\text{L}} \times \dfrac{0.082\ 06 \dfrac{\text{L} \cdot \text{atm}}{\text{mol} \cdot \text{K}} \times 291 \text{ K}}{0.980 \text{ atm}} = 36.3 \text{ g / mol}$

8. First, find the number of moles of O_2, then convert to the number of moles of NO.

$$n = \frac{0.855 \text{ atm} \times 75 \text{ L}}{0.082\ 06 \dfrac{\text{L} \cdot \text{atm}}{\text{mol} \cdot \text{K}} \times 373 \text{ K}} = 2.10 \text{ mol} ; \quad 2.10 \text{ mol O}_2 \times \frac{4 \text{ mol NO}}{5 \text{ mol O}_2} = 1.68 \text{ mol NO}$$

We can now calculate the number of liters of NO.

$$V = \frac{nRT}{P} = \frac{1.68 \text{ mol NO} \times 0.082\ 06 \dfrac{\text{L} \cdot \text{atm}}{\text{mol} \cdot \text{K}} \times 773 \text{ K}}{0.967 \text{ atm}} = 110 \text{ L}$$

9. To solve this problem, we must first calculate the number of moles of NH_3 and O_2 and convert to the number of moles of NO and H_2O. Once we know the <u>total</u> number of moles of product, we can then convert to the total pressure.

$$n = \frac{0.951 \text{ atm} \times 0.175 \text{ L}}{0.082\ 06 \dfrac{\text{L} \cdot \text{atm}}{\text{mol} \cdot \text{K}} \times 303 \text{ K}} = 6.69 \times 10^{-3} \text{ mol O}_2 ;$$

$$n = \frac{0.807 \text{ atm} \times 0.275 \text{ L}}{0.082\ 06 \dfrac{\text{L} \cdot \text{atm}}{\text{mol} \cdot \text{K}} \times 323 \text{ K}} = 8.37 \times 10^{-3} \text{ mol NH}_3$$

From the balanced equation, we find that the oxygen is the limiting reactant.

$$6.69 \times 10^{-3} \text{ mol O}_2 \times \frac{4 \text{ mol NH}_3}{5 \text{ mol O}_2} = 5.35 \times 10^{-3} \text{ mol NH}_3$$

We need to calculate the number of moles of NO and H_2O based on the number of moles of O_2.

$$6.69 \times 10^{-3} \text{ mol O}_2 \times \frac{4 \text{ mol NO}}{5 \text{ mol O}_2} = 5.35 \times 10^{-3} \text{ mol NO} ;$$

$$6.69 \times 10^{-3} \text{ mol O}_2 \times \frac{6 \text{ mol H}_2\text{O}}{5 \text{ mol O}_2} = 8.03 \times 10^{-3} \text{ mol H}_2\text{O}$$

total number of moles produced = $(5.35 \times 10^{-3}) + (8.03 \times 10^{-3}) = (1.34 \times 10^{-2})$

mol NH_3 remaining = $(8.37 \times 10^{-3}) - (5.35 \times 10^{-3}) = 3.02 \times 10^{-3}$

total number of moles present = $(1.34 \times 10^{-2}) + (3.02 \times 10^{-3}) = 1.64 \times 10^{-2}$ mol

We can now solve for the total pressure of the products by using the total number of moles present along with the information given.

$$P = \frac{1.64 \times 10^{-2} \text{ mol} \times 0.082\ 06 \dfrac{\text{L} \cdot \text{atm}}{\text{mol} \cdot \text{K}} \times 473 \text{ K}}{0.500 \text{ L}} = 1.27 \text{ atm} ;$$

$$1.27 \text{ atm} \times \frac{760 \text{ mm Hg}}{1 \text{ atm}} = 965 \text{ mm Hg}$$

10. The balanced equation is:

$$C_3H_8 \ (g) + \ 5 \ O_2 \ (g) \ \rightarrow \ 3 \ CO_2 \ (g) \ + \ 4 \ H_2O \ (g)$$

$$8.92 \text{ g } C_3H_8 \times \frac{1 \text{ mol } C_3H_8}{44.0 \text{ g } C_3H_8} = 0.203 \text{ mol } C_3H_8 \ ;$$

$$0.203 \text{ mol } C_3H_8 \times \frac{3 \text{ mol } CO_2}{1 \text{ mol } C_3H_8} = 0.609 \text{ mol } CO_2$$

$$V = \frac{0.609 \text{ mol} \times 0.082 \ 06 \ \dfrac{L \cdot atm}{mol \cdot K} \times 273 \text{ K}}{1 \text{ atm}} = 13.6 \text{ L}$$

11. The number of moles and pressure of the gas is being held constant, so we can use the equation

$$\frac{nR}{P} = \left(\frac{V}{T}\right)_{initial} = \left(\frac{V}{T}\right)_{final} \ ; \quad \frac{575 \text{ mL}}{296 \text{ K}} = \frac{350 \text{ mL}}{T_{final}} \ ; \quad T_{final} = \frac{350 \text{ mL} \times 296 \text{ K}}{575 \text{ mL}} = 180 \text{ K} \ ;$$

$$180 - 273.15 = -93^\circ C$$

12. $X_{CH_4} = \dfrac{0.68 \text{ atm}}{2.0 \text{ atm}} = 0.34 \ ; \quad X_{C_3H_8} = \dfrac{1.05 \text{ atm}}{2.0 \text{ atm}} = 0.52 ; \quad X_{C_4H_{10}} = \dfrac{0.27 \text{ atm}}{2.0 \text{ atm}} = 0.14$

13. a. Since the total pressure depends on the total molar amount of gas present, we can write

$$P_{total} = (n_1 + n_2 + n_3)\left(\frac{RT}{V}\right) \ ; \quad \text{We can determine the total number of moles of gas present by}$$

solving for $(n_1 + n_2 + n_3)$.

$$(n_1 + n_2 + n_3) = \frac{1.25 \text{ atm} \times 15.0 \text{ L}}{\left(0.08206 \dfrac{atm \cdot L}{mol \cdot K}\right) 301 \text{ K}} = 0.759 \text{ mol}$$

b. To determine the mole fraction of each gas, we need to know the number of moles of each gas present.

$$4.0 \text{ g } O_2 \times \frac{1 \text{ mol } O_2}{32.0 \text{ g } O_2} = 0.125 \text{ mol } O_2 \ ; \quad X_{O_2} = \frac{0.125 \text{ mol } O_2}{0.759 \text{ total mol}} = 0.165$$

$$3.0 \text{ g } CO_2 \times \frac{1 \text{ mol } CO_2}{44.0 \text{ g } CO_2} = 0.068 \text{ mol } CO_2 \ ; \quad X_{CO_2} = \frac{0.068 \text{ mol } CO_2}{0.759 \text{ total mol}} = 0.090$$

$$0.759 \text{ total mol} - 0.125 \text{ mol } O_2 - 0.068 \text{ mol } CO_2 = 0.566 \text{ mol } N_2$$

$$X_{N_2} = \frac{0.566 \text{ mol N}_2}{0.759 \text{ total mol}} = 0.746$$

c. $P_{O_2} = 0.165 \times 1.25 \text{ atm} = 0.206 \text{ atm}$; $P_{CO_2} = 0.090 \times 1.25 \text{ atm} = 0.113 \text{ atm}$;

$P_{N_2} = 0.746 \times 1.25 \text{ atm} = 0.932 \text{ atm}$

d. $0.566 \text{ mol N}_2 \times \dfrac{28.0 \text{ g N}_2}{1 \text{ mol N}_2} = 12.3 \text{ g N}_2$

14. Gas pressure is created when the gas particles collide with the walls of the container. If the volume of the container is decreased while the temperature and number of moles of gas is constant, there will be more collisions between the gas particles and the walls of the container, causing an increase in the pressure.

15. $\dfrac{\text{Rate of effusion of He}}{\text{Rate of effusion of Ar}} = \sqrt{\dfrac{39.9 \text{ g}/\text{mL}}{4.00 \text{ g}/\text{mL}}} = 3.16$

CHAPTER 10

LIQUIDS, SOLIDS, AND CHANGES OF STATE

Chapter Learning Goals

1⚛ Using only VSEPR geometries and electronegativity trends, determine whether a molecule is expected to be polar.

2⚛ Identify the major type of intermolecular force present in substances and determine which of two substances exhibits the stronger intermolecular force.

3⚛ For a phase change, determine whether enthalpy is increasing or decreasing and whether entropy is increasing or decreasing.

4⚛ Use the $\Delta G = \Delta H - T\Delta S$ equation to calculate the entropy change for a phase change or the temperature (boiling point, melting point, sublimation point) at which the phase change occurs.

5⚛ Use the Clasius-Clapeyron equation to calculate vapor pressure or heat of vaporization.

6⚛ For metals crystallizing in one of the three cubic unit cells, determine the number of atoms, mass, volume, density, atomic radius, and packing efficiency.

7⚛ Sketch a phase diagram, labeling the axes and each of the regions, and locate the triple point, critical point, normal melting point, and the normal boiling point. Use the phase diagram to describe physical changes.

Chapter In Brief

Unlike gases, liquids and solids have strong attractive forces between the particles. In this chapter, you will examine the nature of these attractive forces and learn how they arise from the polarity of the particles as well as how these forces affect the properties of particular liquids and solids. You will also learn about the relationship between these attractive forces and the transitions between the three states of matter. You will examine the relationship between the transitions that occur between gases, liquids, and solids, and the ΔG that accompanies these changes along with the effect that temperature and pressure has on these transitions. You will discover the different types of solids and how you can use information on the structure of the unit cell to calculate both the radius and density of metals. You end this chapter by taking a look at the description of ionic and covalent network solids.

Polar Covalent Bonds and Dipole Moments

    A. Bond dipole - a bond that has a partial-positive (δ^+) and partial-negative (δ^-) end due to the difference in electronegativity of the atoms in the bond.

        1. Represented by ↤→; indicates direction of electron displacement.
            a. point of arrow represents δ^- end of dipole
            b. crossed end represents δ^+ end of dipole

1⚛   B. Polar molecules - due to the net sum of individual bond polarities and lone-pair contributions in the molecule.

        1. Molecular dipoles - center of mass of positive charge (nuclei) doesn't coincide with the center of mass of all negative charges (electrons).
            a. leads to a net polarity in molecule

    C. Dipole moment (μ) - the magnitude of the charge Q at either end of the molecular dipole times the distance r between the charges.

2☒   **Intermolecular Forces**
    A. Intermolecular forces - attractive forces between molecules that hold them together at certain temperatures.
       1. Van der Waals forces.
       2. Divided into categories.
          a. ion-dipole
          b. dipole-dipole
          c. London dispersion forces
          d. hydrogen bonding
       3. Electrical in nature.
    B. Ion-dipole forces - result of electrical interactions between an ion and the partial charges on a polar molecule.
       1. Interaction energy $E = z{\cdot}\mu/r^2$.
       2. Important in aqueous solutions of ionic substances where dipolar water molecules surround the ions.
    C. Dipole-dipole forces - result from electrical interactions among dipoles on neighboring molecules.
       1. Generally weak.
          a. strength depends on the sizes of the dipole moments involved
       2. Significant only when molecules are in close contact.
       3. Correlation between dipole moment and boiling point.
          a. high dipole moment corresponds to strong intermolecular forces
          b. substance must overcome intermolecular forces to boil
          c. stronger intermolecular forces require higher boiling points
    D. London Dispersion Forces - result from the motion of electrons around atoms.
       1. Any given instance, electron distribution may be unsymmetrical.
          a. creates a short-lived dipole moment (instantaneous dipole)
       2. Instantaneous dipole induces a temporary dipole on neighboring atoms.
       3. Weak attractive forces.
       4. Polarizability - ease with which a molecule's electron cloud can be distorted by a nearby electric field.
          a. smaller molecules and lighter atoms with fewer electrons
             i. relatively nonpolarizable
             ii. smaller dispersion forces
          b. larger molecules and heavier atoms with more electrons
             i. more polarizable
             ii. larger dispersion forces
          c. molecules with more spread-out shapes
             i. maximize molecular surface area
             ii. greater contact between molecules
             iii. give rise to higher dispersion forces
    E. Hydrogen Bonds - an attractive interaction between a hydrogen atom bonded to an electronegative O, N, or F atom and an unshared electron pair on another nearby electronegative atom.
       1. Quite strong.
       2. Responsible for water's remarkable properties.
       3. Give rise to higher boiling points than might be expected.
       4. Combination of forces.
          a. dipole-dipole interactions
             i. H-F, H-O, H-N bonds are highly polar
          b. hydrogen can be approached very closely
             i. no core electrons and small size

EXAMPLE:

Order the following molecules by increasing strength of intermolecular forces: C_2H_5OH, PH_3, SF_6.

SOLUTION: The order is $SF_6 < PH_3 < C_2H_5OH$. SF_6 has an octahedral geometry and no lone pairs; therefore it is a nonpolar molecule. The only intermolecular forces present are London forces. PH_3 is a polar compound. The strongest type of intermolecular forces present are dipole-dipole forces. C_2H_5OH has an hydrogen atom bonded to oxygen. The strongest type of intermolecular forces present is hydrogen bonding.

Some Properties of Liquids
  A. Viscosity - the measure of a liquid's resistance to flow.
    1. SI unit - $N \cdot s/m^2$.
    2. Ease with which molecules move around in the liquid.
    3. Related to intermolecular forces.
      a. stronger the forces the greater the surface tension
  B. Surface tension - the resistance of a liquid to spreading out and increasing its surface area.
    1. Due to the difference in intermolecular forces felt by the molecules on the surface of the liquid and the molecules in the interior of the liquid.
    2. Related to intermolecular forces.
      a. stronger the forces the greater the viscosity
  C. Properties are temperature dependent.
    1. An increase in temperature corresponds to an increase in kinetic energy.
      a. molecules with high kinetic energies can more easily overcome intermolecular forces

Phase Changes
  A. Phase changes (changes of state) - the physical form but not the chemical identity of a substance changes.
    1. Sublimation - solid changes directly into a gas.
3⊠ B. Phase change is associated with a free-energy change, ΔG.
    1. $\Delta G = \Delta H - T\Delta S$.
    2. Enthalpy part - energy change associated with making or breaking the intermolecular attractions that hold liquids and solids together.
    3. Entropy part - associated with the change in disorder between various states.
    4. Solid → liquid, solid → gas, liquid → gas: ΔH and ΔS are both positive.
    5. Gas → liquid, gas → solid, liquid → solid: ΔH and ΔS are both negative.
4⊠   6. Can calculate temperature at which two phases are in equilibrium knowing ΔH and ΔS for a phase transition.
      a. at equilibrium: $\Delta G = 0$; $T = \Delta H/\Delta S$

EXAMPLE:

The boiling point for NH_3 is $-33.4°$ C, and the $\Delta H°_{vap} = 23.4$ kJ/mol. Calculate $\Delta S°_{vap}$ for NH_3.

SOLUTION:
  When substances boil, $\Delta G°_{vap} = 0$. Therefore, $T_b = \Delta H°_{vap}/\Delta S°_{vap}$. Rearranging this equation and solving for $\Delta S°_{vap}$ gives

$$\Delta S^o_{vap} = \frac{23.4 \text{ kJ} / \text{mol}}{240 \text{ K}} = 0.0976 \ \frac{\text{kJ}}{\text{mol} \cdot \text{K}} = 97.6 \ \frac{\text{J}}{\text{mol} \cdot \text{K}}$$

155

C. Heating curve - graphically displays the results of adding heat to a sample (see Fig. 10.12, page 391 in your text).
   1. Melting point - temperature at which solid and liquid coexist in equilibrium as molecules break free from their position in the crystal and enter the liquid phase.
      a. heat of fusion (ΔH_{fusion}) - the amount of energy required for overcoming enough intermolecular forces to convert a solid into a liquid
   2. Boiling point - temperature at which liquid and vapor coexist in equilibrium as molecules break free from the surface of the liquid and enter the gas phase.
      a. heat of vaporization (ΔH_{vap}) the amount of energy necessary to convert a liquid into a gas
   3. $\Delta H_{vap} \gg \Delta H_{fusion}$
      a. for vaporization, must overcome all intermolecular forces in compound
      b. for fusion, must overcome fewer intermolecular forces

Evaporation, Vapor Pressure, and Boiling Point
A. Evaporation - the escape of molecules from the surface of a liquid.
B. Vapor pressure - the pressure exerted by the molecules in a vapor over a liquid in a closed container.
   1. Reaches a dynamic equilibrium.
      a. number of molecules escaping the liquid = number of molecules returning to the liquid
      b. total number of molecules in both liquid and vapor phases are steady
      c. individual molecules are constantly passing back and forth from one phase to another
C. Processes explained by kinetic molecular theory.
   1. An increase in temperature results in a higher fraction of molecules with sufficient kinetic energy to overcome the surface tension and escape into the vapor.
D. Value of vapor pressure related to intermolecular forces and temperature.
   1. The smaller the intermolecular forces, the higher the vapor pressure.
      a. molecules are loosely held in liquid and can easily escape
   2. The higher the temperature, the higher the kinetic energy.
      a. molecules will have sufficient kinetic energy to escape the liquid
E. Clausius-Clapeyron Equation - relates the vapor pressure of a liquid to the inverse of its temperature.

   1. $\log\left(P_{vap}\right) = \left(\dfrac{-\Delta H_{vap}}{2.303RT}\right) + C$.

   2. Can calculate ΔH_{vap} of a liquid knowing the vapor pressure at several temperatures.
F. Boiling point - the temperature at which the vapor pressure of a liquid is equal to the external pressure pushing on the surface and all of the liquid is able to change into the vapor phase.
   1. Normal boiling point - external pressure = 1 atm.
   2. External pressure < 1 atm; liquid boils at a lower temperature.
   3. External pressure > 1 atm; liquid boils at a higher temperature.

Kinds of Solids
A. Crystalline solids - solids whose atoms, ions, or molecules have an ordered arrangement extending over a long range.
   1. Seen on visible level - have flat faces and sharp angles.
   2. Ionic solids - constituent particles are ions. (example: NaCl).
      a. ordered into a 3-D arrangement and held together by ionic bonds
   3. Molecular solids - constituent particles are molecules held together by intermolecular forces. (example: sucrose or ice)
   4. Covalent network solids - atoms are linked together by covalent bonds into a giant 3-D array. (example: diamond or quartz)

     5. Metallic solids - comparable to network solids but they consist of metal atoms. (example: Ag or Fe)
        a. have metallic properties
   B. Amorphous solids - constituent particles are randomly arranged and have no ordered long-range structure.

Probing the Structure of Solids: X-Ray Crystallography
   A. Diffraction - a beam of electromagnetic radiation is scattered by an object containing regularly spaced lines or points (*i.e.* atoms in a crystal).
     1. Spacing must be comparable to the λ of the radiation.
     2. Due to interference between two overlapping λ's.
        a. constructive interference - waves are in-phase - (peak-to-peak and trough-to-trough)
           i. increases intensity of wave
        b. destructive interference - waves are out-of-phase - waves cancel
   B. Bragg analysis - Xrays are diffracted by different layers of atoms in the crystal, leading to constructive and destructive interference.
     1. Bragg equation: $n\lambda = 2\mathrm{d} \times \sin\theta, \quad \mathrm{d} = \dfrac{n\lambda}{2\sin\theta}.$
        a. λ - known; $\sin\theta$ - angle at which incoming rays are reflected (can be measured); n - an integer (usually 1)

Unit Cells in Crystalline Solids
   A. Unit cells - small repeating units found in crystals.
     1. Symmetrical geometries.
     2. Stack to minimize space.
     3. Fourteen different geometries.
        a. parallelepipeds (six-sided geometric solids whose faces are parallelograms)
        b. differ in lengths of cell edges and the angles between the edges
6⊠   B. Cubic cells - all edges are equal in length and all angles are 90°.
     1. Primitive-cubic unit cell for metals - an atom at each of the eight corners.
        a. each atom is shared with seven other neighboring cubes that come together at the same point
        b. one-eighth of each corner atom is in any one cube
     2. Body-centered cubic unit cell - additional atom in the center of the cube.
     3. Face-centered cubic unit cell - additional atom on each of its faces.
        a. shared with one other neighboring cube
        b. one-half of each face atom belongs to any one cube

Packing of Spheres and the Structures of Metals
   A. Particles pack together in crystals so that they can be as close together as possible to maximize intermolecular attractions.
   B. Simple cubic packing - the spheres in one layer sit directly on top of those in the previous layer.
     1. All layers are identical.
     2. Primitive-cubic unit cell.
     3. Coordination number = 6: sphere touches four neighbors in the same layer, one above and 1 below.
     4. Uses only 52% of available volume.
   C. Body-centered cubic packing - spheres are in alternate layers in an *a-b-a-b* arrangements where the spheres in the *b* layers fit into the small depressions between spheres in the neighboring *a* layers.
     1. Body-centered cubic cell.

   2. Coordination number = 8; four neighbors above and four neighbors below.
   3. Occupies 68% of the available volume.
 D. Hexagonal closest-packed - noncubic unit cell with two alternating layers (*a-b-a-b*).
   1. Hexagonal arrangement of touching spheres.
   2. Spheres in a *b* layer fit into the small triangular depressions between spheres in an *a* layer.
   3. Coordination number = 12.
     a. six neighbors in the same layer, three above and three below
 E. Cubic closest-packed - face-centered cubic unit cell with three alternating layers, *a-b-c-a-b-c*.
   1. *a-b* layers identical to hexagonal closest-packed.
   2. third layer is offset from both *a* and *b*.
   3. Coordination number = 12.

Structure of Some Ionic Solids
 A. Spheres are not all the same size.
   1. Anions are larger than cations.
 B. Adopt a variety of different unit cells depending on the size and charge of the particles.
   1. Face-centered cubic - NaCl, KCl.
   2. Other common ionic unit cells - see Fig. 10.25, page 406 in the text.

Structure of Some Covalent Network Solids
 A. Carbon:
   1. Diamond - each carbon atom is sp^3 hybridized and covalently bonded with tetrahedral geometries.
   2. Allotropes - different structural forms of the same element which differ in physical and chemical properties.
   3. More than 40 amorphous forms of carbon.
   4. Graphite - 2-dimensional sheets of fused six-membered rings; each carbon is sp^2 hybridized.
   5. Fullerene - a spherical C_{60} molecule with the shape of a soccer ball.
 B. Silica (SiO_2) - four single bonds between silicon and four oxygens in a covalent network structure.
   1. Quartz glass - result of heating silica above 1600° C and then cooling the viscous liquid.
     a. Si-O bonds re-form in a random arrangement
     b. amorphous solid
     c. mix in additives - prepare a wide variety of glass
       i. window glass - add $CaCO_3$ and Na_2CO_3
       ii. colored glass - add transition metal ions
       iii. borosilicate glass (Pyrex) - add B_2O_3; resistant to thermal shock because it doesn't expand much on heating

7⊠ **Phase Diagrams**
 A. Change any one state of matter spontaneously into either of the other two depending on the temperature and pressure.
 B. Phase diagram - a graphical method of illustrating the pressure and temperature dependencies of a pure substance in a closed system.
   1. Boundary line - points on this line represent pressure/temperature combinations at which the two phases are in equilibrium.
   2. Triple point - a unique combination of pressure and temperature at which all three phases coexist in equilibrium.
   3. Critical temperature - the temperature beyond which a gas cannot be liquefied.
   4. Critical pressure - the pressure required to liquefy a gas at the critical point.
   5. Critical point - a point defined by the critical temperature and critical pressure.

6. Supercritical fluid - a substance that is neither a liquid nor a gas.
   a. pressure of a gas at the critical point is so high and the molecules are so close together that it is hard to distinguish between the gas and a liquid
   b. temperature of a liquid at the critical point is so high that it is hard to distinguish between the liquid and the gas
7. Effect of pressure on the slope of solid/liquid boundary line depends on the relative densities of the solid and liquid phases.

Self-Test

This section is intended to test your knowledge of the material covered in this chapter. Think through these problems and make certain you understand what is going on. Ask yourself if your answer makes sense. Many of these questions are linked to the chapter learning goals. Therefore, successful completion of these problems indicates you have mastered the learning goals for this chapter. You will receive the greatest benefit from this section if you use it as a mock exam. You will then discover which topics you have mastered and which topics you need to study in more detail.

True/False

1. If a molecule has polar bonds, then the molecule is polar.

2. The different types of intermolecular forces are all electrical in nature.

3. Smaller molecules and lighter atoms with fewer electrons are easily polarizable.

4. Hydrogen bonding is one of the strongest types of intermolecular forces.

5. Strong intermolecular forces result in a large surface area of a liquid.

6. For the phase change from a solid to liquid, there is a decrease in enthalpy and an increase in entropy.

7. The value of the vapor pressure of a given liquid is related to the intermolecular forces and temperature.

8. Constructive interference of two waves leads to cancellation of the waves.

9. The structure of silica (SiO_2) is quite similar to the structure of CO_2.

10. A supercritical fluid is a liquid at very high pressures.

Multiple Choice

1. Interactions which are the result of the attraction between the δ^- end of one molecule with the δ^+ end of another molecule are:
   a. ion-dipole forces
   b. dipole-dipole forces
   c. London dispersion forces
   d. hydrogen bonding

2. The ease with which a molecule's electron cloud can be distorted by a nearby electric field is referred to as:
   a. van der Waals forces
   b. London dispersion forces
   c. polarizability
   d. viscosity

3. The strongest type of intermolecular force present in C_2H_5OH is:
   a. ion-dipole forces
   b. dipole-dipole forces
   c. London dispersion forces
   d. hydrogen bonding

4. Which of the following liquids will have the higher boiling point?
   a. He
   b. H_2Se
   c. CH_3OH
   d. $HgCl_2$

5. Which of the following liquids has the highest vapor pressure?

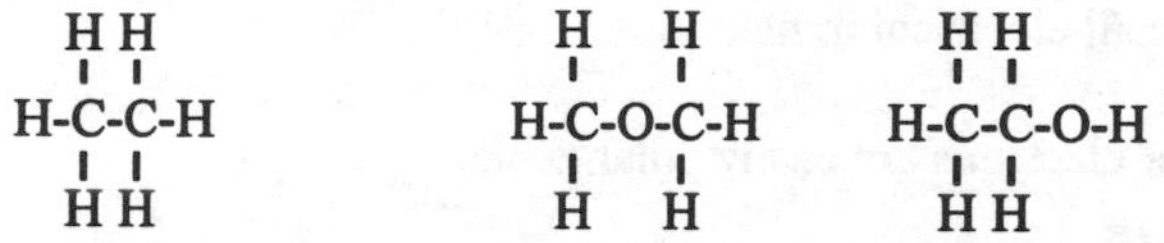

```
   H H                 H   H              H H
   I I                 I   I              I I
H-C-C-H             H-C-O-C-H          H-C-C-O-H
   I I                 I   I              I I
   H H                 H   H              H H
   (A)                 (B)                (C)
```

   a. A
   b. B
   c. C
   d. both B and C will have comparable vapor pressures

6. If the external pressure is >1 atm, then a liquid boils at:
   a. the normal boiling point
   b. a temperature below the normal boiling point
   c. a temperature above the normal boiling point

7. A solid whose constituent particles are molecules held together by intermolecular forces is a(n):
   a. ionic solid
   b. molecular solid
   c. covalent network solid
   d. metallic solid

8. The type of packing present when the spheres in a cubic unit cell are in alternate layers in an *a-b-a-b* arrangement is:
   a. simple cubic packing
   b. body-centered cubic packing
   c. hexagonal closest packed
   d. cubic closest packed

9. Fullerene is an allotrope of:
   a. carbon
   b. diamond
   c. silica
   d. phosphorus

10. The effect of pressure on the slope of the solid/liquid boundary line in a phase diagram depends on
    a. the triple point
    b. the critical pressure
    c. the critical temperature
    d. the relative densities of the solid and liquid phases

Fill-in-the-Blank

1. When the electron distribution in a molecule is temporarily uneven, an _____

   is created which _____ on a

   neighboring atom.

2. The type of intermolecular force described in the above question is a _____

   _____.

3. The _____ is the resistance of a liquid to spreading out and

   increasing its surface area.

4. Both ΔH and ΔS are negative during the following phase changes:_____

   _____.

5. $\Delta H_{vap} \gg \Delta H_{fusion}$ because _____.

6. The value of the vapor pressure is related to the temperature because _____

   _____.

7. A solid whose constituent particles are randomly arranged and have no ordered long-range

   structure is a(n) _____.

8. The type of packing present in a noncubic unit cell with 2 alternating layers is referred to as

   _____.

9. In diamond, each carbon is _____ hybridized and covalently bonded with _____

   geometry.

10. A supercritical fluid exists at the critical point because _____

    _____

    _____.

Problems

1. Determine whether the following molecules are polar or nonpolar:
   a. PCl_5 b. XeF_4 c. SF_4

2. Which compound has the stronger intermolecular forces: C_2H_4 or N_2H_4?

1. For sodium, ΔH_{vap} = 98.0 kJ/mol and ΔS_{vap} = 84.8 J/K·mol. What is the boiling point of sodium?

4. The normal boiling point of propylene glycol is 188.2° C and ΔH_{vap} = 56.8 kJ/mol. What is the vapor pressure of propylene glycol at a temperature of 150° C?

5. Calculate the reflection angle of X-rays with a wavelength equal to 172 pm when they strike a crystal with planes spaced: a) 350 pm apart; and b) 975 pm apart. Assume that n = 1.

6. Calculate the spacing between planes (in picometers) that correspond to reflections of θ = 15.0°, 25°, and 35° by X-rays with a wavelength of 175 pm.

7. The edge of the unit cell of palladium is 389 pm. The density of palladium is 12.02 g/cm³. If palladium has a cubic crystal structure, how many palladium atoms are in a unit cell? Identify the cubic unit cell.

8. Nickel crystallizes in a face-centered arrangement with the edge of the unit cell being 352 pm long. What is the radius of a nickel atom?

Solutions

True/False
1. F. The polarity of a molecule is due to the net sum of the individual bond polarities and lone-pair contributions. If that net sum is 0, then it is possible to have a molecule with polar bonds be nonpolar.
2. T
3. F. Smaller molecules and lighter atoms with fewer electrons are relatively nonpolarizable.
4. T
5. F. Strong intermolecular forces result in a higher surface tension which leads to a decrease in surface area.
6. F. For solid → liquid, there is an increase in both ΔH and ΔS.
7. T
8. F. Constructive interference leads to an increase in the intensity of the wave.
9. F. CO_2 is a linear molecule with double bonds between the C and O's; SiO_2 has four single bonds between Si and four O's in a covalent network structure.
10. F. A supercritical fluid is neither a liquid or solid.

Multiple Choice
1. b
2. c
3. d
4. d
5. a
6. c
7. b
8. b
9. a
10. d

Fill-in-the-Blank
1. instantaneous dipole; induces a temporary dipole
2. London dispersion force

3. surface tension
4. gas → liquid, gas → solid; liquid → solid
5. all intermolecular forces must be overcome to change a liquid to a solid
6. an increase in temperature leads to an increase in the kinetic energy of the molecules and allows them to escape the liquid
7. amorphous solid
8. hexagonal closest packing
9. sp^3; tetrahedral
10. the pressure of a gas at the critical point is so high and the molecules are so close together that it is hard to distinguish between the gas and a liquid (that the temperature of a liquid at the critical point is so high that it is hard to distinguish between the liquid and the gas)

Problems

1. To determine the polarity of the molecules, you must first determine the Lewis structures of the molecules.

a. PCl_5 - The bonds in this molecule are polar; however, the net sum of the bond polarities is equal to zero. Therefore, the molecule is nonpolar.

b. XeF_4 - The bonds in this molecule are also polar, plus there are two lone pairs present. However, once again, the net sum of the bond polarities is equal to zero; also the dipoles created by the lone pairs cancel because they are separated by $180°$. Therefore, the molecule is nonpolar.

c. SF_4 - The bond polarities in this molecule do not cancel; and this molecule is therefore, polar.

2. The intermolecular forces in N_2H_4 are stronger than the intermolecular forces in C_2H_4. The forces present in N_2H_4 are hydrogen bonds while the forces present in C_2H_4 are London dispersive forces. (C_2H_4 is a nonpolar molecule.)

3. $T = \dfrac{\Delta H_{vap}}{\Delta S_{vap}}$; $T = \dfrac{98{,}000 \text{ J} / \text{mol}}{84.8 \text{ J} / \text{K} \cdot \text{mol}} = 1156 \text{ K} = 883° \text{ C}$

2. $\log P_2 = \log P_1 + \dfrac{\Delta H_{vap}}{2.303R}\left(\dfrac{1}{T_1} - \dfrac{1}{T_2}\right)$;

$\log P_2 = \log(760 \text{ mm Hg}) + \dfrac{56{,}800 \dfrac{\text{J}}{\text{mol}}}{(2.303)\left(8.3145 \dfrac{\text{J}}{\text{mol} \cdot \text{K}}\right)}\left(\dfrac{1}{461.4 \text{ K}} - \dfrac{1}{423.2 \text{ K}}\right) = 2.30$;

$P_2 = 199.5 \text{ mm Hg}$

5. a. $172 \text{ pm} = 2(350 \text{ pm}) \times \sin \theta$; $\sin \theta = \dfrac{172 \text{ pm}}{(2 \times 350 \text{ pm})} = 0.246$; $\theta = 14.2°$

   b. $\sin \theta = \dfrac{172 \text{ pm}}{2 \times 975 \text{ pm}} = 0.0882$; $\theta = 5.06°$

6. $d = \dfrac{n\lambda}{2\sin\theta}$;

    a. $d = \dfrac{175 \text{ pm}}{2 \times 0.259} = 338 \text{ pm}$

    b. $d = \dfrac{175 \text{ pm}}{2 \times 0.423} = 207 \text{ pm}$

    c. $d = \dfrac{175 \text{ pm}}{2 \times 0.574} = 152 \text{ pm}$

7. volume of cube = d^3 = $(3.89 \times 10^{-8} \text{ cm})^3$ = $5.89 \times 10^{-23} \text{ cm}^3$;

    Density of unit cell = $\dfrac{\text{mass of unit cell}}{\text{volume of unit cell}}$;

    mass of unit cell $= \left(12.02\, \dfrac{\text{g}}{\text{cm}^3}\right) \times \left(5.89 \times 10^{-23} \text{ cm}^3\right) = 7.08 \times 10^{-22} \text{ g}$

    mass of unit cell =# atoms$\left(\dfrac{\text{molar mass of Pd}}{6.022 \times 10^{23}}\right)$;

    # atoms $= \dfrac{(7.08 \times 10^{-22} \text{ g}) \times (6.022 \times 10^{23} \text{ atoms} / \text{mol})}{106.42 \text{ g} / \text{mol}} = 4 \text{ atoms}$

    Each unit cell has four atoms. To determine what type of cubic unit cell this is, we must first count the number of atoms in a simple cube. Each cube has an atom at the corner, but each of these atoms are shared by eight cubes. Therefore, the total number of corner atoms = 1/8 x 8 = 1. This leaves three atoms unaccounted for. If the unit cell were body-centered, we would have only one more atom for a total of two. Since we know we have four atoms, we can discount this type of cell. For a face-centered cell, there are six faces each with one atom which is shared by two faces. This gives a total of 1/2 x 6 = 3 atoms. If we add these three face atoms to the one corner atom, we have a total of four atoms per unit cell. Therefore, the type of unit cell is face-centered.

8. diagonal of the cell = 4r; From the Pythagorean theorem we know that $d^2 + d^2 = (4r)^2$; $2d^2 = 16r^2$; $2(352 \text{ pm})^2 = 16r^2$; r = 124 pm

CHAPTER 11

SOLUTIONS AND THEIR PROPERTIES

Chapter Learning Goals

1⊠ Explain the rule of thumb: "like dissolves like" by analyzing the solution process in terms of forces overcome in the solute and solvent and forces formed between solute and solvent particles.

2⊠ Define solution density, molarity, mole fraction, weight percent, parts per million, parts per billion, and molality, and perform calculations using these quantities.

3⊠ Perform calculations using Henry's law.

4⊠ Use Raoult's law to calculate the vapor pressure over a solution containing a nonvolatile solute and a solution containing two volatile liquids.

5⊠ Perform calculations involving freezing point depression and boiling point and determine the molar mass of the solute.

6⊠ Use the equation $\Delta G = \Delta H - T\Delta S$ to calculate normal boiling point, heat of vaporization, or entropy of vaporization, given the other two.

7⊠ Perform calculations involving the osmotic pressure equation, and determine the molar mass of the solute.

8⊠ Describe fractional distillation with the aid of a liquid/vapor phase diagram.

Chapter in Brief

This chapter concentrates on the topic of homogeneous mixtures with particular emphasis given to solutions. You begin this study by examining the solution process and the energy changes that occur. This is followed by an in-depth description of the concentration units used to indicate the amounts of solute and solvent in a solution. You will also learn how to perform calculations to determine the concentration of a solution and how to interconvert between the different units. Next, you will examine the meaning of solubility and how it is affected by both temperature and pressure. Finally, you will explore four types of colligative properties, the free energy change that is associated with these properties, and some useful applications of these properties

Solutions

A. Heterogeneous mixtures - mixtures in which the mixing of components is visually nonuniform.

B. Homogeneous mixtures (solutions) - mixtures in which the mixing of components is visually uniform.

C. Suspensions - contain particles that are greater than about 1000 nm in diameter and are visible with a low-power microscope.

D. Colloids - contain particles with diameters in the range 2-1000 nm.

E. Solutions - contain particles the size of a typical ion or covalent molecule.

F. Any one state of matter can form a solution with any other state.

    1. Seven different kinds of solutions (Table 11.2, page 422 in your text).

    2. Gas or solid dissolved in a liquid.

        a. solute - the dissolved substance

        b. solvent - the liquid

    3. Liquid dissolved in a liquid.

        a. solute - the minor component

        b. solvent - the major component

Energy Changes and the Solution Process
    A. Most solutions involve condensed phases.
        1. Intermolecular forces are important for explaining the properties of solutions.
    B. Three types of interactions among particles to take into account.
        1. Solvent-solvent interactions.
        2. Solute-solute interactions.
        3. Solvent-solute interactions.

1⊠   C. "Like dissolves like".
        1. Solutions can form when the three types of interactions are similar in kind and in magnitude.
            a. ionic solids dissolve in polar solvents
            b. nonpolar organic substances dissolve in nonpolar organic solvents
    D. Solvated - ions are surrounded and stabilized by a shell of solvent molecules.
        1. Hydrated - water is the solvent.
    E. Free energy change when a solution is formed.
        1. ΔG is negative - spontaneous process and the substance dissolves.
        2. ΔG is positive - nonspontaneous process and the substance does not dissolve.
        3. ΔH_{soln} - the enthalpy of solution.
            a. difficult to predict - can be either negative (exothermic) or positive (endothermic)
        4. ΔS_{soln} - the entropy of solution.
            a. usually positive - increase in molecular randomness during dissolution
    F. Variations in heats of solution due to the interplay of the three kinds of interactions.
        1. Solvent-solvent interactions - positive ΔH.
            a. need to overcome intermolecular forces between solvent molecules
            b. separate molecules to make room for solute molecules
        2. Solute-solute interactions - positive ΔH.
            a. need to overcome intermolecular forces between solute molecules
            b. lattice energy for ionic solids
        3. Solvent-solute interactions - negative ΔH.
            a. solvent molecules cluster around solute particles and solvate them
            b. ionic substances in water
               i. increase in hydration energy with a decrease in cation size
               ii. increase in hydration energy with an increase in the charge on the ion
        4. Sum of three interactions determines whether ΔH_{soln} is endothermic or exothermic.

2⊠ **Units of Concentration**
    A. Concentration of a solution - the exact amount of solute dissolved in a given amount of solvent.
    B. Molarity $= \dfrac{\text{Moles of solute}}{\text{Liter of solution}}$.
        1. Advantages:
            a. simplifies stoichiometry calculations - uses moles instead of mass
            b. amounts of solution are measured by volume rather than by mass
               i. simplifies volumetric titrations
        2. Disadvantages:
            a. exact concentration depends on the temperature
               i. volume changes as temperature changes
            b. need to know the density of the solution to determine the amount of solvent present

C. Mole fraction $(X) = \dfrac{\text{Moles of component}}{\text{Total moles in the solution}}$.

    1. Advantages:
        a. independent of temperature
        b. useful for calculations involving gas mixtures
    2. Disadvantages:
        a. not convenient for liquid solutions

D. Weight percent (wt %) $= \dfrac{\text{Mass of component}}{\text{Total mass of solution}} \times 100$.

    1. For very dilute solutions:
        a. parts per million (ppm) $= \dfrac{\text{Mass of component}}{\text{Total mass of solution}} \times 10^6$

        b. parts per billion (ppb) $= \dfrac{\text{Mass of component}}{\text{Total mass of solution}} \times 10^9$

    2. Advantage - values are independent of temperature.
        a. masses don't change when temperature changes
    3. Disadvantages:
        a. difficult to measure mass of a liquid solution
        b. need to know the density of a solution to convert to molarity

E. Molality $(m) = \dfrac{\text{Moles of solute}}{\text{Mass of solvent (kg)}}$.

    1. Advantages:
        a. temperature independent
        b. well suited for calculating certain properties of solutions
    2. Disadvantages:
        a. difficult to measure mass of a liquid solution
        b. need to know the density of a solution to convert to molarity

EXAMPLE:

A solution is prepared by dissolving 17.84 grams of glucose ($C_6H_{12}O_6$) in 250 g of water. The density of this solution is 1.16 g/mL. Calculate the molality, weight percent, and molarity of the solution.

SOLUTION:

Molality is the number of moles of solute per kg of solvent. We first need to calculate the moles of glucose. The molecular weight of glucose is 180.2 g/mol.

$$17.84 \text{ g } C_6H_{12}O_6 \times \frac{1 \text{ mol } C_6H_{12}O_6}{180.2 \text{ g } C_6H_{12}O_6} = 0.099\ 00 \text{ mol } C_6H_{12}O_6$$

$$m = \frac{0.099\ 00 \text{ mol } C_6H_{12}O_6}{0.250 \text{ kg } H_2O} = 0.396\ m$$

To determine weight percent, we need to know the total mass of the solution.

mass of solution = mass of solute + mass of solvent

mass of solution = 17.84 g + 250 g = 268 g

$$\text{wt \%} = \frac{17.84 \text{ g}}{268 \text{ g}} \times 100\% = 6.66\%$$

To determine the molarity of the solution, we need to know the volume of the solution. This can be calculated from the grams of solution and the density.

$$\text{Volume} = \frac{268 \text{ g}}{1.16 \frac{\text{g}}{\text{mL}}} = 231 \text{ mL}; \quad M = \frac{0.099 \; 00 \text{ mol } C_6H_{12}O_6}{0.231 \text{ L soln}} = 0.428 \text{ M}$$

EXAMPLE:

A one-liter sample of water is analyzed and found to contain 5.68 ppb of lead. How many grams of lead are present? What is the molar concentration of this sample?

SOLUTION:

A concentration of 1 ppb means that each liter of an aqueous solution contains 0.001 mg of solute. Therefore, the number of grams of lead in this solution is 0.005 68 mg. To calculate the molarity of this solution, we need to convert milligrams to moles.

$$0.005 \; 68 \text{ mg Pb} \times \frac{1 \text{ g}}{1 \times 10^3 \text{ mg}} \times \frac{1 \text{ mol}}{207.19 \text{ g Pb}} = 2.74 \times 10^{-8} \text{ mol}$$

Since we have one liter of solution, the molarity is equal to 2.74×10^{-8}. You can see why the unit of ppb is more useful for solutions with trace impurities.

Some Factors Affecting Solubility
  A. Saturated Solution - a solution in which the number of ions leaving a crystal to go into solution is equal to the number of ions returning from solution to the crystal.
    1. At equilibrium with undissolved solid.
  B. Supersaturated solution - contain a greater-than-equilibrium amount of solute.
  C. Solubility - the amount of solute per unit of solvent needed to form a saturated solution.
    1. Physical property characteristic of a particular substance.
    2. Temperature dependent.
  D. Effect of temperature on solubility.
    1. Gases become less soluble as the temperature is increased.
    2. More difficult to predict effect on the solubility of solids.
  E. Effect of pressure on solubility.
    1. None on solids and liquids.
    2. Henry's law: solubility $= k \cdot P$ (k = Henry's law constant and P = partial pressure of gas).
    3. Increase in pressure leads to an increase in solubility.
      a. due to change in position of the equilibrium between dissolved and undissolved gas
      b. more gas particles are forced into solution

Physical Behavior of Solutions: Colligative Properties
  A. Colligative properties - properties that depend on the amount of dissolved solute but not on the chemical identity of the solute.
    1. Boiling point elevation.
    2. Freezing point depression.
    3. Vapor pressure of a solution.

    4. Osmosis - the migration of solvent and other small molecules through a semipermeable membrane.

4⊠   **Vapor-Pressure Lowering of Solutions: Raoult's Law**

    A. Solutions with a nonvolatile solute - lower vapor pressure than pure solvent.
        1. Raoult's law: $P_{soln} = P_{solv} \times X_{solv}$.
        2. For ionic substances, calculate mole fractions based on the total number of solute particles rather than on the number of formula units.
    B. Free energy change accompanies vapor pressure lowering.
        1. More negative ΔG, the easier the vaporization process.
    C. Liquid $\rightarrow$ gas:
        1. ΔH is positive (need energy to overcome intermolecular forces).
          a. similar value for both solvent and solution
        2. ΔS is positive (molecular disorder increases).
          a. ΔS smaller for solution than for pure solvent
        3. Larger ΔG for solution (subtracting a smaller $T\Delta S$ from ΔH).
          a. vaporization is more difficult
          b. lower vapor pressure
    D. Raoult's law only applies to ideal solutions.
        1. Solute concentrations are low.
        2. Solute and solvent particles have similar intermolecular forces.
    E. Solute-solvent intermolecular forces $<$ solvent intermolecular forces; vapor pressure higher than predicted by Raoult's law.
    F. Solute-solvent intermolecular forces $>$ solvent intermolecular forces; vapor pressure less than predicted by Raoult's law.
    G. Solutions with a volatile solute - vapor pressure of the solution is always intermediate between the vapor pressures of the two pure substances.
        1. $P_{total} = P_A + P_B$.
        2. P_A and P_B are calculated from Raoult's law.
          a. $P_A = X_A(P_A^0);\quad P_B = X_B P_B$
        3. $P_{total} = X_A(P_A^0) + X_B(P_B^0)$.

EXAMPLE:
    A student needs to prepare an aqueous solution of sucrose at a temperature of 20° C with a vapor pressure of 15.0 mm Hg. How many grams of sucrose does she need if she uses 375 g H_2O? (The vapor pressure of water at 20° C is 17.5 mm Hg.)

SOLUTION: This solution consists of a nonvolatile solute in a volatile solvent. The form of Raoult's law that we will use is $P_{soln} = P_{solv} \times X_{solv}$. We are given both P_{soln} and P_{solv}. We can solve for X_{solv} and then determine the mass of sucrose needed for this solution.

$$X_{solv} = \frac{15.0 \text{ mm Hg}}{17.5 \text{ mm Hg}} = 0.857$$

$$X_{solv} = \frac{\text{mol } H_2O}{\text{mol sucrose } + \text{ mol } H_2O};$$

Rearranging this equation gives:

$$\text{mol sucrose} = \frac{\text{mol H}_2\text{O} - X_{solv} \cdot \text{mol H}_2\text{O}}{X_{solv}}$$

$$\text{mol H}_2\text{O} = 375 \text{ g H}_2\text{O} \times \frac{1 \text{ mol H}_2\text{O}}{18.0 \text{ g H}_2\text{O}} = 20.8 \text{ mol H}_2\text{O}$$

$$\text{mol sucrose} = \frac{20.8 - (0.875 \times 20.8)}{0.857} = 3.03 \text{ mol sucrose}$$

$$3.03 \text{ mol sucrose} \times \frac{342 \text{ g sucrose}}{1 \text{ mol sucrose}} = 1040 \text{ g sucrose}$$

5⊠ **Boiling-Point Elevation and Freezing-Point Depression of Solutions**

    A. Boiling-point elevation and freezing-point depression of a solution relative to that of a pure solvent depends upon the number of solute particles.
1. $\Delta T_b = K_b \cdot m$.
2. $\Delta T_f = K_f \cdot m$.
3. Use molality as the concentration unit because the numbers of solute and solvent particles are independent of temperature.

6⊠   B. Due to an entropy difference between pure solvent and solvent in a solution.
1. $T_b = \dfrac{\Delta H_{vap}}{\Delta S_{vap}}$ at equilibrium.
2. ΔS_{vap} is smaller for a solution, therefore T_b is larger.
3. $T_f = \dfrac{\Delta H_{fusion}}{\Delta S_{fusion}}$ at equilibrium.
4. ΔS_{fusion} is larger for a solution, therefore T_f is smaller.

EXAMPLE:

    What will be the freezing point and boiling point of an aqueous solution containing 55.0 g of glycerol, $C_3H_5(OH)_3$, and 250 g of water? $K_b(H_2O) = 0.51°$ C/m and $K_f = 1.86°$ C/m.

SOLUTION: To determine the freezing point and boiling point of this solution, we need to first calculate the molality of the solution. This requires that we determine the number of moles of glycerol present.

$$55.0 \text{ g glycerol} \times \frac{1 \text{ mol glycerol}}{92.0 \text{ g glycerol}} = 0.598 \text{ mol glycerol}; \qquad \frac{0.598 \text{ mol glycerol}}{0.250 \text{ kg H}_2\text{O}} = 2.39 \ m$$

$$\Delta T_f = \left(1.86 \frac{°C}{m}\right) \times 2.39 \ m = 4.45° \text{ C}; \qquad \text{F. pt. Solution} = 0° - 4.45° = -4.45°C$$

$$\Delta T_b = \left(0.51 \frac{°C}{m}\right) \times 2.39 \ m = 1.22°C; \qquad \text{B. pt. Solution} = 100° + 1.22° = 101.22°C$$

7⊠ **Osmosis and Osmotic Pressure**
   A. Semipermeable membrane - membranes that allow water or other small molecules to pass through, but block the passage of large solute molecules or ions.
   B. Osmosis - the migration of solvent and other small molecules through a semipermeable membrane.
   C. Osmotic pressure of a solution (Π) - the pressure needed to prevent the osmotic flow of solvent through a semipermeable membrane.
      1. $\Pi = MRT$.
      2. Can use molarity since measurements are made at the temperature specified in the equation.
      3. Due to an increase in entropy when pure solvent passes through the membrane and mixes with the solution.

EXAMPLE:
   Determine the osmotic pressure of a 0.075 M solution of aspartic acid at 18.5° C.

SOLUTION:

$$0.075\frac{\text{mol}}{\text{L}} \times 0.0821\frac{\text{L} \cdot \text{am}}{\text{mol} \cdot \text{K}} \times 291.7 \text{ K} = 1.80 \text{ atm}$$

Some Uses of Colligative Properties
   A. Molar mass determinations.
      1. Can use any four colligative properties.
      2. Most accurate is osmotic pressure - magnitude of osmosis effect is so great.

EXAMPLE:
   A solution is prepared from 25.0 g of benzene, C_6H_6, and 2.50 g of a compound with an empirical formula of C_6H_5P. The freezing point of this solution is 4.3° C. Determine the molar mass and molecular formula of the compound.

SOLUTION: To solve this problem, we need to first determine the molality of the solution. This is accomplished by rearranging the equation for freezing point depression. (The freezing point of benzene is 5.5* C.)

$$m = \frac{1.2°\text{C}}{5.12°\text{C}/m} = 0.234 \ m$$

This molality represents the number of moles of the unknown in 1 kg of benzene. The actual number of moles in this solution can be calculated using the mass of benzene in the solution.

$$\frac{0.234 \text{ mol unknown}}{1 \text{ kg benzene}} \times 0.025 \text{ kg benzene} = 0.005 \ 85 \text{ mol unknown}$$

Knowing the mass and number of moles of the unknown, we can now calculate the molar mass of the sample.

$$\frac{2.50 \text{ g unknown}}{0.005 \ 85 \text{ mol unknown}} = 427 \ \text{g}\!/\!\text{mol}$$

To determine the molecular formula, we divide the molar mass of the unknown by the molar mass of the empirical formula.

$$\frac{427 \text{ g/mol}}{108 \text{ g/mol}} = 4$$

Multiplying the coefficients of the empirical formula by 4 gives $C_{36}H_{20}P_4$.

8⊠ **Fractional Distillation of Liquid Mixtures**

    A. Fractional distillation - a mixture of volatile liquids is boiled and the vapors are condensed.
       1. Vapor is enriched in the more volatile component.
       2. Condensed vapor is also enriched in more volatile component.
       3. Repeat boil/condense cycle many times - can have complete purification of the more volatile liquid component.

Self-Test

This section is intended to test your knowledge of the material covered in this chapter. Think through these problems and make certain you understand what is going on. Ask yourself if your answer makes sense. Many of these questions are linked to the chapter learning goals. Therefore, successful completion of these problems indicates you have mastered the learning goals for this chapter. You will receive the greatest benefit from this section if you use it as a mock exam. You will then discover which topics you have mastered and which topics you need to study in more detail.

True-False

1. A mixture in which the mixing of components is visually uniform is referred to as a solution.

2. Polar solutes can be dissolved in either nonpolar or polar solvents.

3. The entropy of solution is usually negative.

4. The molality of a solution is determined by dividing the number of moles of solute by the mass (in kg) of the solvent.

5. A solution in which the number of ions leaving a crystal to go into solution is equal to the number of ions returning from solution to the crystal is said to be supersaturated.

6. According to Henry's law, an increase in pressure leads to an increase in solubility.

7. Solutions with a nonvolatile solute will have a higher vapor pressure than the pure solvent.

8. If the solute-solvent intermolecular forces are greater than the intermolecular forces of the pure solvent, the vapor pressure of the solution will be higher than the vapor pressure which is predicted by Raoult's law.

9. ΔS_{vap} is smaller for a solution than a pure solvent; therefore the T_b will be higher.

10. The colligative property which is considered to be the most accurate for determining the molar mass of a compound is freezing point depression.

Multiple Choice

1. When a gas or solid is dissolved in a liquid, the solute is
   a. the major component
   b. the liquid
   c. the dissolved substance
   d. the minor component

2. ΔH_{soln} will be exothermic if
   a. ΔH for solvent-solvent interactions is negative
   b. ΔH for solute-solute interactions is negative
   c. ΔH for solute-solvent interactions is negative
   d. the sum of the three types of interactions leads to a negative ΔH

3. Benzene, a nonpolar organic compound, is most likely to dissolve in
   a. CCl_4
   b. CH_3CH_2OH
   c. water
   d. NH_3

4. The disadvantage of using mole fraction to express the concentration of a solution is
   a. the exact concentration depends on the temperature
   b. you need to know the density of the solution to determine the amount of solvent present
   c. it is not convenient for liquid solutions
   d. it is difficult to measure the mass of a liquid solution

5. The vapor pressure of a solution will be higher than that predicted by Raoult's law when
   a. solute-solvent interactions > solute interactions
   b. solute-solvent interactions < solute interactions
   c. solute-solvent interactions > solvent interactions
   d. solute-solvent interactions < solvent interactions

6. The ionic substance with the highest hydration energy is
   a. NaCl
   b. $BaCl_2$
   c. $AlCl_3$
   d. $PbCl_4$

7. Which of the following is more likely to form a solution?
   a. an ionic solid is mixed with a nonpolar solvent
   b. a nonpolar solute is mixed with water
   c. a nonpolar solute is mixed with NH_3
   d. a nonpolar solute is mixed with a nonpolar solvent

8. The ionic substance with the lowest hydration energy is
   a. $BaSO_4$
   b. $SrSO_4$
   c. $CaSO_4$
   d. $MgSO_4$

9. If you are performing a volumetric titration in the laboratory, the most likely concentration unit you would use is
   a. molality
   b. molarity
   c. ppb
   d. mole fraction

10. A mixture which contains particles large enough to be visible with a low-power microscope is a
    a. heterogeneous mixture
    b. solution
    c. suspension
    d. colloid

Fill-in-the-Blank

1. When a liquid is dissolved in a liquid, the solute is _____ and the solvent is _____.

2. ΔS_{soln} is usually _____ due to _____
   _____.

3. One disadvantage of using molarity as a concentration unit is that it is dependent upon
   _____ . This dependence is due to the change in _____ as the
   _____ changes.

4. Gases become _____ soluble as the temperature increases.

5. Colligative properties are properties that depend on _____
   but not on the _____ of the substance.

6. The value of ΔS_{vap} is _____ for a solution than for a pure solvent.

7. Dissolved ions that are surrounded and stabilized by a shell of solvent molecules are said to be
   _____ . If water is the solvent, the term used is _____.

8. The freezing point of solution is lower because _____.

9. Osmosis is _____
   _____.

10. Molarity can be used to calculate osmotic pressure because _____
    _____.

Matching

Colloids   a. the amount of solute per unit of solvent needed to form a saturated solution.

Supersaturated solution   b. the pressure needed to prevent the osmotic flow of solvent through a semipermeable membrane.

174

Solubility

c. mixtures which contain particles with diameters in the rage of 2 to 1000 nm.

Semipermeable membrane

d. a process in which a mixture of volatile liquids is boiled and the vapors are condensed.

Osmotic pressure

e. a solution which contains a greater than equilibrium amount of solute.

Fractional distillation

f. membrane that allows water or other small molecules to pass through, but block the passage of large solute molecules or ions.

Problems

1. Identify the intermolecular forces present in both the solute and solvent, and predict whether a solution will form between the two.

   a. CCl_4 and Br_2

   b. CH_3OH and Br_2

   c. KCl and NH_3

   d. NH_3 and H_2O

2. The density of a NaOH solution is 1.109 g/mL and has a weight percent of 9.99 %. Calculate the molality and molarity of the solution.

3. A 0.838 M acetic acid, CH_3COOH, solution has a density of 1.0055 g/mL. Calculate the molality and weight percent of this solution.

4. A 8.0 m̲ solution of NH_3 has a density equal to 0.950 g/mL. Calculate both the molarity and weight percent of this solution.

5. A 50 mL sample of water was found to have 32.5 ppb of Hg. Calculate the number of grams and molarity of Hg in this sample.

6. The Henry's law constant for methane is 3.34 x 10^{-3} mol/L·atm. Calculate the solubility of methane at 852 mm Hg.

7. The solubility of N_2 gas at 25° C and 650 mm Hg is 5.85×10^{-4} mol/L. What is the solubility of N_2 at 725 mm Hg?

8. The vapor pressure of water at 35° C is 42.175 mm Hg. Calculate the vapor pressure of an aqueous solution that contains 15.8 g NaCl and 72.1 g of water at this temperature.

9. The vapor pressure of benzene, C_6H_6, at 25° C is 93.4 mm Hg. The vapor pressure of toluene, $C_6H_5CH_3$, is 26.9 mm Hg at 25° C. Calculate the vapor pressure of a solution prepared from 8.0 g of benzene and 15.0 g of toluene.

10. Heptane, C_7H_{16}, has a vapor pressure of 791 mm Hg at 100° C. 250 g of hexane were mixed with 150 g of an unknown substance whose vapor pressure is 352 mm Hg at 100° C. The vapor pressure of the resulting solution is 639 mm Hg. Calculate the molar mass of the unknown substance.

11. Calculate the boiling point and freezing point of a solution prepared by mixing 10.0 g $CaCl_2$ with 90.0 g of water.

12. A solution was prepared by mixing 50.0 g of benzene (freezing point = 5.5^o C) and 2.50 g of an unknown substance. The freezing point of the solution was 3.5^o C. Calculate the molar mass of the unknown substance.

13. 2.5 g of ethanol (CH_3CH_2OH) were mixed with 10.0 g of water. Calculate the freezing point of the solution and ΔS_{fusion} of the solution. ΔH_{fusion} for water = 6.01 kJ/mol.

14. Calculate the osmotic pressure of a solution at 15° C containing 8.0 g of glucose ($C_6H_{12}O_6$) in 1 L of solution.

15. Calculate the molar mass of 0.250 g of a tripeptide in 1 L of water at 15° C that has an osmotic pressure of 16.4 mm Hg.

Solutions

True-False
1. T
2. F. "Like dissolves like." Polar solutes can be dissolved in polar solvents.
3. F. Disorder increases when a solution is formed; therefore, the entropy of solution is positive.
4. T
5. F. The solution described is a saturated solution.
6. T
7. F. Solutions with a nonvolatile solute will have a lower vapor pressure than the pure solvent.
8. F. The vapor pressure of the solution will be less than that predicted by Raoult's law.
9. T
10. F. The colligative property considered to be the most accurate for determining molar mass is osmotic pressure.

Multiple Choice
1. c
2. d
3. a
4. c
5. d
6. d
7. d
8. a
9. b
10. c

Fill-in-the-Blank
1. the minor component; the major component
2. positive; due to the increase in molecular randomness upon dissolution.
3. temperature; volume; temperature

4. less
5. the amount of dissolved solute; chemical identity of the solute
6. smaller
7. solvated; hydrated
8. ΔS_{fusion} is greater for a solution than pure solvent and $T_f = \dfrac{\Delta H_f}{\Delta S_f}$
9. the migration of solvent and other small molecules through a semipermeable membrane
10. because measurements are made at the temperature specified in the equation

Matching

Colloids - c
Supersaturated solution - e
Solubility - a
Semipermeable membrane - f
Osmotic pressure - b
Fractional distillation - d

Problems

1. a. CCl_4 - nonpolar substance; London dispersion forces; Br_2 - nonpolar substance; London
   dispersion forces; will form a solution
   b. CH_3OH - polar substance; hydrogen bonding; Br_2 - nonpolar substance; London dispersion
   forces; will not form a solution
   c. KCl - ionic substance; ionic attractions; NH_3 - polar substance; hydrogen bonding; will form a
   solution
   d. NH_3 - polar substance; hydrogen bonding; H_2O - polar substance; hydrogen bonding; will form
   a solution

2. Assume 100 g of solution. We now have 9.99 g of NaOH and 90.01 g of water. To calculate both
   molality and molarity, we need to convert grams of NaOH to moles of NaOH.

$$9.99 \text{ g NaOH} \times \frac{1 \text{ mol NaOH}}{40.0 \text{ g NaOH}} = 0.250 \text{ mol NaOH}$$

$$\frac{0.250 \text{ mol NaOH}}{0.09001 \text{ kg}} = 2.78 \text{ } m$$

To calculate molarity, we also need to know the volume of the solution which we can obtain from
the mass and density of the solution.

$$\frac{100 \text{ g soln}}{1.109 \text{ g soln / mL}} = 90.2 \text{ mL}; \qquad \frac{0.250 \text{ mol NaOH}}{0.0902 \text{ L soln}} = 2.77 \text{ M}$$

3. Assume that you have 1 L of solution; therefore, you also have 0.838 moles of acetic acid. To
   calculate molality, you need to know the number of kg of solvent. This can be calculated from the
   grams of solution and grams of solute. (g soln = g solvent + g solute)

$$0.838 \text{ mol CH}_3\text{COOH} \times \frac{60.0 \text{ g CH}_3\text{COOH}}{1 \text{ mole CH}_3\text{COOH}} = 50.3 \text{ g CH}_3\text{COOH}$$

$$1000 \text{ mL soln} \times \frac{1.0055 \text{ g soln}}{1 \text{ mL soln}} = 1,005.5 \text{ g soln}; \quad 1,005.5 \text{ g soln} - 50.3 \text{ g solute} = 955.2 \text{ g solvent}$$

$$\frac{0.838 \text{ mol CH}_3\text{COOH}}{0.9552 \text{ kg H}_2\text{O}} = 0.877 \ m.$$

To calculate % weight, we simply divide the grams of acetic acid by the grams of solution.

$$\frac{50.3 \text{ g CH}_3\text{COOH}}{1,005.5 \text{ g soln}} \times 100 = 5.00\%$$

4. By assuming we have 1 kg of solvent, we know that we have 8.0 moles of NH_3. We need to know the grams of NH_3 for both calculations.

$$8.0 \text{ mol NH}_3 \times \frac{17.0 \text{ g NH}_3}{1 \text{ mol NH}_3} = 136 \text{ g NH}_3$$

We also need to know the mass of the solution for both calculations.

$$136 \text{ g NH}_3 + 1000 \text{ g H}_2\text{O} = 1136 \text{ g soln}$$

To calculate molarity, we need to determine the volume of the solution from the mass and density of the solution.

$$\frac{1,136 \text{ g soln}}{0.950 \text{ g soln/mL}} = 1,200 \text{ mL soln}; \qquad \frac{8 \text{ mol NH}_3}{1.2 \text{ L soln}} = 6.67 \text{ M}$$

To calculate % weight, we divide the mass of NH_3 by the mass of the solution.

$$\frac{136 \text{ g NH}_3}{1,136 \text{ g soln}} \times 100 = 12.0\%$$

5. From the definition for ppb, we know that this solution contains 0.0325 mg of Hg in a 1 L sample. The number of grams of Hg in a 50 mL sample is

$$\frac{0.0325 \text{ mg Hg}}{1 \text{ L sample}} \times 0.050 \text{ L} \times \frac{1 \times 10^{-3} \text{ g}}{1 \text{ mg}} = 1.6 \times 10^{-6} \text{ g Hg}$$

To calculate molarity, we need to convert grams of Hg to moles.

$$1.63 \times 10^{-6} \text{ g Hg} \times \frac{1 \text{ mol Hg}}{200.6 \text{ g Hg}} = 8.0 \times 10^{-9} \text{ mol Hg}; \qquad \frac{8.0 \times 10^{-9} \text{ mol Hg}}{0.050 \text{ L}} = 1.6 \times 10^{-7} \text{ M}$$

6. $\text{Solubility} = 3.34 \times 10^{-3} \ \dfrac{\text{mol}}{\text{L} \cdot \text{atm}} \times \left(852 \text{ mm Hg} \times \dfrac{1 \text{ atm}}{760 \text{ mm Hg}} \right) = 3.74 \times 10^{-3} \ \dfrac{\text{mol}}{\text{L}}$

7. We can use the first set of data to calculate the Henry's law constant.

$$5.85 \times 10^{-4} \frac{\text{mol}}{\text{L}} = k \times \left(725 \text{ mm Hg} \times \frac{1 \text{ atm}}{760 \text{ mm Hg}}\right); \quad k = 6.84 \times 10^{-4} \frac{\text{mol}}{\text{L} \cdot \text{atm}}$$

8. To calculate the vapor pressure of the solution, we need to know the mole fraction of water. To calculate the mole fraction of water, we need to know the moles of water and the total number of moles of particles in the solution.

$$72.1 \text{ g H}_2\text{O} \times \frac{1 \text{ mol H}_2\text{O}}{18.0 \text{ g H}_2\text{O}} = 4.01 \text{ mol H}_2\text{O}$$

$$15.8 \text{ g NaCl} \times \frac{1 \text{ mol NaCl}}{58.5 \text{ g NaCl}} = 0.270 \text{ mol NaCl};$$

Because we are using an ionic solid, we must take into account the number of moles of particles, not just the number of moles of formula units of NaCl. There are 0.270 mol of Na^+ and 0.270 mol of Cl^- in 0.270 mol of NaCl. This gives 0.540 mol of ions in the solution. This amount will be used to calculate the total number of moles of particles present in the solution.

$$X_{H_2O} = \frac{4.01 \text{ mol H}_2\text{O}}{4.55 \text{ mol particles}} = 0.88 ; \qquad P_{soln} = 0.881 \times 42.175 \text{ mm Hg} = 37.2 \text{ mm Hg}$$

9. $$8.0 \text{ g C}_6\text{H}_6 \times \frac{1 \text{ mol C}_6\text{H}_6}{78.0 \text{ g C}_6\text{H}_6} = 0.103 \text{ mol C}_6\text{H}_6;$$

$$15.0 \text{ g C}_6\text{H}_5\text{CH}_3 \times \frac{1 \text{ mol C}_6\text{H}_5\text{CH}_3}{92.0 \text{ g C}_6\text{H}_5\text{CH}_3} = 0.163 \text{ mol C}_6\text{H}_5\text{CH}_3$$

$$X_{C_6H_5CH_3} = \frac{0.163 \text{ mol C}_6\text{H}_5\text{CH}_3}{0.266 \text{ mol particles}} = 0.613; \qquad X_{C_6H_6} = \frac{0.103 \text{ mol C}_6\text{H}_6}{0.266 \text{ mol particles}} = 0.387$$

$$P_{soln} = (0.387 \times 93.4 \text{ mm Hg}) + (0.613 \times 26.9 \text{ mm Hg}) = 52.6 \text{ mm Hg}$$

10. To calculate the molar mass of the unknown, we need to know both the mass and number of moles of unknown. The mass of the unknown was given. We can find the number of moles of unknown by determining the mole fraction of heptane and the unknown. Remember that the mole fraction of heptane is the number of moles of heptane divided by the number of moles of heptane and the unknown,

$$X_{heptane} = \frac{\text{mol heptane}}{\text{mol heptane} + \text{mol unknown}}$$

Let's start by calculating the moles of heptane.

$$250 \text{ g C}_7\text{H}_{16} \times \frac{1 \text{ mol C}_7\text{H}_{16}}{100.0 \text{ g C}_7\text{H}_{16}} = 2.50 \text{ mol C}_7\text{H}_{16}$$

Let x be the number of moles of unknown. The mole fraction for heptane and the unknown are:

179

$$X_{C_7H_{16}} = \frac{2.50 \text{ mol } C_7H_{16}}{2.50 \text{ mol } C_7H_{16} + x \text{ mol unk}}; \qquad X_{unk} = \frac{x \text{ mol unk}}{2.50 \text{ mol } C_7H_{16} + x \text{ mol unk}}$$

Substituting these mole fractions into the equation for the vapor pressure of a solution gives:

$$639 \text{ mm Hg} = \left(\frac{2.50 \text{ mol } C_7H_{16}}{2.50 \text{ mol } C_7H_{16} + x \text{ mol unk}}\right)(791 \text{ mm Hg}) + \left(\frac{x \text{ mol unk}}{2.50 \text{ mol } C_7H_{16} + x \text{ mol unk}}\right)(352 \text{ mm Hg})$$

$$639 \text{ mm Hg} = \frac{1}{2.50 \text{ mol } C_7H_{16} + x \text{ mol unk}}\left[(2.50 \text{ mol } C_7H_{16} \times 791 \text{ mm Hg}) + (x \text{ mol unk} \times 352 \text{ mm Hg})\right]$$

For simplification, let's drop the units out of the calculations, keeping in mind that the units for x are moles of unknown.

$$639(2.50 + x) = (2.50 \times 791) + (x \times 352);$$

$$(1.60 \times 10^3) + 639x = (1.98 \times 10^3) + 352x \quad 287x = 380; \quad x = 1.32$$

Now that we know the number of moles of unknown, we can calculate the molar mass of the unknown.

$$\frac{150 \text{ g unk}}{1.32 \text{ mol unk}} = 114 \text{ g/mol}$$

11. To calculate both the boiling point and freezing point of a solution, we need to know the molality of the solution. Remember, when calculating the molality of an ionic substance, we need to use the number of moles of solute particles, not formula units.

$$10.0 \text{ g CaCl}_2 \times \frac{1 \text{ mol CaCl}_2}{111.1 \text{ g CaCl}_2} = 9.00 \times 10^{-2} \text{ mol CaCl}_2;$$

There are three moles of ions for every one mole of $CaCl_2$, so the number of moles of solute particles is 0.270. The molality of the solution is

$$\frac{0.270 \text{ mol ions}}{0.090 \text{ kg H}_2\text{O}} = 3.00 \ m.$$

To calculate the boiling point and freezing point of the solution, we need to calculate ΔT_b and ΔT_f. K_b for water = $0.51°$ C/m and K_f for water = $1.86°$ C/m..

$$\Delta T_b = (0.51° \text{ C/}m)(3.00 \ m) = 1.53°\text{C} \ ; \qquad \Delta T_f = (1.86° \text{ C/}m)(3.00 \ m) = 5.58°\text{C}$$

b. pt. soln. = $100.00°$C + $1.53°$ C = $101.53°$ C   f. pt. soln. = $0.00°$C – $5.58°$C = $-5.58°$C

12. To calculate molar mass, we again need to find the number of moles of solute. This time, we will do that by finding the molality of the solution from the change in freezing point. From the definition of molality, we can then determine the exact number of moles in the solution.

$\Delta T_f = 5.5° \text{ C} - 3.5° \text{ C} = 2.0° \text{ C}; \ \ K_f = 5.12° \text{ C}/m; \ \ m = \dfrac{2.0° \text{ C}}{5.12° \text{ C}/m} = 0.391 \ m$

$\dfrac{0.391 \text{ mol solute}}{1 \text{ kg benzene}} \times 0.050 \text{ kg benzene} = 0.0196 \text{ mol solute}; \ \ \ \dfrac{2.50 \text{ g solute}}{0.0196 \text{ mol solute}} = 128 \text{ g}/\text{mol}$

13. $2.5 \text{ g ethanol} \times \dfrac{1 \text{ mol ethanol}}{46.0 \text{ g ethanol}} = 0.0543 \text{ mol ethanol}; \ \ \ \dfrac{0.0543 \text{ mol ethanol}}{0.010 \text{ kg water}} = 5.43 \ m;$

$K_f = 1.86° \text{ C}/m \ \ \ \ \ \ \ \ \Delta T_f = \left(1.86° \text{ C}/m\right)\left(5.43 \ m\right) = 10.1° \text{ C} \ ;$

f. pt. soln. $= 0.0° \text{ C} - 10.1° \text{ C} = -10.1° \text{ C}$

Since ΔH_{fusion} for the solvent is the same as ΔH_{fusion} for the solution, we now have all the information we need to calculate ΔS_{fusion} from the equation $\Delta H_{fusion} = T\Delta S_{fusion}$.

$\dfrac{6.01 \text{ kJ}/\text{mol}}{\left(273-10.1\right)} \times \dfrac{1000 \text{ J}}{1 \text{ kJ}} = 221.9 \text{ J}/\text{mol}\cdot\text{K}$

14. $8.0 \text{ g } C_6H_{12}O_6 \times \dfrac{1 \text{ mol } C_6H_{12}O_6}{180.0 \text{ g } C_6H_{12}O_6} = 0.044 \text{ mol } C_6H_{12}O_6; \ \ \dfrac{0.044 \text{ mol } C_6H_{12}O_6}{1 \text{ L soln}} = 0.044 \text{ M}$

$\Pi = 0.044 \dfrac{\text{mol}}{\text{L}} \times 0.08206 \dfrac{\text{L}\cdot\text{atm}}{\text{mol}\cdot\text{K}} \times 288 \text{ K} = 1.04 \text{ atm}$

15. $\left(16.4 \text{ mm Hg} \times \dfrac{1 \text{ atm}}{760 \text{ mm Hg}}\right) = \text{M} \times 0.08206 \dfrac{\text{atm}\cdot\text{L}}{\text{mol}\cdot\text{K}} \times 288 \text{ K}; \ \ \ \text{M} = 9.13 \times 10^{-4}$

Since we have 1 L of solution, we know that we have 9.13×10^{-4} moles of the tripeptide. We can now calculate the molar mass.

$\dfrac{0.250 \text{ g}}{9.13 \times 10^{-4} \text{ mol}} = 274 \text{ g}/\text{mol}$

CHAPTER 12

CHEMICAL KINETICS

Chapter Learning Goals

1⊠ Use a table of concentration *versus* time data to calculate an average rate of reaction over a period of time.

2⊠ From the coefficients of a balanced chemical equation, express the relative rates of consumption of reactants and formation of products.

3⊠ From a table of initial concentrations of reactants and initial rates, determine the order of reaction with respect to each reactant, the overall order of reaction the rate law, the rate constant, and the initial rate for any other set of initial concentrations.

4⊠ Use integrated first- and second-order rate laws to find the value of one variable, given values of the other variables.

5⊠ From plots of log concentrations *versus* time and 1/concentration *versus* time, determine the order of reaction.

6⊠ Use the expression for half-life of a first- or second-order reaction to determine $t_{1/2}$ from k, or vice versa.

7⊠ From a plot of concentration *versus* time, estimate the half-life of a first-order reaction.

8⊠ Given a reaction mechanism and an experimental rate law, identify the reaction intermediates, determine the molecularity of each elementary reaction, and determine if the mechanism is consistent with the experimental rate law.

9⊠ Prepare an Arrhenius plot and determine the activation energy from the slope of the line.

10⊠ Solve the Arrhenius equation for any variable given the others.

11⊠ Sketch a potential energy profile showing the activation energies for the forward and reverse reactions and showing how they are affected by the addition of a catalyst.

Chapter in Brief

Chemical kinetics is the area of chemistry concerned with reaction rates and the sequence of steps by which reactions occur. In this chapter you will discover how to describe reaction rates and examine how they are affected by variables such as reactant concentrations and temperatures. You will learn how to determine reaction rates from plots of concentration *versus* time and how to relate the rates of disappearance or appearance of the individual reactants and products. You will examine experimental rate laws and the order of reaction and learn how to obtain the experimental rate law from initial rate data. You will study integrated rate laws and explore how these rate laws can be used to calculate the concentration of a reactant at any time *t*, the fraction of reactant that remains at any time, or the time required for the initial concentration of a reactant to drop to any particular value or fraction of its initial concentration. You will also learn how to use plots of concentration *versus* time to determine the reaction order. You will examine how kinetics allows us to postulate a reaction mechanism, how rate constants depend on temperature, and how collision theory leads to the Arrhenius equation. Finally, you will gain an understanding of the effect of a catalyst on the rate of a reaction, and the difference between a homogeneous and heterogeneous catalyst.

Reaction Rates

    A. Rate of a reaction - how fast the concentration of a reactant or a product changes per unit time.

        1. $\text{Rate} = \dfrac{\Delta(\text{concentration})}{\Delta\,(\text{time})}$

           a. increase in the concentration of a product per unit time

b. decrease in the concentration of a reactant per unit time
2. Units - M/s or mol/(L·s).
   a. allows rate to be independent of the scale of the reaction
   b. use a minus sign in calculating the rate of disappearance of a reactant

2⊠   B. Relative rates of product formation and reactant consumption depend on the coefficients in the balanced equation.
   1. Specify the reactant or product when quoting a rate.
   C. Rate changes as the reaction proceeds.
   1. Specify the time.
   2. Reaction rates decrease as the reaction mixture runs out of reactants.

1⊠   D. Plot concentration (y axis) *versus* time (x axis).
   1. Δ(concentration) and Δ(time) represent vertical and horizontal sides of a right triangle.
   2. Slope of hypotenuse of triangle is the average rate during that time period.
   E. Instantaneous rate at time t - the slope of the tangent to a concentration-versus-time curve at time t.
   F. Initial rate - the instantaneous rate at the beginning of a reaction ($t = 0$).

EXAMPLE:

It was found that the rate of formation of $N_2 (g)$ in the following reaction

$$4\,NH_3 (g) + 3\,O_2 (g) \rightarrow 2\,N_2 (g) + 6\,H_2O (g)$$

is 0.52 M·s^{-1} at a particular point in time. Determine the rate of disappearance of NH_3.

SOLUTION: Knowing that the rate of appearance is $0.52 \dfrac{mol\,N_2}{L \cdot s}$, we can use the stoichiometry of the balanced equation to determine the rate of disappearance of NH_3.

$$0.52\ \frac{mol\,N_2}{L \cdot s} \times \frac{4\ mol\,NH_3}{2\ mol\,N_2} = 1.04\ \frac{mol\,NH_3}{L \cdot s}$$

This rate should be reported as -1.04 M·s^{-1} because we are reporting the rate of *disappearance* of NH_3.

Rate Laws and Reaction Order
   A. Rate law - states the dependence of the reaction rate on concentration.
   1. Equation that tells how the rate depends on the concentration of each reactant.
   2. For the reaction $a\,A + b\,B \rightarrow$ products, the rate law is

$$Rate = -\frac{\Delta[A]}{\Delta t} = k[A]^m[B]^n$$

   a. k = proportionality constant called the rate constant
   B. Reaction order - determined by the values of the exponents.
   1. Values of m and n indicate the reaction order with respect to A and B.
     a. exponent = 1; first order
     b. exponent = 2; second order
     c. exponent = 3; third order
   2. Overall reaction order = $m + n$.
   3. Indicates how the change in concentration can affect the rate.
   4. Unrelated to the coefficients in the balanced equation.
   5. Usually small positive integers but can be negative, zero, or even fractions.
     a. exponent = 1; rate depends linearly on the concentration of the corresponding reactant

       b. exponent = 0; the rate is independent of the concentration of the corresponding reactant

       c. exponent < 1; the rate decreases as the concentration of the corresponding reactant increases

      6. ***The values of the exponents in a rate law must be determined by experiment; they can't be deduced from the stoichiometry of the reaction.***

3⊠ Experimental Determination of a Rate Law

    A. To determine the reaction order (values of the exponents in a rate law) - measure the initial rate of a reaction as a function of different sets of initial concentrations.

       1. Design pairs of experiments to investigate the effect of the initial concentration of a single reactant on the inital rate of change.

       2. If, by doubling the concentration of a reactant, the rate also doubles, then the reaction is first order with respect to that reactant.

       3. If, by doubling the concentration of a reactant, the rate increases by a factor of $2^2 = 4$, the reaction is second order with respect to that reactant.

       4. If, by doubling the concentration of a reactant, the rate of the reaction increases by a factor of $2^3 = 8$, the reaction is third order with respect to that reactant.

       5. Use initial rates to avoid complications from the reverse reaction.

          a. initial rates measure only the rate of the forward reaction

          b. only reactants and catalysts appear in the rate law

    B. Can determine the value of k from the rate law.

       1. Value is characteristic of a reaction.

       2. Depends on temperature.

       3. Does not depend on concentration.

       4. Units depend on the number of concentration terms in the rate law and on the values of the exponents.

EXAMPLE:

The following data was collected for the reaction

$$2\,NO\,(g)\ +\ H_2\,(g)\ \rightarrow\ N_2O\,(g)\ +H_2O\,(g)$$

$[NO]_I$	$[H_2]_I$	Rate $(M\cdot s^{-1})$
0.15	0.15	8.54×10^{-6}
0.30	0.15	3.42×10^{-6}
0.45	0.15	7.68×10^{-6}
0.15	0.30	1.71×10^{-5}
0.15	0.45	2.56×10^{-5}

Determine the rate law from this data. What is the order of the reaction with respect to each reactant? What is the overall order of the reaction? Calculate the value of k.

SOLUTION: In the first three solutions, the concentration of NO is changing while the concentration of H_2 remains constant. Therefore, we know that any changes which occur in the rate are a consequence of the change in concentration of NO. When the concentration of NO is doubled in the first two experiments, the rate increases by a factor of four. When the concentration of NO is tripled (exp. 1 and 3) the rate increases by a factor of nine. We know that $2^2 = 4$ and that $3^2 = 9$. Therefore, the rate of reaction with respect to NO depends on $[NO]^2$. In comparing experiments 1,4, and 5, the concentration of NO remains constant while the concentration of H_2 changes. Therefore, we know that any changes which occur in the rate are a consequence of the change in concentration of H_2. When the concentration of H_2 is doubled in experiments 1 and 4, the rate is doubled. When the concentration of H_2 is tripled in experiments

1 and 5, the rate is tripled. We know that $2^1 = 2$ and $3^1 = 3$. Therefore, the rate of reaction with respect to H_2 depends on $[H_2]$. We can now write our rate law.

$Rate = k[NO]^2[H_2]$

We can calculate the value of k using the data from any one of the five experiments. Using the data in experiment four gives:

$$1.71 \times 10^{-5} \frac{mol}{L \cdot s} = k\left(0.15 \frac{mol}{L}\right)^2\left(0.30 \frac{mol}{L}\right); \qquad k = \frac{1.71 \times 10^{-5} \frac{mol}{L \cdot s}}{6.75 \times 10^{-3} \frac{mol^3}{L^3}} = 2.53 \times 10^{-3} \frac{L^2}{mol^2 \cdot s}$$

Integrated Rate Law for a First-order Reaction
A. Integrated Rate law - a concentration-time equation that allows us to calculate the concentration of a reactant at any time t or the fraction of a reactant that remains at any time t.
1. Can be used to calculate the time required for the initial concentration of a reactant to drop to any particular value or to any particular fraction of its initial concentration.
B. For the reaction $a A \rightarrow$ products, the integrated rate law is
$$\log\frac{[A]_t}{[A]_0} = \frac{-kt}{2.303}$$
1. Can rearrange the equation to give
$$\log[A]_t = \frac{-kt}{2.303} + \log[A]_0$$
2. Plot of log [A] *versus* time gives a straight line if the reaction is first order in A.
3. $k = -2.303$(slope).

EXAMPLE:
When sucrose reacts with water, glucose is formed according to the reaction:

$C_{12}H_{22}O_{11} + H_2O \rightarrow 2 C_6H_{12}O_6$

This reaction follows first order kinetics with respect to the sucrose. Calculate the value of k, if it takes 9.70 hours for the concentration of sucrose to decrease from 0.00375 M to 0.00252 M. Determine the amount of time required for the reaction to go to 80% completion.

SOLUTION: To calculate the value of k, we simply substitute the data given into the first-order integrated rate equation.

$$\log\frac{0.00252}{0.00375} = \frac{-k(9.70 \text{ h})}{2.303}; \qquad k = 4.10 \times 10^{-2} \text{ h}^{-1}$$

We can now calculate the amount of time required for the reaction to be 80% complete. To determine the concentration at this time, we multiply the initial concentration by 0.80.

$0.80 \times 0.00375 = 0.00300$

This value represents the amount of sucrose that has reacted. The amount of sucrose remaining after the reaction is 80% complete is $0.00375 - 0.00300 = 0.00075$. We now have the initial concentration and the concentration at time t, which can be substituted into the first-order integrated rate law, along with the value of k we calculated in the first part of the problem.

185

$$\log \frac{0.00075}{0.00375} = \frac{-(4.10 \times 10^{-2} \text{ h}^{-1})t}{2.303}; \qquad t = 39.3 \text{ h}$$

Half-Life of a First-order Reaction

6⊠   A. Half-life ($t_{1/2}$) - the time required for the reactant concentration to drop to one-half of its initial value.

7⊠   1. $t_{1/2} = \dfrac{0.693}{k}$

         2. For first-order reaction, half-life is a constant.

           a. depends only on the rate constant

EXAMPLE:

Determine the half-life for the reaction of sucrose with water.

SOLUTION: We can calculate the half-life by substituting the rate constant determined in the previous example into the equation for the first-order half-life.

$$t_{1/2} = \frac{0.693}{4.10 \times 10^{-2} \text{ h}^{-1}} = 16.9 \text{ h}$$

Second-order Reactions

4⊠   A. Integrated rate law for a second-order reaction.

$$\frac{1}{[A]_t} = kt + \frac{1}{[A]_0}$$

5⊠   1. Plot of $\dfrac{1}{[A]_t}$ *versus* time gives a straight line if the reaction is second-order.

           a. slope of line = k; intercept = $\dfrac{1}{[A]_0}$

7⊠   B. Half-life for a second-order reaction

         1. Depends on both the rate constant and the initial concentration.

$$t_{1/2} = \frac{1}{k[A]_0}$$

EXAMPLE:

The reaction 2 NOBr (g) → 2 NO (g) + Br$_2$ (g) is a second order reaction with respect to NOBr. The rate constant for this reaction is $k = 0.810$ M$^{-1} \cdot$s^{-1} when the reaction is carried out at a temperature of 10° C. If the initial concentration of NOBr = 7.5×10^{-3} M, how much NOBr will be left after a reaction time of 10 minutes? Determine the half-life of this reaction.

SOLUTION: We can solve for the amount of NOBr after 10 minutes by substituting the given data into the integrated rate law for a second-order reaction.

$$\frac{1}{[NOBr]_t} = (0.810 \text{ M}^{-1} \cdot \text{s}^{-1}) \times (600 \text{ s}) + \frac{1}{7.5 \times 10^{-3} \text{ M}};$$

$$\frac{1}{[NOBr]_t} = 6.19 \times 10^2 \text{ M}^{-1}; \qquad [NOBr]_t = 1.6 \times 10^{-3} \text{ M}$$

To determine the half-life for this reaction, we substitute the intitial concentration of NOBr and the rate constant for the reaction into the equation for the half-life of a second-order reaction.

$$t_{1/2} = \frac{1}{0.810 \text{ M}^{-1} \cdot \text{s}^{-1} (7.5 \times 10^{-3} \text{ M})} = 160 \text{ s}$$

Reaction Mechanisms

A. Reaction mechanism - the sequence of molecular events, or reaction steps, that defines the pathway from reactants to products.
   1. Reaction steps - involve the breaking of chemical bonds and/or the making of new bonds.
   2. Knowing reaction mechanisms allows for better control of known reactions and predictions of new reactions.
B. Elementary step - a single step in a reaction mechanism.
   1. Description of an individual molecular event (collisions of individual molecules).
   2. Describe the reaction mechanism.
   3. Classified on the basis of their molecularity.
      a. molecularity - the number of molecules on the reactant side of the chemical equation
      b. unimolecular reaction - elementary reaction that involves a single reactant molecule
      c. bimolecular reaction - elementary reaction that results from energetic collisions between two reactant molecules
      d. termolecular reaction - involve three atoms or molecules; rare
   4. Reaction intermediate - a species that is formed in one step of a reaction mechanism and consumed in a subsequent step.
      a. do not appear in the net equation for the overall reaction
      b. presence is only noticed in the elementary steps
C. Balanced equation for an overall reaction.
   1. Provides no information about how the reaction occurs.
   2. Describes reaction stoichiometry.

8⊠ **Rate Laws and Reaction Mechanisms**

A. Rate law for overall reaction - determined by experimentation.
B. Rate law for an elementary reaction - determined from its molecularity.
   1. Contains the concentration of each reactant raised to an exponent equal to its coefficient in the chemical equation for the elementary reaction.
   2. *Only applies to elementary reactions, not overall reactions.*
   3. Rate of a unimolecular reaction is first order in the concentration of the reactant molecule.
   4. Overall reaction order for an elementary reaction is equal to its molecularity. (see Table 12.5, page 487)
C. Experimentally observed rate law for an overall reaction depends on the reaction mechanism.
   1. Overall reaction occurs in a single elementary step.
      a. experimental rate law = rate law for the elementary step
   2. Overall reaction occurs in two or more steps.
      a. rate-determining step - the slowest step in a reaction mechanism
         i. limits the rate at which reactants can be converted to products
      b. overall reaction can occur no faster than the speed of the rate-determining step
D. Two criteria for an acceptable reaction mechanism.
   1. The elementary steps must sum to give the overall reaction.
   2. The mechanism must be consistent with the observed rate law for the overall reaction.
E. Procedure used for establishing a reaction mechanism.
   1. Determine the overall rate law experimentally.
   2. Devise a series of elementary steps.
   3. Predict the rate law based on the reaction mechanism.

4. If observed and predicted rate laws agree, the proposed mechanism is a plausible pathway for the reaction.

5. Easy to disprove a mechanism; impossible to "prove" a mechanism.

Reaction Rates and Temperature; The Arrhenius Equation

A. Reaction rates tend to double when the temperature is increased by $10°C$.

B. Collision theory model - a bimolecular reaction occurs when two properly oriented reactant molecules come together in a sufficiently energetic collision.

C. For the reaction: $A + BC \rightarrow AB + C$.

1. For a single step reaction, a new bond, A-B, develops at the same time as the old bond, B-C, breaks.

2. Nuclei pass through a configuration in which all three atoms are weakly linked together.

    a. $A + B\text{-}C \rightarrow A\text{--}B\text{--}C \rightarrow A\text{-}B + C$

3. Need energy to overcome repulsions.

    a. comes from kinetic energy of the colliding particles

    b. stored as potential energy in A--B--C.

       i. A--B--C has more potential energy than either the reactants or products

4. Potential energy barrier that must be surmounted before reactants can be converted to products.

    a. potential energy profile - plot of potential energy *versus* reaction progress

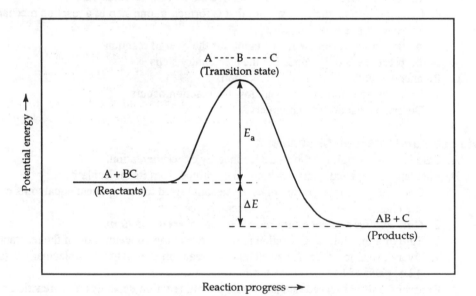

    b. activation energy, E_a, - the height of the barrier

    c. transition state (or activated complex) -configuration of atoms at the maximum in the potential energy profile

5. All the energy needed to climb the potential energy barrier must come from the kinetic energy of the colliding molecules.

D. Comparison of collision rates and reaction rates leads to experimental evidence for the idea of an activation energy barrier.

1. Only a small fraction of collisions lead to reaction.

    a. very few collisions occur with a kinetic energy as large as the activation energy

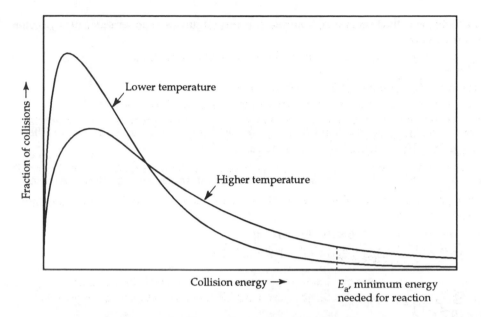

b. area under the curve to the right of E_a represents the fraction of the collision with an energy equal to or greater than the activation energy

    i. fraction, $f = e^{-E_a/RT}$

2. Distribution of collision energies broadens and shifts to higher energies with an increase in temperature.

    a. exponential increase in the fraction of collision that lead to products

3. Accounts for exponential dependence of reaction rates on temperature.

4. Explains why reaction rates are so much lower than collision rates.

E. Orientation requirement also reduces the fraction of collisions that lead to products.

    1. Reactants won't react unless the orientation of the reaction partners is correct for formation of the transition state.

    2. Steric factor, p, - fraction of collisions having proper orientation.

F. For a bimolecular collision between A and B.

    1. Collision rate = Z[A][B].

        a. Z is a constant related to collision frequency

    2. Reaction rate < collision rate by a factor of $p \times f$.

        a. Reaction rate = $p\,fZ$[A][B]

    3. Reaction rate = k[A][B].

        a. $k = pfZ = pZe^{-E_a/RT}$

G. Arrhenius equation.

    1. $k = Ae^{-E_a/RT}$.

    2. A = frequency factor = pZ.

Using the Arrhenius Equation

A. Can determine the activation energy if values of the rate constant are known at different temperatures.

B. Rearrange the equation to get

$$\log k = \left(\frac{-E_a}{2.303R}\right)\left(\frac{1}{T}\right) + \log A$$

9 ⊠

    1. Plot log k *versus* $1/T$ gives a straight line.

        a. slope = $-E_a/2.303R$

10⊠   C. Can estimate the activation energy from rate constants at just two temperatures using another form of the equation.

$$\log\left(\frac{k_2}{k_1}\right) = \left(\frac{-E_a}{2.303R}\right)\left(\frac{1}{T_2} - \frac{1}{T_1}\right)$$

EXAMPLE:

The activation energy for the reaction $ClO_2F\ (g) \rightarrow ClOF\ (g) + O\ (g)$ is 186 kJ/mol. If the value of k is $6.76 \times 10^{-4}\ s^{-1}$ at 322° C, what is the value of k at 50° C?

SOLUTION: To solve this problem, we simply substitute the information given into the Arrhenius equation.

$$\log\left(\frac{k_2}{6.76 \times 10^{-4}\ s^{-1}}\right) = \left(\frac{-1.86 \times 10^5\ J/mol}{2.303 \cdot 8.314\ J/mol \cdot K}\right)\left(\frac{1}{323\ K} - \frac{1}{595\ K}\right)$$

$$\log\left(\frac{k_2}{6.76 \times 10^{-4}\ s^{-1}}\right) = -13.7$$

To determine the value of k_2, we need to take the antilog of both sides of the equation.

$$\frac{k_2}{6.76 \times 10^{-4}\ s^{-1}} = 2.00 \times 10^{-14}; \qquad k_2 = 1.35 \times 10^{-17}\ s^{-1}$$

Catalysis
A. Catalyst - a substance that increases the rate of a reaction without being consumed in the reaction.
   1. Important in chemical industry and in living organisms.
   2. Chemical industry - favor formation of specific products and lower reaction temperatures.
   3. Living organisms - enzymes (catalysts) facilitate specific reactions of crucial biological importance.

11⊠   B. Accelerates the rate of a reaction by making a new and more efficient pathway available for the conversion of reactants to products.
   1. Speeds up reaction in two ways:
      a. increases the frequency factor A
      b. decreases the activation energy
         i. rate constant is more sensitive to the E_a and a catalyst usually functions by lowering the activation energy

Homogeneous and Heterogeneous Catalysts
A. Homogeneous catalyst - one that exists in the same phase as the reactants.
B. Heterogeneous catalyst - exists in a different phase from the reactants.
   1. Mechanism is complex and not well understood.
   2. Important steps:
      a. adsorption of reactants onto the surface of the catalyst
      b. conversion of reactants to products on the surface
      c. desorption of products from the surface
   3. Adsorption step:
      a. chemical bonding of reactants to the highly reactive metal atoms on the surface
      b. breaking or weakening of bonds in the reactants

4. Industrial chemical processes use mostly heterogeneous catalysts due to the ease of separation of the catalyst from the reaction products.
5. Used in automobile catalytic converters.

Self-Test

This section is intended to test your knowledge of the material covered in this chapter. Think through these problems and make certain you understand what is going on. Ask yourself if your answer makes sense. Many of these questions are linked to the chapter learning goals. Therefore, successful completion of these problems indicates you have mastered the learning goals for this chapter. You will receive the greatest benefit from this section if you use it as a mock exam. You will then discover which topics you have mastered and which topics you need to study in more detail.

True/False

1. The relative rates of product formation and reactant consumption depend on the coefficients in the balanced equation.

2. The rate of a reaction stays the same as the reaction proceeds.

3. If the rate law for the reaction a A + b B → products is rate = $k[A]^m[B]^n$, the values of m and n are equal to the coefficients a and b, respectively, in the chemical equation.

4. For the second order reaction, A → products, doubling the concentration of A will cause the rate to also double.

5. The units for the rate constant, k, depend on the number of concentration terms in the rate law and on the values of the exponents.

6. For a first-order reaction, the half life depends only on the rate constant.

7. The molecularity of a reaction is the number of molecules which participate in the reaction.

8. The balanced equation for an overall reaction describes the reaction mechanism.

9. The experimentally observed rate law for an overall reaction can be determined from the slow step in the reaction mechanism.

10. Industrial chemical processes use mostly homogeneous catalysts due to the ease of recovery.

Multiple Choice

1. The overall reaction order for the rate law, rate = $k[A]^2[B]^3$ is
   a. two
   b. three
   c. five
   d. six

2. A reaction has the rate law, rate = $k[A]^3$. If the initial rate for this reaction is 0.15 mol/L· sec, and the concentration of A is doubled, the new initial rate for this reaction will be
   a. 0.45 mol/L·sec
   b. 1.20 mol/L·sec
   c. 0.30 mol/L·sec
   d. 1.35 mol/L·sec

3. If a plot of $\dfrac{1}{[A]_t}$ versus time gives a straight line for a particular reaction, that reaction is
   a. first-order
   b. 1/2 order
   c. second-order
   d. zero order

4. The reaction order for the elementary reaction step, $2\,A + B \rightarrow$ products, is
   a. second-order
   b. third-order
   c. cannot be determined from the balanced equation
   d. first-order

5. For every collision that occurs in a reaction
   a. the products are formed
   b. the particles have enough kinetic energy to overcome the potential energy barrier
   c. the fraction of molecules with enough kinetic energy to overcome the potential energy barrier is
      $$f = e^{-E_a/RT}$$
   d. none of the above.

6. The exponents in a rate law indicate the
   a. dependence of the rate on concentration
   b. sum of the coefficients in a balanced equation
   c. time required for a certain amount of reactant to disappear
   d. none of the above

7. The half-life of a first order reaction with $k = 1.65 \times 10^{-3}$ s^{-1}
   a. cannot be determined without knowing the initial concentration of the reactant
   b. equals 1.14×10^{-3} s
   c. equals 4.06×10^{-2} s
   d. 4.2×10^2 s

8. The half-life of a second order reaction depends on
   a. both the rate constant and the initial concentration
   b. only the rate constant
   c. only the initial concentration
   d. none of the above

9. Reaction rates tend to double when the temperature is
   a. doubled
   b. increased by 100°
   c. multiplied by 2
   d. increased by 10°

10. The activated complex has
    a. more energy than either the reactants or products
    b. is the configuration of the atoms at the minimum of the potential energy barrier
    c. less energy than the reactants but more energy than the products
    d. the same configuration as an intermediate in an elementary step in a reaction mechanism

Matching

Kinetics

Integrated rate law

Half-life

Reaction mechanism

Elementary step

Molecularity

Bimolecular reaction

Potential energy profile

Transition state

Catalyst

Homogeneous catalyst

a. a single step in a reaction mechanism

b. a plot of potential energy *versus* reaction progress.

c. configuration of atoms at the maximum in the potential energy profile

d. a substance that increases the rate of a reaction without being consumed in the reaction.

e. the area of chemistry concerned with reaction rates and the sequence of steps by which reactions occur.

f. the time required for the reactant concentration to drop to one-half of its initial value.

g. the sequence of molecular events, or reaction steps, that defines the pathway from reactants to products.

h. a concentration-time equation that allows us to calculate the concentration of a reactant at any time t or the fraction of a reactant that remains at any time t.

i. one that exists in the same phase as the reactants.

j. the number of molecules on the reactant side of the chemical equation for an elementary step.

k. elementary reaction that results from energetic collisions between two reactant molecules.

Fill-in-the-Blank

1. The relative rates of product formation and reactant consumption depend on _____
   _____ .

2. The reaction order indicates _____
   _____ .

3. The values of the exponents in a rate law are determined _____ .

4. A reaction is third order if, by _____ the concentration of a reactant, the rate
   increases by a factor of _____ .

5. If a plot of log [A] *versus* time gives a straight line, the reaction is _____ with
   respect to A.

6. An elementary step in a reaction mechanism gives a description of an _____
   _____ .

7. The rate law for an elementary reaction is determined from _____ .

8. If an overall reaction occurs in two or more steps, the rate-determining step is the _____
   _____ .

9. Reaction rates tend to double when the temperature is increased by _____ .

10. Reactants won't react unless the _____ of the reaction partners is correct for
    formation of the _____ .

Problems

1. The following data for the reaction A + B → C was collected. Calculate the average rate for the
   formation of C between 45 and 75 seconds.

[C] (mol/L)	time (sec)
0	0
0.005	15
0.010	30
0.015	45
0.017	60
0.019	75
0.020	90
0.021	105
0.022	120

2. The rate of disappearance of PH_3 for the reaction

   $$4 PH_3(g) \rightarrow P_4(g) + 6 H_2 (g)$$

   is 1.47×10^{-3} mol/L·sec. What is the rate of appearance of H_2?

194

3. The following data was collected for the reaction

$2 NO (g) + Cl_2 (g) \rightarrow 2 NOCl (g)$

[NO] (mol/L)	[Cl$_2$] (mol/L)	Initial Rate mol/(L·sec)
0.025	0.025	39.1
0.050	0.025	156.3
0.075	0.025	351.6
0.025	0.050	78.1
0.025	0.075	117.2

Determine the rate law from this data. What is the order of the reaction with respect to each reactant? What is the overall order of the reaction? Calculate the value of k.

4. The following data was collected for the reaction

$3 A + 2 B \rightarrow 2 D$

[A] (mol/L)	[B] (mol/L)	Initial Rate mol/(L·sec)
0.150	0.150	1.56×10^{-2}
0.300	0.150	0.125
0.450	0.150	0.421
0.150	0.300	1.56×10^{-2}
0.150	0.450	1.56×10^{-2}

Determine the rate law from this data. What is the order of the reaction with respect to each reactant? What is the overall order of the reaction? Calculate the value of k. If the initial concentration of A = 0.639 M and the initial concentration of B = 0.75 M, what would the initial rate be?

5. The conversion of cyclopropane to propene obeys first order kinetics. If the initial concentration of cyclopropane is 0.500 M, how long will it take for the reaction to reach 35% completion? 70% completion? $(k = 1.16 \times 10^{-6} s^{-1})$

6. The following data was collected for the first order reaction

$4 PH_3 (g) \rightarrow P_4 (g) + 6 H_2 (g)$

[PH$_3$] (mol/L)	t (sec)
0.450	0
0.248	30.0
0.137	60.0
0.0757	90.0
0.418	120.0

Determine the half-life for this reaction from the plotted data. Calculate the value of k. From the calculated value of k, calculate the half-life of this reaction. Do your two answers for half-life agree?

7. The rate law for the decomposition of O_3 to O_2 is second-order in ozone. The rate constant for this reaction equals 1.40×10^{-2} L/(mol·sec). If the initial concentration of ozone is 2.75 M, how much O_3 will be present after 24 hours?

8. The following data was collected for the reaction

$2 \, HI \, (g) \rightarrow H_2 \, (g) + I_2 \, (g)$

[HI] (mol/L)	time (min)
2.50	0
1.45	3
1.02	6
0.788	9
0.641	12
0.541	15
0.468	18
0.412	21
0.368	24
0.332	27
0.303	30

Determine the order of this reaction. Calculate the value of k and the half-life at the beginning of the reaction.

9. The proposed reaction mechanism for the overall reaction

$2 \, NO_2 \, (g) + F_2 \, (g) \rightarrow 2 \, NO_2F \, (g)$

is

$NO_2 \, (g) + F_2 \, (g) \rightarrow NO_2F \, (g) + F \, (g)$

$F \, (g) + NO_2 \, (g) \rightarrow NO_2F \, (g)$

What is the molecularity of each step? If the first step is the slow step in the mechanism, what would the rate law be?

10. The reaction mechanism for the decomposition of hydrogen peroxide is

$H_2O_2 \rightarrow 2 \, OH$

$H_2O_2 + OH \rightarrow H_2O + HO_2$

$HO_2 + OH \rightarrow H_2O + O_2$

What is the overall reaction? What are the reaction intermediates? What is the molecularity of each step? Which step is the slowest step if the experimental rate law is: Rate $= k[H_2O_2]$.

11. The following data were collected for the reaction

    A + B → product

k (sec^{-1})	T ($^{\circ}$C)
1.69 x 10^{-4}	50
2.89 x 10^{-4}	75
5.29 x 10^{-4}	100
1.04 x 10^{-3}	125
2.25 x 10^{-3}	150
5.43 x 10^{-3}	175
1.5 x 10^{-2}	200

    Prepare an Arrhenius plot and determine the activation energy from the slope of the line.

12. If the rate constant for a certain reaction is $k = 1.67 \times 10^{-4}$ L/mol·sec at a temperature of 68° C and the activation energy for the reaction is 111.8 kJ/mol, what is the value of A?

13. The activation energy is E_a = 100.25 kJ/mol for the reaction

    2 NOCl → 2 NO + Cl$_2$

    If $k = 9.3 \times 10^{-6}$ sec^{-1} at 77° C, at what temperature will $k = 2.8 \times 10^{-3}$ sec^{-1}?

14. Draw a potential energy profile for the endothermic reaction A + B → products. Indicate the change in activation energy that will occur when a catalyst is added to this reaction.

Solutions

True/False
1. T
2. F. The reaction rate usually decreases as the reaction runs out of reactants.
3. F. The exponents m and n can only be determined from experiments.
4. F. Doubling the concentration of A will cause the rate to increase by a factor of 4 for a second order reaction.
5. T
6. T
7. F. Molecularity is the number of molecules only on the reactant side of an elementary reaction.
8. F. The balanced equation for an overall reaction only describes the stoichiometry of the reaction.
9. T. Only if the reaction mechanism is consistent with the observed rate law for the reaction.
10. F. Industrial processes use heterogeneous catalysts.

Multiple Choice
1. c
2. b
3. c
4. b
5. c
6. a
7. d
8. a

9. d
10. a

Matching
Kinetics - e
Integrated rate law - h
Half-life - f
Reaction mechanism - g
Elementary step - a
Molecularity - j
Bimolecular reaction - k
Potential energy profile - b
Transition state - c
Catalyst - d
Homogeneous catalyst - i

Fill-in-the-Blank
1. the coefficients in the balanced equation
2. how a change in concentration can affect the rate
3. experimentally
4. doubling; eight
5. first-order
6. individual molecular event
7. the molecularity of the reaction
8. slowest step in the reaction mechanism
9. $10°C$
10. orientation; transition state

Problems

1. $\dfrac{\Delta[C]}{\Delta t}$ can be calculated by first plotting the data and then determining the slope of the right triangle created from $\Delta[C]$ and Δt.

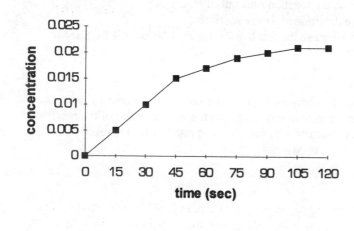

$$\frac{\Delta[C]}{\Delta t} = \frac{0.019 - 0.015}{75 - 45} = 1.3 \times 10^{-4} \ M/s$$

2. Rate of appearance of $H_2 =$

$$\left[1.47 \times 10^{-3} \text{ mol PH}_3 /(\text{L} \cdot \text{s})\right] \times \frac{6 \text{ mol H}_2}{4 \text{ mol PH}_3} = 2.20 \times 10^{-3} \text{ mol H}_2 /(\text{L} \cdot \text{s})$$

3. In the first three experiments, the concentration of NO changes while that of Cl_2 remains constant. Therefore, the change in rate for these three experiments is dependent only on the change in concentration of NO. When the concentration of NO is doubled from 0.025 M to 0.050 M, the rate increases by a factor of 4. When the concentration of NO is tripled from 0.025 M to 0.075 M, the rate increases by a factor of 9. We know that $2^2 = 4$ and that $3^2 = 9$. Therefore, the rate of the reaction with respect to NO depends on $[NO]^2$. In the first, fourth, and fifth experiments, the concentration of NO remains constant while that of the Cl_2 is changed. Therefore, the change in rate for these three experiments is dependent only on the change in concentration of Cl_2. When the concentration of Cl_2 is doubled from 0.025 M to 0.050 M, the rate increases by a factor of 2. When the concentration of Cl_2 is tripled from 0.025 M to 0.075 M, the rate increases by a factor of 3. We know that $2^1 = 2$ and that $3^1 = 3$. Therefore, the rate of reaction with respect to Cl_2 depends on $[Cl_2]$. The overall rate law is: rate $= k[NO]^2[Cl_2]$. The overall order of the reaction is third order. k can be calculated from any one of the five experiments.

$$117.2 \text{ mol}/(\text{L} \cdot \text{s}) = k(0.025 \text{ mol}/\text{L})^2 (0.075 \text{ mol}/\text{L});$$

$$k = \frac{117.2 \text{ mol}/(\text{L} \cdot \text{s})}{4.69 \times 10^{-5} \text{ mol}^3/\text{L}^3} = 2.50 \times 10^6 \text{ L}^2/(\text{mol}^2 \cdot \text{s})$$

4. In the first three experiments, the concentration of B is held constant, while the concentration of A is changing. Therefore, the change in rate for these three experiments is dependent only on the change in concentration of A. When the concentration of A is doubled, the rate increases by a factor of 8. When the concentration of A is tripled, the rate increases by a factor of 27. We know that $2^3 = 8$ and that $3^3 = 27$. Therefore, the rate of reaction with respect to A is proportional to $[A]^3$. In the first, fourth, and fifth experiments, the concentration of A is held constant, while the concentration of B is changing. Therefore, the rate of reaction in these three experiments is dependent only upon the concentration of B. When the concentration of B is doubled, the rate remains the same. When the concentration of B is tripled, the rate remains the same. (This is equivalent to multiplying the rate by 1.) We know that $2^0 = 1$ and that $3^0 = 1$. Therefore, the rate of reaction with respect to B is proportional to $[B]^0$ or 1. The overall rate law is: rate $= k[A]^3$. The overall order of the reaction is third order. k can be calculated from any one of the five experiments.

$$1.56 \times 10^{-2} \text{ mol}/(\text{L} \cdot \text{s}) = k(0.150 \text{ mol}/\text{L})^3$$

$$k = \frac{1.56 \times 10^{-2} \text{ mol}/(\text{L} \cdot \text{s})}{\left(3.38 \times 10^{-3} \text{ mol}^3/\text{L}^3\right)} = 4.62 \text{ L}^2/(\text{mol}^2 \cdot \text{s})$$

$$\text{Rate} = 4.62 \text{ L}^2/(\text{mol}^2 \cdot \text{s}) \times (0.639 \text{ mol}/\text{L})^3 = 1.21 \text{ mol}/(\text{L} \cdot \text{s})$$

5. When the reaction has reached 35% completion, then 0.500×0.35 or 0.175 M has reacted. This means that the concentration of cyclopropane after 35% completion is $0.500 - 0.175$ or 0.325 M. We can now use the integrated rate law for a first order reaction to calculate t.

$$\log\left(\frac{0.325 \text{ M}}{0.500 \text{ M}}\right) = \frac{-\left(1.16 \times 10^{-6} \text{ s}^{-1}\right)t}{2.303}; \quad t = \frac{-0.187 \times 2.303}{-1.16 \times 10^{-6} \text{ s}^{-1}} = 3.71 \times 10^5 \text{ s}$$

When the reaction has reached 70% completion, the concentration of cyclopropane is 0.150 M.

$$\log\left(\frac{0.150}{0.500}\right) = \frac{-\left(1.16\times10^{-6}\,s^{-1}\right)t}{2.303} \;; \qquad t = \frac{-0.523\times2.303}{-1.16\times10^{-6}\,s^{-1}} = 1.04\times10^{6}\,s$$

6. The half-life can be determined from the plotted data by finding the time when the concentration of PH_3 is 0.225 M.

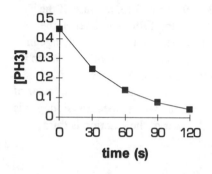

The estimated half-life from the plot is 35 s. k can be calculated by using the integrated rate law for a first order reaction.

$$\log\frac{0.248}{0.450} = \frac{-k(30.0\,s)}{2.303} \;; \qquad k = -\left(\frac{-0.259\times2.303}{30.0\,s}\right) = 1.99\times10^{-2}\;s^{-1}$$

The calculated half-life is

$$t_{\frac{1}{2}} = \frac{0.693}{1.99\times10^{-2}\;s^{-1}} = 34.8\;s$$

7. $\dfrac{1}{[O_3]} = \left[1.40\times10^{-2}\,L/(mol\cdot s)\right]\left(24\,h\times\dfrac{3600\,s}{1\,h}\right) + \dfrac{1}{2.75\,M} = 1.21\times10^{3}\,L/mol\;;$

$$[O_3] = 8.26\times10^{-4}\;M$$

8. We can determine the order of the reaction by plotting log{HI} *versus* t. If we obtain a straight line, we know we have a first-order reaction. If we don't obtain a straight line, then we need to plot 1/[HI] *versus* t. If we obtain a straight line, we know we have a second-order reaction.

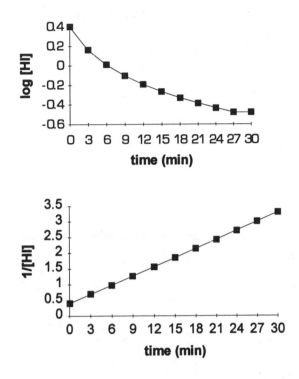

From the two plots, we can see that the reaction is second-order. We can use the integrated rate law for a second order reaction to calculate the value of k and the half-life.

$$\frac{1}{1.45\,M} = k(3\,\text{min}) + \frac{1}{2.50\,M} \;;\quad k = \frac{\dfrac{1}{1.45\,M} - \dfrac{1}{2.50\,M}}{3\,\text{min}} = 9.66 \times 10^{-2}\ \text{L}/(\text{mol}\cdot\text{min})$$

$$t_{\frac{1}{2}} = \frac{1}{\left[9.66 \times 10^{-2}\ \text{L}/(\text{mol}\cdot\text{min})\right] 2.50\,\text{mol}/\text{L}} = 4.14\,\text{min}$$

9. Both steps are bimolecular. Rate = $k[NO_2][F_2]$.

10. Overall reaction: $2\,H_2O_2 \rightarrow 2\,H_2O + O_2$; OH^- and HO_2 are reaction intermediates; first step - unimolecular; second and third step - bimolecular; first step is the slowest step.

11.

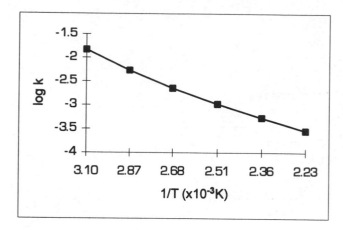

From the plot, we can see that we have a straight line. To obtain the slope we calculate $\dfrac{\Delta y}{\Delta x}$ for the two farthest points. (Don't forget that you are plotting **log** k *versus* $1/T$ in kelvin.)

$$\text{Slope} = \frac{-3.78 - (-1.82)}{\left(3.10 \times 10^{-3}\right) - \left(2.11 \times 10^{-3}\right)} = -1.98 \times 10^{3}$$

Remember that the slope $= \dfrac{-E_a}{2.303R}$; therefore

$$E_a = \left(1.98 \times 10^{3}\ \text{K}\right)\left(2.303\right)\left(8.314\ \text{J / mol}\cdot\text{K}\right) = 3.79 \times 10^{4}\ \text{J / mol}$$

12. $\log\!\left(1.67 \times 10^{-4}\right) = \left(\dfrac{-1.118 \times 10^{5}\ \text{J / mol}}{(2.303)(8.314\ \text{J / mol}\cdot\text{K})}\right)\!\left(\dfrac{1}{341\ \text{K}}\right) + \log A\,; \qquad \log A = 13.3\,;$

$\qquad A = 2.22 \times 10^{13}$

13. $\log\!\left(\dfrac{2.8 \times 10^{-3}}{9.3 \times 10^{-6}}\right) = \left(\dfrac{-1.0025 \times 10^{5}\ \text{J / mol}}{2.303 \times 8.314\ \text{J / (mol}\cdot\text{K)}}\right)\!\left(\dfrac{1}{T_2} - \dfrac{1}{350\ \text{K}}\right)$

$\quad 2.48 = -5.236 \times 10^{3}\ \text{K}\!\left(\dfrac{1}{T_2} - \dfrac{1}{350\ \text{K}}\right); \quad -4.74 \times 10^{-4}\ \text{K}^{-1} = \dfrac{1}{T_2} - \dfrac{1}{350\ \text{K}}\,;$

$\quad \dfrac{1}{T_2} = 2.38 \times 10^{-3}\ \text{K}^{-1}; \quad T_2 = 420\ \text{K}$

14.

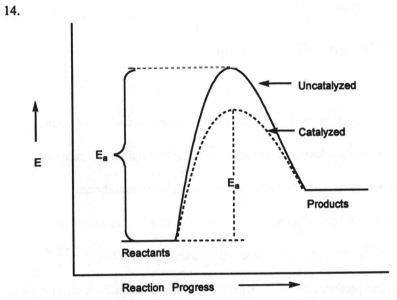

CHAPTER 13

CHEMICAL EQUILIBRIUM

Chapter Learning Goals

1⊠ Given any balanced chemical equation representing a homogenous or heterogeneous equilibrium, write the equilibrium equation.

2⊠ From the equilibrium concentrations of products and reactants, calculate the equilibrium constant K_c.

3⊠ Given the equilibrium partial pressures of reactants and products, calculate the equilibrium constant K_p.

4⊠ From a value of K_c and a balanced equation, calculate K_p. From a value of K_p and a balanced equation, calculate K_c.

5⊠ From the value of K_c or K_p, determine whether mainly products or mainly reactants exist at equilibrium.

6⊠ For a given mixture of reactants and products, determine whether a system is at equilibrium. If it is not, determine the direction in which the reaction must go to achieve equilibrium.

7⊠ Given K_c and initial concentrations of reactants and/or products, calculate the final concentrations of reactants and/or products.

8⊠ Determine the reaction direction when a system at equilibrium reacts to a stress applied to the system, including changes in concentrations, pressure and volume, or temperature.

9⊠ Describe the effect of adding a catalyst to a system at equilibrium.

10⊠ Describe the relationship between the equilibrium constant and the ratio of the rate constants for the forward and reverse reactions. Solve problems involving this relationship.

Chapter in Brief

In this chapter, you begin the study of chemical equilibria. The concepts found in this chapter will be applied to various chemical systems in Chapters 15 and 16. A state of chemical equilibrium is achieved when, with time, the concentrations of both the reactants and products in a reversible reaction level off at constant, equilibrium values. Your study of chemical equilibria will explore several aspects of equilibrium mixtures. You will study the relationship between the concentration of the reactants and products as well as how to determine the equilibrium concentrations knowing the initial concentrations. You will also study which factors can be employed to change the composition of an equilibrium mixture. Finally, you will examine the link between chemical equilibrium and kinetics.

The Equilibrium State

A. Reversible reaction - reaction in which the product molecules can recombine to form reactant molecules.
1. Reactants - substances on the left side of the equation.
2. Products - substances on the right side of the equation.
3. Use a double arrow to indicate a reaction that is at equilibrium.

   a. reactants $\rightleftarrows$ products

B. With time, the concentration of reactant decreases and the concentration of product increases until both concentrations level off at constant, equilibrium values.
   1. Rates of the forward and reverse reactions are equal.
   2. State of chemical equilibrium.
   3. Equilibrium mixture - a mixture of reactants and products in the equilibrium state.
   4. Dynamic state - forward and reverse reactions continue at equal rates.
      a. no net conversion of reactants to products

1,2⊠ **The Equilibrium Constant K_c**
   A. Equilibrium mixture obeys an equilibrium equation (or law of mass action).
      1. For the reaction: $a\,A + b\,B \rightleftarrows c\,C + d\,D$.
         a. $K_c = \dfrac{[C]^c[D]^d}{[A]^a[B]^b}$
         b. K_c = equilibrium constant
            i. units depend on the particular equilibrium expression
            ii. units usually omitted when quoting values of equilibrium constants
         c. equilibrium constant expression - expression on right side of equilibrium equation
      2. For the reverse reaction.
         a. equilibrium expression = reciprocal of the original expression
         b. equilibrium constant $K_c' = \dfrac{1}{K_c}$

EXAMPLE:
For the reaction CO (g) + 2 H$_2$ (g) $\rightleftarrows$ CH$_3$OH (g), the equilibrium concentrations are: [CO]$_e$ = 2.58 M; [H$_2$]$_e$ = 0.280 M; [CH$_3$OH]$_e$ = 2.93 M at a temperature of 210°C. Calculate the equilibrium constant, K_c, for this reaction.

SOLUTION: Before we calculate K_c for this reaction, we need to write an equilibrium equation for the balanced chemical reaction.

$$K_c = \frac{[CH_3OH]}{[CO][H_2]^2}$$

Once we have the equilibrium equation, we can substitute the equilibrium concentrations and solve for K_c.

$$K_c = \frac{(2.93)}{(2.58)(0.280)^2} = 14.5$$

The Equilibrium Constant K_p
3⊠   A. For gas-phase reactions use partial pressures rather than molar concentrations in equilibrium equations.
      1. K_p - equilibrium constant is defined using partial pressures.
4⊠   B. Relationship between K_c and K_p.
      1. $K_p = K_c(RT)^{\Delta n}$.
         a. Δn = the sum of the coefficients of the gaseous products minus the sum of the coefficients of the gaseous reactants

EXAMPLE:

Calculate K_p for the reaction in the previous example.

SOLUTION: We first need to determine Δn for the reaction. The product side of the equation has 1 mol of CH_3OH while the reactant side of the equation has 1 mol CO and 2 mol H_2. Therefore, $\Delta n = -2$. We can now calculate the value of K_p from the value of Δn and the value of K_c determined in the previous example.

$$K_p = (14.5)\left[\left(0.0821\frac{L \cdot atm}{mol \cdot K}\right)(483K)\right]^{-2} = 9.22 \times 10^{-3}$$

Heterogeneous Equilibria

A. Homogeneous equilibria - reactants and products are in a single phase.
B. Heterogeneous equilibria - reactants and products are present in more than one phase.
   1. Solids - molar concentrations are constants
      a. can be calculated from the densities and molar masses.
      b. independent of its amount
   2. Concentrations of pure solids or pure liquids are not included when writing the equilibrium equation for any heterogeneous equilibrium.

Using the Equilibrium Constant

A. Judging the extent of reaction.
   1. Large value of K_c ($> 10^3$)
      a. products predominate over reactants
      b. reaction proceeds almost to completion
      c. position of reaction is to the right
   2. Small value of K_c ($< 10^{-3}$)
      a. reactants predominate over products
      b. reaction hardly proceeds
      c. position of the reaction is to the left
   3. $10^{-3} < K_c < 10^3$
      a. appreciable amounts of both reactants and products present.
B. Predicting the direction of reaction.
   1. Reaction quotient, Q_c, same as the equilibrium constant expression except that the concentrations are not necessarily equilibrium values.
   2. Predict direction of reaction by comparing the value of Q_c to K_c.
      a. $Q_c < K_c$; reaction goes form left to right
      b. $Q_c > K_c$; reaction goes from right to left
      c. $Q_c = K_c$; reaction is at equilibrium

EXAMPLE:

$K_c = 4.18 \times 10^{-9}$ at $425°$ C for the following reaction:

$$2\ HBr\ (g) \rightleftarrows H_2\ (g) + Br_2\ (g)$$

What is the position of equilibrium? If the concentrations of all species present are: [HBr] = 0.75 M; $[H_2] = [Br_2] = 2.5 \times 10^{-4}$ M is the reaction mixture at equilibrium? In which direction will the reaction proceed?

SOLUTION: We can determine the position of equilibrium by looking at the value of K_p. Since K_p $< 10^{-3}$, we know that reactants predominate over products and that the position of equilibrium is to the left. To determine if the reaction mixture is at equilibrium, we need to calculate Q_c and compare that value to K_c.

$$Q_c = \frac{[H_2][Br_2]}{[HBr]^2} = \frac{(2.5 \times 10^{-4})(2.5 \times 10^{-4})}{(0.75)^2} = 1.1 \times 10^{-7}$$

Since $Q_c > K_c$, the reaction is not at equilibrium and will proceed from right to left.

7⊠

C. Calculating equilibrium concentrations.

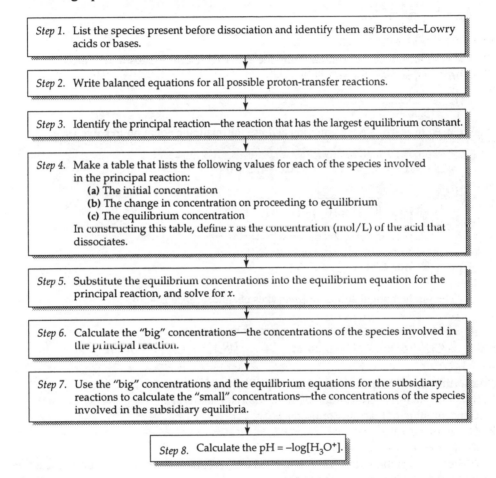

Step 1. List the species present before dissociation and identify them as Bronsted–Lowry acids or bases.

Step 2. Write balanced equations for all possible proton-transfer reactions.

Step 3. Identify the principal reaction—the reaction that has the largest equilibrium constant.

Step 4. Make a table that lists the following values for each of the species involved in the principal reaction:
    (a) The initial concentration
    (b) The change in concentration on proceeding to equilibrium
    (c) The equilibrium concentration
In constructing this table, define x as the concentration (mol/L) of the acid that dissociates.

Step 5. Substitute the equilibrium concentrations into the equilibrium equation for the principal reaction, and solve for x.

Step 6. Calculate the "big" concentrations—the concentrations of the species involved in the principal reaction.

Step 7. Use the "big" concentrations and the equilibrium equations for the subsidiary reactions to calculate the "small" concentrations—the concentrations of the species involved in the subsidiary equilibria.

Step 8. Calculate the pH $= -\log[H_3O^+]$.

EXAMPLE:

The reaction $PCl_5(g) \rightleftarrows PCl_3(g) + Cl_2(g)$ has $K_c = 85.0$ at a temperature of $760°C$. Calculate the equilibrium concentrations of PCl_5, PCl_3, and Cl_2 if the initial concentration of PCl_5 is 5.00 M. Assume a volume of 1.00 L

To solve this problem, follow the steps which are outlined in Fig. 13.6 in your text.

1. The balanced equation is given.

2. The initial concentration of PCl_5 is 5.00 M. Let x be the concentration of PCl_5 that reacts on going to the equilibrium state. Since the mole to mole ratios are 1:1:1, if x amount of PCl_5 reacts, then x mol/L of PCl_3 and Cl_2 will be present at equilibrium. This can be summarized in the following table.

	$PCl_5(g)$ $\rightleftharpoons$	$PCl_3(g)$ +	$Cl_2(g)$
Initial Concentration (M)	5.00	0	0
Change (M)	-x	+x	+x
Equilibrium Concentration (M)	5.00 - x	x	x

3. The equilibrium expression for the reaction is:

$$K_c = \frac{[PCl_3][Cl_2]}{[PCl_5]}.$$

The equilibrium concentrations from the above table can be substituted into this expression, and we can solve for x.

$$85.0 = \frac{(x)(x)}{(5.00 - x)}$$

We need to rearrange the equation and use the quadratic equation to solve for x.

$$425 - 85.0x = x^2; \qquad x^2 + 85.0x - 425 = 0$$

$$x = \frac{-85.0 \pm \sqrt{(7.23 \times 10^3) + (1.7 \times 10^3)}}{2}$$

$$x = -89.7 \text{ or } 4.74.$$

The mathematical solution that makes chemical sense is **4.74**.

4. The equilibrium concentrations can now be calculated.

$$[PCl_5] = 5.00 - 4.74 = 0.26 \text{ M}; \qquad [PCl_3] = [Cl_2] = 4.74 \text{ M}$$

Factors that Alter the Composition of an Equilibrium Mixture

A. Factors that alter the composition of an equilibrium mixture:
   1. Concentration of reactants or products.
   2. Pressure and volume.
   3. Temperature.
B. Le Châtelier's Principle: If a stress is applied to a reaction mixture at equilibrium, reaction occurs in the direction that relieves the stress.
   1. Predicts changes in the composition of an equilibrium mixture if above changes occur.

Altering an Equilibrium Mixture: Changes in Concentration

8⊠
A. Disturb an equilibrium by adding or removing a reactant or product.
   1. Stress of an added reactant or product is relieved by reaction in the direction that consumes the added substance.
      a. add reactant - reaction shifts right (toward product)
      b. add product - reaction shifts left (toward reactant)
   2. Stress of removing reactant or product is relieved by reaction in the direction that replenishes the removed substance.
      a. remove reactant - reaction shifts left
      b. remove product - reaction shifts right

       B. Changes occur due to change in value of Q_c.
          1. Add reactant - denominator in Q_c expression becomes larger.
            a. $Q_c < K_c$
            b. to return to equilibrium, Q_c must increase
               i. numerator of Q_c expression must increase and the denominator must decrease
               ii. implies net conversion of reactants to products
               iii. reaction shifts right
          2. Remove reactant - denominator in Q_c expression becomes smaller.
            a. $Q_c > K_c$
            b. to return to equilibrium, Q_c must decrease
               i. numerator of Q_c expression must decrease and the denominator must increase
               ii. implies net conversion of products to reactants
               iii. reaction shifts left

Altering an Equilibrium Mixture: Changes in Pressure and Volume

8⊠   A. Number of moles of gaseous reactants in the balanced equation is different from the number of moles of gaseous products.
          1. Change in pressure (due to changing volume) changes composition of equilibrium mixture.

8⊠   B. Increase in pressure (due to decrease in volume) results in reaction in the direction of fewer number of moles of gas.
       C. Decrease in pressure (due to increase in volume) results in reaction in the direction of greater number of moles of gas.
       D. Changes occur due to change in value of Q_c.
          1. Decrease volume - molarity ($= n/V$) increases.
          2. If reactant side has more moles of gas.
            a. increase in denominator is greater than increase in numerator
            b. $Q_c < K_c$
            c. to return to equilibrium, Q_c must increase
               i. numerator of Q_c expression must increase and the denominator must decrease
               ii. implies net conversion of reactants to products (shifts towards fewer moles of gas)
          3. If product side has more moles of gas.
            a. increase in numerator is greater than increase in denominator
            b. $Q_c > K_c$
            c. to return to equilibrium, Q_c must decrease
               i. denominator of Q_c expression must decrease and the numerator must increase
               ii. implies net conversion of products to reactants (shifts towards fewer moles of gas)
       E. Reaction involves no change in the number of moles of gas.
          1. No effect on composition of equilibrium mixture.
       F. For heterogeneous equilibrium.
          1. Effect of pressure changes on solids and liquids can be ignored.
            a. volume is nearly independent of pressure
       G. Change in pressure due to addition of an inert gas.
          1. No change in the molar concentrations of reactants or products.
          2. No effect on composition of equilibrium mixture.

Altering an Equilibrium Mixture: Changes in Temperature

       A. Disturb an equilibrium by a change in concentration, pressure, or volume.
          1. $Q_c \neq K_c$.
          2. K_c remains constant with constant T.
8⊠   B. Change in temperature always changes equilibrium constant.
          1. Depends on sign of ΔH^o for the reaction.

2. Exothermic reaction (ΔH^o) - equilibrium constant decreases as the temperature increases.
3. Endothermic reaction (ΔH^o) - equilibrium constant increases as the temperature increases.
C. Le Châtelier's principle - add heat to an equilibrium mixture, net reaction occurs in the direction that relieves the stress of the added heat.
   1. Endothermic reaction - heat is absorbed by reaction in the forward direction.
      a. equilibrium mixture contains more product than reactant
      b. K_c increases with increasing temperature
   2. Exothermic reaction - heat is absorbed by reaction in the reverse direction.
      a. equilibrium mixture contains more reactant than product
      b. K_c decreases with decreasing temperature

EXAMPLE:

How will the following changes alter the equilibrium for the reaction:

$$3\,Fe\,(s)\ +\ 4\,H_2O\,(g)\ \rightleftarrows\ Fe_3O_4\,(s)\ +\ 4\,H_2\,(g) \qquad \Delta H^o = \text{-}150\ kJ$$

a. H_2O is removed from the system.
b. H_2 is removed from the system.
c. The volume of the container is increased.
d. Fe_3O_4 is added to the system.
e. The temperature is raised.

SOLUTION:

a. When H_2O is removed from the system, the equilibrium shifts left.
b. When H_2 is removed from the system, the equilibrium shifts right.
c. Increasing the volume of the container decreases the pressure in the container. However, the number of moles of gas is equal on both sides of the reaction. Therefore, a change in volume and pressure does not affect the equilibrium.
d. Adding a solid to the system does not affect the equilibrium since solids are not included in the equilibrium expression.
e. Raising the temperature causes the equilibrium to shift left.

9⊠ The Effect of a Catalyst

A. Catalyst increases the rate of a chemical reaction.
   1. Provides a new, lower energy pathway.
   2. Forward and reverse reactions pass through the same transition state.
      a. rate for forward and reverse reactions increases by the same factor
   3. Does not affect the composition of the equilibrium mixture.
      a. does not alter the equilibrium constant or equilibrium concentrations
      b. does not appear in the balanced chemical equation
   4. Can influence choice of optimum conditions for a reaction.

10⊠ The Link Between Chemical Equilibrium and Chemical Kinetics

A. For the reaction: $A + B \rightleftarrows C + D$.
   1. Rate of forward reaction $= k_f[A][B]$.
   2. Rate of reverse reaction $= k_r[C][D]$.
   3. At equilibrium: $k_f[A][B] = k_r[C][D]$ or $\dfrac{k_f}{k_r} = \dfrac{[C][D]}{[A][B]} = K_c$.
B. Relative values of k_f and k_r determine the composition of the equilibrium mixture.
   1. $k_f \gg k_r$; K_c is very large; reaction goes to completion.
      a. irreversible reaction

  b. reverse reaction is too slow to be detected
2. $k_f \approx k_r$; $K_c \approx 1$; reactants and products are present at equilibrium.
  a. reversible reaction

EXAMPLE:

Calculate the equilibrium constant K_c for the reaction:

$$H_2O\ (l) \rightleftarrows H^+\ (aq)\ +\ OH^-\ (aq)$$

given that $k_f = 2.4 \times 10^{-5}\ s^{-1}$ and $k_r = 1.3 \times 10^{11}\ M^{-1} \cdot s^{-1}$ and that K is for the expression

$$K = \frac{[H^+][OH^-]}{[H_2O]}$$

SOLUTION: To calculate K_c, we use the equation

$$K_c = \frac{2.4 \times 10^{-5}\ s^{-1}}{1.3 \times 10^{11}\ M^{-1} \cdot s^{-1}} = 1.8 \times 10^{-16}$$

Self-Test

This section is intended to test your knowledge of the material covered in this chapter. Think through these problems and make certain you understand what is going on. Ask yourself if your answer makes sense. Many of these questions are linked to the chapter learning goals. Therefore, successful completion of these problems indicates you have mastered the learning goals for this chapter. You will receive the greatest benefit from this section if you use it as a mock exam. You will then discover which topics you have mastered and which topics you need to study in more detail.

True/False

1. The reaction quotient Q_c is the same as the equilibrium constant expression.

2. Le Châtelier's principle predicts the changes in the composition of an equilibrium mixture if a stress is applied to a system at equilibrium.

3. If Q_c, is greater than K_c, then the reaction proceeds from left to right.

4. If a product is added to a system at equilibrium, the reaction will shift to the right.

5. If the pressure of a system at equilibrium is increased, the reaction will shift in the direction of the fewer number of moles of gas.

6. If an inert gas is added to a system at equilibrium, the reaction will shift in the direction of the fewer number of moles of gas.

7. A change in the temperature of a reaction will change the value of the equilibrium constant.

8. Adding a catalyst to a system at equilibrium will change the composition of an equilibrium mixture.

9. The relative values of k_f and k_r determine the composition of the equilibrium mixture.

10. For a large value of K_c, the position of the reaction is to the right.

Multiple Choice

1. A state of chemical equilibrium is reached when:
   a. the rate of the forward reaction is greater than the rate of the reverse reaction
   b. the concentration of the products and reactants are equal
   c. more product is present than reactant
   d. the concentrations of the products and reactants have reached constant value

2. If $Q_c > K_c$, the reaction:
   a. proceeds from left to right
   b. proceeds from right to left
   c. is at equilibrium

3. When a reactant is added to a system at equilibrium:
   a. the reaction shifts towards the left
   b. Q_c must increase because the denominator in the Q_c expression becomes larger
   c. there is a net conversion of products to reactants
   d. the value of Q_c does not change

4. When a system at equilibrium which has more moles of gas on the reactant side experiences a decrease in pressure:
   a. the molarity of the species present increases
   b. $Q_c < K_c$
   c. Q_c must decrease for the system to return to equilibrium
   d. the numerator of Q_c expression must increase and the denominator must decrease

5. When the temperature is increased for a system at equilibrium:
   a. the value of K_c remains the same
   b. the effect is independent of the value of ΔH
   c. the value of K_c decreases as the temperature increases for an exothermic reaction
   d. the value of K_c decreases as the temperature decreases for an exothermic reaction

6. When a catalyst is added to a system at equilibrium:
   a. the value of the equilibrium constant changes
   b. the equilibrium concentrations change
   c. a new, lower energy pathway is established
   d. the rate of the forward and reverse reactions are no longer equal

7. When $K_c > 10^3$:
   a. the reaction hardly proceeds
   b. appreciable amounts of both reactants and products are present
   c. the reactants predominate over the products
   d. the position of the reaction is to the right

8. K_p:
   a. is the same as K_c.
   b. is the reciprocal of K_c.
   c. is defined as the equilibrium constant using partial pressures.
   d. includes the concentration of pure liquids in the equilibrium expression.

9. The reaction quotient, Q_c, is:
   a. calculated in the same manner as K_c
   b. less than K_c at equilibrium
   c. greater than K_c at equilibrium
   d. equal to K_p

10. When a reactant is removed from a system at equilibrium:
   a. $Q_c < K_c$
   b. the reaction shifts left
   c. the reactants are converted to products
   d. the denominator in the K_c expression becomes larger

Matching

Chemical equilibrium

a. state in which the forward and reverse reactions continue at equal rates

Equilibrium constant, K_c

b. If a stress is applied to a reaction mixture at equilibrium, reaction occurs in the direction that relieves the stress.

Homogeneous equilibria

c. reactants and products are present in more than one phase

Heterogeneous equilibria

d. reaction in which the product molecules can recombine to form reactant molecules

Reaction quotient, Q_c

e. reactants and products are present in the same phase

Reversible reaction

f. a dynamic state in which the concentrations of reactants and products remain constant because the rates of the forward and reverse reactions are equal

Dynamic state

g. the number obtained when equilibrium concentrations (in mol/L) are substituted into the equilibrium constant expression

Le Châtelier's principle

h. the number obtained when the concentration of all species (not necessarily the equilibrium concentration) are substituted into the equilibrium constant expression

Fill-in-the-Blank

1. When approaching equilibrium from the reactant side of a reaction, the concentration of reactant

   _____ and the concentration of product _____ until both

   concentrations _____.

2. The equilibrium constant, K_c', for the reverse reaction is equal to the _____

   _____.

3. The concentrations of pure solids or pure liquids are _____

   when writing the equilibrium equation for any heterogeneous equilibrium.

4. If $K_c < 10^{-3}$, the reaction _____.

5. The stress of an added reactant or product to a system at equilibrium is relieved by reaction

   _____.

6. A decrease in pressure to a system at equilibrium results in reaction _____

   _____.

7. For heterogeneous equilibrium, the effect of pressure changes on solids and liquids can be

   _____.

8. There is _____ on the composition of the equilibrium mixture when

   an inert gas is added to a system at equilibrium because _____

   _____.

9. For an exothermic reaction, K_c _____ with decreasing temperature.

10. The composition of the equilibrium mixture is determined by the relative values of _____.

Problems

1. Write equilibrium expressions for the following balanced equations.

   a. $7 H_2 (g) + 2 NO_2 (g) \rightleftarrows 2 NH_3 (g) + 4 H_2O (g)$

   b. $3 Cl_2 (g) + CS_2 (l) \rightleftarrows CCl_4 (l) + S_2Cl_2 (l)$

   c. $Br_2 (g) + 2 HI (g) \rightleftarrows I_2 (g) + 2 HBr (g)$

   d. $2 NaHCO_3 (s) \rightleftarrows Na_2CO_3 (s) + CO_2 (g) + H_2O (g)$

2. The partial equilibrium pressures for N_2, O_2, and NO in the reaction

   $N_2 (g) + O_2 (g) \rightleftarrows 2 NO (g)$

   are $p(N_2) = 0.25$ atm, $p(O_2) = 0.198$ atm, and $p(NO) = 0.050$ atm. Calculate K_p for this reaction.

3. Calculate K_c for the reaction in Question 2.

4. From the value of K_c, determine whether mainly products or mainly reactants exist at equilibrium for the following balanced equations:

   a. $N_2 (g) + O_2 (g) \rightleftarrows 2 NO (g)$ $K = 1 \times 10^{238}$

b. $SO_2 (g) + NO_2 (g) \rightleftarrows NO (g) + SO_3 (g)$ $K = 85.0$

c. $2 CO_2 (g) \rightleftarrows 2 CO (g) + O_2 (g)$ $K = 6.4 \times 10^{-7}$

5. For the reaction

$PCl_5 (g) \rightleftarrows PCl_3 (g) + Cl_2 (g)$

$K_c = 33.3$ at $760°$ C.   Is this system at equilibrium if $[PCl_5] = 1.25 \times 10^{-3}$ M, $[PCl_3] = [Cl_2] = 0.75$ M? If not, determine the direction the reaction must go to reach equilibrium.

6. Nitrogen dioxide is produced from the reaction of dinitrogen oxide and oxygen according to the equation

$2 N_2O (g) + 3 O_2 (g) \rightleftarrows 4 NO_2 (g)$

At $25°$ C, 0.0292 moles of N_2O and 0.0713 moles of O_2 are placed in a 1.00 L container and allowed to react. The $[NO_2]$ at equilibrium is 0.0284 moles. What are the equilibrium concentrations of N_2O and O_2? What is the value of K_c?

7. The equilibrium constant for the reaction

$SO_2 (g) + NO_2 (g) \rightleftarrows NO (g) + SO_3 (g)$

is $K_c = 85.0$ at $460°$ C. Calculate the concentrations of all species at equilibrium when the initial reaction mixture contains 0.0650 M SO_2 and 0.0650 M NO_2.

8. Consider the reaction

$2 SO_3 (g) \rightleftarrows 2 SO_2 (g) + O_2 (g)$ $\Delta H° = 197$ kJ

What will be the direction of the reaction when the following stress is applied:

a. addition of SO_3
b. decrease in temperature
c. increase in volume
d. addition of N_2 gas
e. addition of SO_2 gas

9. For the reaction $A + B \rightleftarrows 2 C$, $k_f = 7.5 \times 10^{-7}$ s^{-1} and $k_r = 3.2 \times 10^{-2}$ s^{-1}. Calculate K.

10. Calculate the equilibrium concentration for all species present in the reaction:

$HCONH_2 (g) \rightleftarrows NH_3 (g) + CO (g)$

if the initial concentration of formamide ($HCONH_2$) is 0.250 M. $K_c = 4.84$ at $127°$ C.

11. For the reaction:

$2 H_2S (g) + CH_4 (g) \rightleftarrows 4 H_2 (g) + CS_2 (g)$

Chapter 13 – Chemical Equilibrium

$K_c = 5.27 \times 10^{-8}$ at 700° C. What is the position of equilibrium for this reaction? Is the reaction mixture at equilibrium when the concentrations of the reactants and products are $[CH_4] = 1.50$ M, $[H_2S] = 3.00$ M, $[H_2] = 0.300$ M, and $[CS_2] = 0.075$ M. If not, in which direction will the reaction proceed?

Solutions

True/False
1. F. The concentrations in the reaction quotient expression are not necessarily equilibrium values.
2. T
3. F. If $Q_c > K_c$, the reaction proceeds from right to left.
4. F. The reaction will shift to the left.
5. T
6. F. When an inert gas is added to a reaction mixture there is no change in the molar concentrations of reactants and products and therefore, no effect on the composition of the equilibrium mixture.
7. T
8. F. The rates of the forward and reverse reaction increase by the same factor. Therefore, the addition of a catalyst does not affect the composition of the equilibrium mixture.
9. T
10. T

Multiple Choice
1. d
2. b
3. b
4. c
5. c
6. c
7. d
8. c
9. a
10. b

Matching
Chemical equilibrium – f
Equilibrium constant, K_c – g
Homogeneous equilibria – e
Heterogeneous equilibria – c
Reaction quotient, Q_c – h
Reversible reaction – d
Dynamic state – a
Le Châtelier's principle – b

Fill-in-the-Blank
1. decreases; increases; level off at constant equilibrium values
2. reciprocal of the original equilibrium expression.
3. not included
4. hardly proceeds
5. in the direction that consumes the added reactant or product
6. in the direction of greater number of moles of gas
7. ignored

8. no effect; there is no change in the molar concentrations of reactants or products
9. increases
10. k_f and k_r

Problems

1. a. $K_c = \dfrac{[NH_3]^2[H_2O]^4}{[H_2]^7[NO_2]^2}$ b. $K_c = \dfrac{1}{[Cl_2]^3}$ c. $K_c = \dfrac{[I_2][HBr]^2}{[Br_2][HI]^2}$

   d. $K_c = [CO_2][H_2O]$

2. $K_p = \dfrac{(p_{NO})^2}{(p_{N_2})(p_{O_2})}$; $K_p = \dfrac{(0.050)^2}{(0.25)(0.198)} = 5.1 \times 10^{-2}$

3. $K_c = \dfrac{K_p}{(RT)^{\Delta n}}$; $\Delta n = 0$; $K_c = K_p = 5.1 \times 10^{-2}$

4. a. mainly products; b. mainly products; c. mainly reactants

5. To determine if a system is at equilibrium, you must first solve for the value of Q_c and then compare it to the value of K_c.

   $Q_c = \dfrac{[PCl_3][Cl_2]}{[PCl_5]}$; $Q_c = \dfrac{(0.75)(0.75)}{(1.25 \times 10^{-3})} = 4.5 \times 10^2$

   Since $Q_c \gg K_c$, the system is not at equilibrium. The reaction would need to shift from right to left to achieve equilibrium.

6. To solve this problem, we need to use the approach outlined in Fig. 13.6 in your text and on page 205 of this chapter.
   a. The balanced equation is given.
   b. The equilibrium concentrations of N_2O and O_2 can be determined by allowing x to be the concentration of N_2O that reacts on going to the equilibrium state. Since the mole to mole ratios are 2:3:4, if x mol/L of N_2O reacts then $1.5x$ mol/L of O_2 reacts and $2x$ mol/L of NO_2 is produced.

   This can be summarized in the following table.

	2 N_2O (g) +	3 O_2 (g) ⇌	4 NO_2 (g)
Initial Concentration (M)	0.0292	0.0713	0
Change (M)	-x	-3/2x	+2x
Eq. Concentration (M)	0.0292 - x	0.0713 - 3/2 x	+2x

   We know that the equilibrium concentration of NO_2 is 0.0284 M and can therefore, solve for the value of x.

   $2x = 0.0284$; $x = 1.42 \times 10^{-2}$

The value of x can be substituted back into the expressions for the equilibrium concentrations of N_2O and O_2.

$$[N_2O] = 0.0292 - (0.0142) = 0.0150; \qquad [O_2] = 0.0713 - 1.5(0.0142) = 0.0500$$

c. The value of K_c can now be calculated by substituting the equilibrium concentrations of all species into the equilibrium expression.

$$K_c = \frac{[NO_2]^4}{[N_2O]^2[O_2]^3}; \qquad K_c = \frac{(0.0284)^4}{(1.50 \times 10^{-2})^2 (5.00 \times 10^{-2})^3} = 23.1$$

7. To solve this problem, we need to use the approach outlined in Fig. 13.6 in your text and on page 205 of this chapter.
   a. The balanced equation is given.
   b. The initial concentrations of SO_2 and NO_2 are 0.0650 M. Let x be the concentration of SO_2 that reacts on going to the equilibrium state. Since the mole to mole ratios are 1:1:1:1, if x mol/L of SO_2 reacts, then x mol/L of NO_2 also reacts and x mol/L of SO_3 and NO is produced. This can be summarized in the following table.

	SO_2 (g)	+ NO_2 (g)	$\rightleftarrows$ SO_3 (g)	+ NO (g)
Initial Concentration (M)	0.0650	0.0650	0	0
Change	-x	-x	+x	+x
Eq. Concentration (M)	0.0650 - x	0.0650 - x	+x	+x

   c. The equilibrium expression for the reaction is $K_c = \dfrac{[NO][SO_3]}{[SO_2][NO_2]}$. The equilibrium concentrations from the above table can be substituted into this expression, and we can solve for x.

$$85.0 = \frac{(x)(x)}{(0.0650-x)(0.0650-x)} = \left(\frac{x}{0.0650-x}\right)^2.$$

Taking the square root of both sides gives

$$\pm 9.22 = \frac{x}{0.0650-x}.$$

Solving for x, we obtain two solutions. The equation with the positive square root of 85 gives

$$+9.22(0.0650 - x) = x; \qquad +0.599 = x + 9.22x; \qquad x = \frac{0.599}{10.22} = 5.86 \times 10^{-2}$$

The equation with the negative square root of 85 gives

$$-9.22(0.0650 - x) = x; \qquad -0.599 = x - 9.22x$$
$$x = \frac{-0.599}{-8.22} = 7.29 \times 10^{-2}$$

Because the initial concentrations of SO_2 and NO_2 are 0.0650, x cannot exceed 0.0650. Therefore, $x = 7.29 \times 10^{-2}$ should be discarded as chemically unreasonable and $x = 5.86 \times 10^{-2}$

is chosen as the solution. The equilibrium concentrations are:

$$[SO_2] = [NO_2] = 0.0650 - (5.86 \times 10^{-2}) = 0.0064 \ M; \quad [SO_3] = [NO] = 0.0586 \ M$$

8. a. reaction shifts to the right
   b. K_c decreases
   c. shifts to the right (greater number of moles of gas)
   d. no effect
   e. shifts to the left

9. $K_c = \dfrac{k_f}{k_r};$ $\qquad\qquad K_c = \dfrac{7.5 \times 10^{-7}}{3.2 \times 10^{-2}} = 2.3 \times 10^{-5}$

10. To solve this problem, we need to use the approach outlined in Fig. 13.6 in your text and on page 205 in this chapter.
    a. The balanced equation is given
    b. The initial concentration of $HCONH_2$ is 0.250 M. Let x by the concentration of $HCONH_2$ that reacts on going to the equilibrium state. Since the mole to mole ratios are 1:1:1, if x mol/L of $HCONH_2$ reacts, then x mol/L of NH_3 and CO are produced. This can be summarized in the following table:

	$HCONH_2$ (g) $\rightleftarrows$	NH_3 (g) +	CO (g)
Initial Concentration (M)	0.250	0	0
Change	$-x$	$+x$	$+x$
Eq. Concentration (M)	$0.250 - x$	$+x$	$+x$

The equilibrium expression for the reaction is $K_c = \dfrac{[NH_3][CO]}{[HCONH_2]}$. The equilibrium concentrations from the above table can be substituted into this expression

$$4.84 = \frac{(x)(x)}{(0.250 - x)} = \frac{x^2}{0.250 - x}$$

To solve for x, we need to rearrange this expression and use the quadratic equation.

$$4.84(0.250 - x) = x^2; \qquad x^2 + 4.84x - 1.21 = 0;$$

$$x = \frac{-4.84 \pm \sqrt{(4.84)^2 - 4(-1.21)}}{2} \qquad\qquad x = -5.08 \text{ or } 0.238$$

The mathematical solution that makes chemical sense is 0.238.

c. The equilibrium concentrations can now be calculated.

$$[HCONH_2] = 0.250 - 0.238 = 0.012 \ M \qquad\qquad [NH_3] = [CO] = 0.238 \ M$$

11. The position of equilibrium is to the left, given the very small value of K_c. We can calculate the value of Q_c from the concentrations given.

219

$$Q_c = \frac{[H_2]^4[CS_2]}{[H_2S]^2[CH_4]} = \frac{(0.300)^4(0.075)}{(3.00)^2(1.50)} = 4.50 \times 10^{-5}$$

$Q_c > K_c$; therefore, the reaction mixture is not at equilibrium. The reaction will proceed from right to left to reach equilibrium.

CHAPTER 14

HYDROGEN, OXYGEN, AND WATER

Chapter Learning Goals

1⊠ Describe the properties of elemental hydrogen. Include its appearance, molecular structure, occurrence in nature, laboratory synthesis, industrial synthesis, and industrial use.
2⊠ Apply the gas laws to problems involving hydrogen, including calculation of its density and calculation of the number of grams of solid reactants needed to produce a certain volume of hydrogen.
3⊠ Describe the isotopes of hydrogen, and compare and contrast the properties of the isotopes and compounds containing the isotopes.
4⊠ Assign oxidation numbers and identify the oxidizing agent and reducing agent for redox reactions of hydrogen. Balance equations for these reactions using the oxidation-number method.
5⊠ Classify binary hydrides as ionic, covalent, or metallic.
6⊠ Describe the properties of elemental oxygen. Include its appearance, molecular structure, occurrence in nature, laboratory synthesis, industrial synthesis, and industrial use.
7⊠ Apply the gas laws to problems involving oxygen, including the calculation of its density and the calculation of the number of grams of solid reactants needed to produce a certain volume of oxygen.
8⊠ Classify oxides as basic, acidic, or amphoteric.
9⊠ Assign oxidation numbers and identify the oxidizing agent and reducing agent for redox reactions of oxygen. Balance equations for these reactions using the oxidation-number method.
10⊠ Identify compounds as oxides, peroxides, and superoxides.
11⊠ Use structure and bonding concepts to explain the physical and chemical properties of elemental oxygen and ozone.
12⊠ Write balanced net ionic equations for the reaction of water with alkali metals, alkaline earth metals, and halogens.
13⊠ Identify and describe the general properties of a hydrate.
14⊠ Determine the empirical formula of a hydrate.

Chapter in Brief

In this chapter, you will continue your study of descriptive chemistry by examining hydrogen, oxygen, and water. This study will include the properties, synthesis, and use of the elements as well as their redox chemistry. You will investigate the properties of ozone, an allotrope of oxygen, the isotopes of hydrogen, and the hydrides and oxides. Finally, you will examine the reactivity and the properties of water along with the properties of hydrates.

1⊠ **Hydrogen**
    A. Henry Cavendish - isolated hydrogen in pure form.
        1. acid + metal → H_2 (g) + metal ion.
    B. Properties of hydrogen.
        1. Colorless, odorless, and tasteless gas.
        2. Nonpolar, diatomic molecule.
        3. Weak intermolecular forces.
            a. low boiling and melting points

4. Bond dissociation energy = 436 kJ/mol.
C. Natural occurrence.
1. 75% of the mass of the universe.
2. Atmospheric abundance - 0.53 ppm by volume.
3. Ninth most abundant element in terms of mass in the earth's crust and oceans.
D. Industrial use.
1. Haber process.
a. largest single consumer of hydrogen
b. $N_2 (g) + 3 H_2 (g) \rightarrow 2 NH_3 (g)$
2. Synthesis of methanol.
a. $CO (g) + 2 H_2 (g) \rightarrow CH_3OH (l)$

3⊠ **Isotopes of Hydrogen**
A. Three isotopes of hydrogen.
1. Protium - ordinary hydrogen ($_1^1H$).
2. Deuterium - heavy hydrogen ($_1^2H$ or D).
3. Tritium - ($_1^3H$ or T).
B. Similar properties.
1. All three isotopes have the same electron configuration.
a. determines chemical behavior
2. Isotope effects - arise from the differences in the mass of the isotopes.
a. greater for hydrogen than any other element
i. percentage differences between the masses of the isotopes are larger for hydrogen
b. D_2 - higher boiling and melting point and greater heat of dissociation than H_2
C. D_2 and D_2O manufactured by electrolysis of an inert electrolyte in ordinary water.
1. Electrolysis - the process of using an electric current to bring about chemical change.
2. D_2O.
a. higher boiling and melting point than H_2O
b. smaller equilibrium constant for dissociation
c. used as a coolant and a moderator in nuclear reactors

1⊠ **Synthesis of Hydrogen**
A. Electrolysis of water.
1. Requires a large amount of energy.
2. Not economical for large-scale production.
3. $2 H_2O (l) \rightarrow 2 H_2 (g) + O_2 (g)$ $\Delta H^o = + 572$ kJ
B. Laboratory synthesis - reaction of dilute acid with an electropositive metal.
C. Large-scale, industrial methods - reducing agent extracts the oxygen from steam.
1. Steam-hydrocarbon reforming process.
a. first step – produce synthesis gas
i. $H_2O (g) + CH_4 (g) \rightarrow CO (g) + 3 H_2 (g)$
ii. high temperature, moderately high pressure, and a nickel catalyst
b. second step – gas shift reaction
i. shifts the composition of synthesis gas
ii. removes CO and produces more H_2
iii. $CO (g) + H_2O (g) \rightarrow CO_2 (g) + H_2 (g)$
iv. metal oxide catalyst, $T = 400°C$

Reactivity of Hydrogen

A. Hydrogen has properties similar to both the alkali metals and the halogens.
  1. Lose electron - H^+.
    a. $E_i = 1312$ kJ/mol
    b. doesn't completely transfer its valence electron in ordinary chemical reactions
    c. shares electron with a nonmetallic element to produce a covalent compound
    d. gas phase - complete ionization of a hydrogen atom to produce a bare proton
    e. liquids and solids - bare proton is too reactive to exist by itself
  2. Gain electron - H^-.
    a. $E_{ea} = -73$ kJ/mol
    b. will accept an electron from an active metal to give an ionic hydride
B. H_2 - relatively unreactive.
  1. Due to strong H-H bond.

5 ⊠ Binary Hydrides

A. Binary hydride - a compound that contains hydrogen and just one other element.
B. Ionic hydride - formed by the alkali metals and the heavier alkaline-earth metals.
  1. Saltlike, high-melting, white, crystalline compounds.
  2. Prepared by direct reaction of the elements.
  3. Alkali-metal hydrides - face-centered cubic crystal structure.
    a. ionic in the liquid state
      i. molten compounds conduct electricity
  4. H^- - good proton acceptor.
    a. ionic hydrides react with water to give H_2 gas
      i. acid-base reaction
C. Covalent hydrides - hydrogen is attached to another element by a covalent bond.
  1. Most common - hydrides of nonmetallic elements.
  2. Discrete, small molecules with relatively weak intermolecular forces.
    a. gases or volatile liquids at ordinary temperatures
D. Metallic hydrides - lanthanide, actinide and certain d-block transition metals.
  1. Interstitial hydrides - crystal lattice of metal atoms with the smaller hydrogen atoms occupying holes (interstices) between the larger metal atoms.
  2. Nonstoichiometric compounds - atomic composition can't be expressed as a ratio of small whole numbers.
  3. Properties depend on composition.
    a. function of the partial pressure of H_2 gas in the surroundings
  4. Potential hydrogen-storage devices.
    a. can contain large amounts of hydrogen

6 ⊠ Oxygen

A. Most abundant element on the planet's surface.
B. Crucial to human life.
C. Occurrence in nature.
  1. 23% of atmosphere - primarily as O_2.
  2. 46% of lithosphere.
  3. 85% of hydrosphere - in the form of H_2O.
D. Photosynthesis process - green plants use solar energy to produce O_2 and glucose from carbon dioxide and water.
  1. Replaces O_2 removed by combustion and respiration processes.
  2. $6\,CO_2 + 6\,H_2O \rightarrow 6\,O_2 + C_6H_{12}O_6$.
E. Metabolism of carbohydrates.
  1. Reverse of photosynthesis.

6☒ **Preparation and Uses of Oxygen**

    A. Properties of oxygen.
        1. Odorless, tasteless gas.
        2. Supports combustion better than air.
    B. Laboratory preparation.
        1. Electrolysis of water.
        2. Decomposition of aqueous hydrogen peroxide in the presence of a catalyst.
        3. Thermal decomposition of an oxoacid salt.
    C. Industrial synthesis - fractional distillation of liquefied air.
        1. Third ranking industrial chemical produced in the U.S.
    D. Industrial uses.
        1. Steel making - removes impurities from iron by forming oxides.
        2. Sewage treatment - destroys malodorous compounds by oxidation.
        3. Paper bleaching - oxidizes compounds that cause unwanted colors.
        4. Oxyacetylene torches - provides the high temperatures needed for cutting and welding
           metals.
        5. Inexpensive and readily available oxidizing agent.

6☒ **Properties of Oxygen**

    A. Paramagnetic in all three phases.
    B. O_2 - double bonded.
    C. High electronegativity.
    D. Reacts with active metals.
        1. Gains two electrons and achieves a noble gas configuration.
    E. Forms covalent oxides with nonmetals.
        1. Sharing of two additional electrons achieves a noble gas configuration.
        2. Forms two single bonds or one double bond.
           a. double bonds are formed between oxygen and other small atoms
               i. good overlap between oxygen $p\pi$ orbital and the $p\pi$ orbitals of small atoms
    F. Reacts directly with nearly all the elements in the periodic table.
        1. Doesn't react with noble gases or platinum and gold.
        2. Reactions are slow at room temperature.
        3. Extremely reactive at higher temperatures.

8☒ **Oxides**

    A. Classified on the basis of the oxygen's oxidation state.
    B. Oxidation number = -2, oxides.
        1. Categorize in terms of their acid-base character.
        2. Basic oxides.
           a. ionic
           b. formed by the metals on the left side of the periodic table
        3. Acidic oxides.
           a. covalent
           b. formed by the nonmetals on the right side of the periodic table
        4. Amphoteric oxides.
           a. exhibit both acidic and basic properties
           b. formed by elements with intermediate electronegativity
           c. strongly polar covalent character
           d. bonds have intermediate ionic-covalent character
        5. Acid-base properties and ionic-covalent character depend on:
           a. position of element in the periodic table

    i. acidic and covalent character increases left to right across the periodic table

    ii. basic and ionic character increases top to bottom down the periodic table

   b. oxidation state of element

    i. acidic and covalent character increases with increasing oxidation state

  6. As the bonding in oxides changes from ionic to covalent, a change in structure and physical properties occurs.

10⬜ C. Oxidation number = -1, peroxides (O_2^{2-}).

10⬜ D. Oxidation number = $-\frac{1}{2}$, superoxides (O_2^-).

10⬜ **Peroxides and Superoxides**

  A. Formed by heating the heavier group 1A and 2A metals in an excess of air.

  B. Ionic solids.

  C. Peroxide ion.

   1. Diamagnetic.

   2. O-O single bond.

   3. Has basic properties.

    a. reacts with strong acids to form hydrogen peroxide, H_2O_2

  D. Superoxide ion.

   1. One unpaired electron.

   2. Paramagnetic.

   3. Bond order = 1.5.

   4. Evolves oxygen when dissolved in water.

    a. disproportionation reaction - a substance is both oxidized and reduced in a reaction

    b. $2\ KO_2\ (s) + H_2O\ (l) \rightarrow O_2\ (g) + 2\ K^+\ (aq) + HO_2^-\ (aq) + OH^-\ (aq)$

       ↑   ↑   ↑

      -1/2   0   -1

Hydrogen Peroxide

  A. Uses (due to oxidizing properties).

   1. Antiseptic, bleach, starting material for synthesis of other peroxide containing compounds.

  B. Properties.

   1. Colorless, syrupy liquid.

   2. m.p. = $-0.4°C$; b.p. = $150°C$.

   3. Explodes when heated.

   4. Strong hydrogen bonding present in liquid H_2O_2 (indicated by high b.p.).

   5. Weak acid in aqueous solutions.

  C. Both an oxidizing and reducing agent.

   1. Oxidizing agent: $O_2^{2-} \rightarrow O^{2-}$.

   2. Reducing agent: $O_2^{2-} \rightarrow O_2$.

   3. Can oxidize and reduce itself (unstable with respect to disproportionation to water and oxygen).

    a. heat and many catalysts can cause rapid, exothermic and potentially explosive decomposition

Ozone

  A. Two allotropes of oxygen.

   1. Ordinary dioxygen, O_2.

   2. Ozone, O_3.

  B. Ozone.

   1. Toxic, pale blue gas.

   2. Sharp, penetrating odor.

   3. Produced by passing an electric discharge through O_2.

11⊠
    C. Bonding in ozone.
       1. Two resonance structures which are bent.
       2. Central atom surrounded by 2 σ bonds and 1 lone pair of electrons.
       3. π bond is delocalized over all 3 oxygen atoms.
       4. Net bond order = 1.5.
    D. Powerful oxidizing agent.
       1. Used to kill bacteria in drinking water.

Water

    A. Most familiar and abundant compound on earth.
    B. Reacts with alkali metals, heavier alkaline-earth metals and the halogens.
       1. Alkali metals and heavier alkaline-earth metals $+ H_2O \rightarrow H_2 +$ metal hydroxide.
       2. Fluorine $+ H_2O \rightarrow O_2 + HF$.
       3. Chlorine, bromine and iodine disproportionate.
         a. extent of disproportionation decreases down the table
         b. $X_2 + H_2O \rightarrow HOX + X^-$ (X = Cl, Br, I)

13⊠ **Hydrates**

    A. Solid compounds that contain water molecules.
    B. Complex or unknown structures.
       1. Use a dot in the formula to specify the composition.
    C. Formation is common with salts that contain +2 and +3 cations.
       1. Bonding interactions between water and a metal cation increase with increasing charge on the cation.
    D. Hygroscopic compounds - absorb water from the air.
       1. Anhydrous compounds that have a tendency to form hydrates.
       2. Useful as drying agents.

Natural Waters

    A. Seawater.
       1. Major constituent - sodium chloride.
       2. Unlimited source of chemicals.
       3. Three substances obtained from sea water commercially: $NaCl$, Mg, and Br_2
    B. Freshwater lakes.
       1. Purification process.
         a. preliminary filtration
         b. sedimentation
         c. sand filtration
         d. aeration
         e. sterilization

Self-Test

This section is intended to test your knowledge of the material covered in this chapter. Think through these problems and make certain you understand what is going on. Ask yourself if your answer makes sense. Many of these questions are linked to the chapter learning goals. Therefore, successful completion of these problems indicates you have mastered the learning goals for this chapter. You will receive the greatest benefit from this section if you use it as a mock exam. You will then discover which topics you have mastered and which topics you need to study in more detail.

True/False

1. Hydrogen transfers a valence electron to a nonmetallic element to produce an ionic compound.

2. Hydrogen is the most abundant element on the planet's surface.

3. Oxygen is an inexpensive and readily available reducing agent.

4. Oxygen is paramagnetic in all three phases.

5. The largest single consumer of hydrogen is the Haber process.

6. The three isotopes of hydrogen have different chemical behavior.

7. Interstitial hydride compounds are considered to be ionic hydrides.

8. The acidic and covalent character of oxides increases from left to right across the periodic table.

9. The peroxide ion has basic properties

10. Hydrogen peroxide is stable with respect to disproportionation.

11. Ozone is a powerful oxidizing agent.

12. The formation of hydrates is common with salts that contain +2 and +3 cations.

Matching

Isotope effect	a. a compound in which the structure consists of a crystal lattice of metal atoms with the smaller hydrogen atoms occupying holes between the larger metal atoms.
Electrolysis	b. a compound that contains hydrogen and just one other element.
Binary hydride	c. a solid compound that contains water molecules.
Interstitial hydride	d. a reaction in which a substance is both oxidized and reduced.
Nonstoichiometric compound	e. differences in properties due to the differences in the mass of isotopes.
Disproportionation reaction	f. the process of using an electric current to bring about chemical change.

Hydrate   g. compound in which the atomic composition can't be expressed as a ratio of small, whole numbers.

Fill-in-the-Blank

1. The low boiling and melting points of hydrogen are due to _____.

2. Hydrogen is synthesized in the laboratory by _____

_____.

3. The water-gas shift reaction shifts the composition of _____ to remove

_____ and produce more ____.

4. Ionic hydrides are formed between hydrogen and _____ and

are prepared by _____.

5. The properties of interstitial hydrides depend on _____ and are a function

of _____.

6. Oxygen does not react with _____.

7. Amphoteric oxides are formed between oxygen and _____

_____.

8. Peroxides and superoxides can be formed between oxygen and the _____

_____ and are prepared by _____.

9. Ozone is produced by _____.

10. Hygroscopic compounds are _____ compounds that have a tendency to form

_____ and are useful as _____ since they absorb

_____.

Problems

2⊠   1. How many grams of zinc are required to generate 750 mL of hydrogen at 22°C and 740 mm Hg by the reaction between zinc and hydrochloric acid?

2. Give a chemical equation for the largest industrial use of hydrogen.

4⊠   3. Write balanced equations for the following reactions using the half-reaction method. Assign oxidation numbers, and identify the oxidizing agent and reducing agent.

     a. $Ag^+ (aq) + AsH_3 \rightarrow H_3AsO_4 (aq) + Ag (s)$ (acidic solution)

     b. $Al (s) + H_2SO_4 (aq) \rightarrow Al_2(SO_4)_3 (aq) + H_2 (g)$

c. $SnO_2 (s) + H_2 (g) \rightarrow Sn (s) + H_2O (l)$

d. $WO_3 (s) + H_2 (g) \rightarrow W (s) + H_2O (g)$

4. Write the chemical equation for the reaction between LiH and H_2O. Identify the reducing and oxidizing agents in this reaction. This reaction can also be classified as an acid-base reaction. Identify the acid and the base.

5. Using chemical equations, explain why the price of ammonia depends on the price of natural gas.

6. Identify the following hydrides as ionic, covalent or metallic:
   a. MgH_2 b. AsH_3 c. B_2H_6 d. $ZrH_{1.9}$ e. UH_3

7. Oxygen can be prepared in the laboratory by the thermal decomposition of $KClO_3$ in the presence of MnO_2. How many grams of $KClO_3$ are needed to produce 500 mL of O_2 at $22^\circ C$ and 758 mm Hg?

8. Write balanced equations for the following reactions using the half-reaction method. Assign oxidation numbers and identify the oxidizing agent and reducing agent.
   a. $VO^{2+} (aq) + Cr_2O_7^{2-} (aq) \rightarrow Cr^{3+} (aq) + VO_2^{+} (aq)$ (acidic solution)

   b. $O_2 (g) + H_2O (l) + Mg (s) \rightarrow Mg(OH)_2 (s)$

   c. $O_2 (g) + 4 H^{+} (aq) + Cu (s) \rightarrow Cu^{2+} (aq) + H_2O (l)$

9. Classify the following oxides as basic, acidic, or amphoteric.
   a. CaO b. P_4O_{10} c. Al_2O_3 d. SO_3

10. Complete and balance the following equaions:
   a. $CaH_2 (s) + H_2O (l) \rightarrow$

   b. $KH (s) + O_2 (g) \rightarrow$

   c. $Li (s) + O_2 (g) \rightarrow$

   d. $P_4 (s) + O_2 (g) \rightarrow$

   e. $Na_2O (s) + H_2O (l) \rightarrow$

   f. $BaO_2 (s) + H_2SO_4 (aq) \rightarrow$

   g. $N_2O_5 (s) + H_2O (l) \rightarrow$

   h. $CsO_2 (s) + H_2O (l) \rightarrow$

   i. $H_2O_2 (aq) + MnO_4^{-} (aq) \rightarrow$ acidic solution

   j. $O_3 (g) + I^{-} (aq) \rightarrow$ basic solution

11. Calculate the density of H_2 if 0.2348 g of CaH_2 reacts with an excess of H_2O at $0^\circ C$ 1 atm of pressure.

7⊠ 12. Calculate the volume of 0.7500 M $KMnO_4$ needed to react with H_2O_2 to produce 3.000 g of O_2 at 0° C and 1 atm pressure. What volume of O_2 is produced? From this information, calculate the density of O_2.

13. Identify the following as oxides, peroxides, or superoxides:
   a. K_2O_2 b. RbO_2 c. CaO d. SrO_2 e. Al_2O_3

14⊠ 14. When 2.879 g of the hydrate $MgSO_4 \cdot XH_2O$ was heated, 1.406 g of $MgSO_4$ was obtained. What was the formula of the hydrate?

14⊠ 15. A 2.75 g sample of Na_2CO_3 was exposed to moist air. If 7.42 g of a hydrate of Na_2CO_3 was obtained, what was the formula of the hydrate?

Solutions

True/False
1. F. Hydrogen doesn't completely transfer its valence electron in ordinary chemical reactions. Hydrogen shares its electron with a nonmetallic element to produce a covalent compound.
2. F. Oxygen is the most abundant element on the planet's surface.
3. F. Oxygen is an inexpensive and readily available oxidizing agent.
4. T
5. T
6. F. The three isotopes of hydrogen all have the same electronic configuration which determines the chemical behavior of an element. The isotopes differ only in the number of neutrons they contain.
7. F. Interstitial hydride compounds are classified as metallic hydrides.
8. T
9. T
10. F. Hydrogen peroxide disproportionates to give water and oxygen.
11. T
12. T

Matching
   Isotope effect - e
   Electrolysis -f
   Binary hydride - b
   Interstitial hydride - a
   Nonstoichiometric compound - g
   Disproportionation reaction - d
   Hydrate - c

Fill-in-the-Blank
1. weak intermolecular forces
2. reaction of a dilute acid with an active metal
3. synthesis gas; CO, and H_2
4. alkali and heavier alkaline-earth metals; direct reaction of the elements
5. their composition; the pressure of H_2 gas in the surroundings
6. the noble gases, platinum, and gold
7. elements with intermediate electronegativities.
8. heavier group 1A and 2A metals; heating the metals in an excess of air
9. passing an electric discharge through O_2
10. anhydrous; hydrates; drying agents; water from the air

Problems

1. The balanced equation for the reaction is: $Zn\,(s) + 2\,HCl\,(aq) \rightarrow ZnCl_2\,(aq) + H_2\,(g)$
   To determine the mass of Zn needed for this reaction, we need to calculate the number of moles of hydrogen using the ideal gas law.

$$n = \frac{PV}{RT}; \qquad n = \frac{\left(740\text{ mm Hg} \times \dfrac{1\text{ atm}}{760\text{ mm Hg}}\right)(0.750\text{ L})}{\left(0.0821\dfrac{\text{L}\cdot\text{atm}}{\text{mol}\cdot\text{K}}\right)(295\text{ K})} = 3.02 \times 10^{-2}\text{ mol}$$

We now can calculate the mass of zinc needed for this reaction.

$$(3.02 \times 10^{-2}\text{ mol H}_2) \times \frac{1\text{ mol Zn}}{1\text{ mol H}_2} \times \frac{65.4\text{ g Zn}}{1\text{ mol Zn}} = 1.98\text{ g Zn}$$

2. The largest industrial use of hydrogen is the Haber process for the production of NH_3. The balanced chemical equation is $N_2\,(g) + 3\,H_2\,(g) \rightarrow 2\,NH_3\,(g)$

3. a. Ag^+ is reduced to Ag (oxidizing agent) and As^{3-} in AsH_3 is oxidized to As^{5+} (reducing agent) so the two half reactions are:

   $Ag^+ \rightarrow Ag$

   $AsH_3 \rightarrow H_3AsO_4$

   Both half reactions are balanced for elements other than oxygen and hydrogen. Add H_2O for any oxygens that are needed and H^+ for any hydrogens that are needed.

   $Ag^+ \rightarrow Ag$

   $AsH_3 + 4\,H_2O \rightarrow H_3AsO_4 + 8\,H^+$

   Balance each reaction for charge

   $Ag^+ + 1\,e^- \rightarrow Ag$

   $AsH_3 + 4\,H_2O \rightarrow H_3AsO_4 + 8\,H^+ + 8\,e^-$

   Make the electron count the same in both reactions.

   $8 \times (Ag^+ + 1\,e^- \rightarrow Ag)$

   $AsH_3 + 4\,H_2O \rightarrow H_3AsO_4 + 8\,H^+ + 8\,e^-$

   Add the two half reactions together, canceling anything that appears on both sides of the equation.

   $8\,Ag^+ + AsH_3 + 4\,H_2O \rightarrow 8\,Ag + H_3AsO_4 + 8\,H^+$

b. Al is oxidized (reducing agent) and H^+ is reduced (oxidizing agent), so the two half reactions are:

$$Al \rightarrow Al^{3+}$$

$$2H^+ \rightarrow H_2$$

The equation is balanced using the method outlined above (see chapter 4) to give

$$2\,Al\,(s)\,+\,3\,H_2SO_4\,(aq)\,\rightarrow\,Al_2(SO_4)_3\,(aq)\,+\,3\,H_2\,(g)$$

c. Sn^{4+} in SnO_2 is reduced (oxidizing agent) and H_2 is oxidized (reducing agent), so the two half reactions are:

$$Sn^{4+} \rightarrow Sn$$

$$H_2 \rightarrow H^+$$

The equation is balanced using the method outlined above (see chapter 4) to give

$$SnO_2\,+\,2\,H_2\,\rightarrow\,Sn\,+\,2\,H_2O$$

d. W^{6+} in WO_3 is reduced (oxidizing agent) and H_2 is oxidized (reducing agent), so the two half reactions are:

$$WO_3 \rightarrow W$$

$$H_2 \rightarrow H^+$$

The equation is balanced using the method outlined above (see chapter 4) to give

$$WO_3\,+\,3\,H_2\,\rightarrow\,W\,+\,3\,H_2O$$

4. $LiH\,(s)\,+\,H_2O\,(l)\,\rightarrow\,H_2\,(g)\,+\,LiOH\,(aq)$; Reducing agent: H^- in LiH; Oxidizing agent: H^+ in H_2O; Base: H^-; Acid: H_2O

5. Ammonia is synthesized by the Haber process: $N_2\,(g)\,+\,3\,H_2\,(g)\,\rightarrow\,2\,NH_3\,(g)$. The large scale industrial method for synthesizing hydrogen is the steam-hydrocarbon reforming process. The primary source of hydrocarbons is natural gas. The chemical equations for the steam hydrocarbon reforming process are:

$$H_2O\,+\,hydrocarbon \rightarrow CO\,+\,H_2$$
$$CO\,+\,H_2O\,\rightarrow\,CO_2\,+\,H_2$$

Since the primary source for synthesizing hydrogen is fuel oil, it follows that the price of ammonia depends upon the price of fuel oil.

6. a. ionic   b. covalent   c. covalent   d. metallic   e. metallic

7. The balanced equation for the reaction is $2\,KClO_3\,(s)\,\rightarrow\,3\,O_2\,(g)\,+\,2\,KCl\,(s)$. To determine the mass of $KClO_3$ needed for this reaction, we need to calculate the number of moles of oxygen using

the ideal gas law.

a. $$n = \frac{\left(758 \text{ mm Hg} \times \dfrac{1 \text{ atm}}{760 \text{ mm Hg}}\right) \times 0.500 \text{ L}}{\left(0.0821 \dfrac{\text{L} \cdot \text{atm}}{\text{mol} \cdot \text{K}}\right)(295 \text{ K})} = 2.06 \times 10^{-2} \text{ mol O}_2$$

We can now calculate the mass of $KClO_3$ needed for this reaction.

$$2.06 \times 10^{-2} \text{ mol O}_2 \times \frac{2 \text{ mol KClO}_3}{3 \text{ mol O}_2} \times \frac{122.6 \text{ g KClO}_3}{1 \text{ mol KClO}_3} = 1.68 \text{ g KClO}_3$$

8. a. Vanadium is oxidized from V^{4+} to V^{5+} (reducing agent = VO^{2+}) and chromium is reduced from Cr^{6+} to Cr^{3+} (oxidizing agent = $Cr_2O_7^{2-}$). The half reactions are

$$VO^{2+} \rightarrow VO_2^{+}$$

$$Cr_2O_7^{2-} \rightarrow Cr^{3+}$$

Using the half-reaction method (see problem 3 and chapter 4), the balanced equation is

$$6 \text{ VO}^{2+} + 2 \text{ H}^{+} + Cr_2O_7^{2-} \rightarrow 6 \text{ VO}_2^{+} + 2 \text{ Cr}^{3+} + H_2O$$

b. Magnesium is oxidized to Mg^{2+} (reducing agent = Mg) and oxygen is reduced to O^{2-} (oxidizing agent = O_2). The half reactions are

$$Mg \rightarrow Mg(OH)_2 \, (s)$$

$$O_2 + H_2O \rightarrow 4 \text{ OH}^{-}$$

Using the half-reaction method (see Problem 3 and Chapter 4), the balanced equation is

$$O_2 + 2 \text{ H}_2O + 2 \text{ Mg} \rightarrow Mg(OH)_2 \, (s)$$

c. Copper is oxidized to Cu^{2+} (reducing agent = Cu) and oxygen is reduced to O^{2-} (oxidizing agent = O_2). The half reactions are

$$Cu \rightarrow Cu^{2+}$$

$$O_2 + 4 \text{ H}^{+} \rightarrow H_2O$$

Using the half-reaction method (see Problem 3 and Chapter 4), the balanced equation is

$$O_2 + 4 \text{ H}^{+} + 2 \text{ Cu} \rightarrow 2 \text{ H}_2O + 2 \text{ Cu}^{2+}$$

9. a. basic   b. acidic   c. amphoteric   d. acidic

10. Some of these equations can be balanced by inspection. Others are redox reactions which need to be balanced using either the half-reaction method (see problems 3 and 8) or the ion-electron method (see chapter 4).

   a. $CaH_2 (s) + 2 H_2O \rightarrow 2 H_2 (g) + Ca(OH)_2 (s)$

   b. $2 KH (s) + O_2 (g) \rightarrow H_2O (g) + K_2O (s)$

   c. $4 Li (s) + O_2 (g) \rightarrow 2Li_2O (s)$

   d. $P_4 (s) + 5 O_2 (g) \rightarrow P_4O_{10} (s)$

   e. $Na_2O (s) + H_2O (l) \rightarrow 2 NaOH (aq)$

   f. $BaO_2 (s) + H_2SO_4 (aq) \rightarrow BaSO_4 (s) + H_2O_2 (aq)$

   g. $N_2O_5 (s) + H_2O (l) \rightarrow 2 HNO_3 (aq)$

   h. $2 CsO_2 (s) + H_2O (l) \rightarrow O_2 (g) + 2 Cs^+ (aq) + HO_2^- (aq) + OH^- (aq)$

   i. $5 H_2O_2 (aq) + 2 MnO_4^- (aq) + 6 H^+ (aq) \rightarrow 5 O_2 (g) + 2 Mn^{2+} (aq) + 8 H_2O (l)$

   j. $O_3 (g) + 2 I^- (aq) + H_2O (l) \rightarrow O_2 (g) + I_2 (g) + 2 OH^- (aq)$

11. The balanced equation for this reaction is $CaH_2 (s) + 2 H_2O (l) \rightarrow 2 H_2 (g) + Ca(OH)_2 (aq)$. We first need to calculate the number of moles and grams of H_2 from the number of moles of CaH_2.

$$0.2348 \text{ g } CaH_2 \times \frac{1 \text{ mol } CaH_2}{42.1 \text{ g } CaH_2} = 5.577 \times 10^{-3} \text{ mol } CaH_2$$

$$5.577 \times 10^{-3} \text{ mol } CaH_2 \times \frac{2 \text{ mol } H_2}{1 \text{ mol } CaH_2} = 1.115 \times 10^{-2} \text{ mol } H_2$$

$$1.115 \times 10^{-2} \text{ mol } H_2 \times \frac{2.00 \text{ g } H_2}{1 \text{ mol } H_2} = 2.230 \times 10^{-2} \text{ g } H_2$$

Knowing moles of H_2 gas, we can now use the ideal gas law to calculate the volume of the gas.

$$V = \frac{\left(1.115 \times 10^{-2} \text{ mol } H_2\right)\left(0.0821 \frac{L \cdot atm}{mol \cdot K}\right)(273 \text{ K})}{1 \text{ atm}} = 0.2499 \text{ L}$$

We now know both the grams and volume of H_2 produced and can calculate the density.

$$\frac{2.230 \times 10^{-2} \text{ g } H_2}{0.2499 \text{ L } H_2} = 0.08924 \text{ g}/_L$$

12. The balanced equation for this reaction is

$$5\,H_2O_2\,(aq) + 2\,MnO_4^{2-} + 6\,H^+\,(aq) \rightarrow 5\,O_2\,(g) + 2\,Mn^{2+}\,(aq) + 8\,H_2O\,(l)$$

To calculate the volume of $KMnO_4$ needed to produce 3.000 g of O_2, we need to first calculate the number of moles of $KMnO_4$ needed to produce 3.000 g of O_2. This can be accomplished by calculating the number of moles of O_2 in 3.000 g of O_2.

$$3.000\,g\,O_2 \times \frac{1\,mol\,O_2}{32\,g\,O_2} = 9.375 \times 10^{-2}\,mol\,O_2$$

$$9.375 \times 10^{-2}\,mol\,O_2 \times \frac{2\,mol\,MnO_4^-}{5\,O_2} \times \frac{1\,KMnO_4}{1\,MnO_4^-} = 3.750 \times 10^{-3}\,mol\,KMnO_4$$

$$\left(3.750 \times 10^{-3}\,mol\,KMnO_4\right) \times \frac{1\,L}{0.7500\,mol\,KMnO_4} = 5.000 \times 10^{-3}\,L\,KMnO_4 \text{ or } 5.000\,mL$$

To determine the density of O_2, we need to calculate the volume of O_2 using the ideal gas law.

$$V = \frac{\left(9.375 \times 10^{-2}\,mol\,O_2\right) \times \left(0.0821\,\dfrac{L \cdot atm}{mol \cdot K}\right)(273\,K)}{1\,atm} = 2.101\,L$$

$$\frac{3.000\,g\,O_2}{2.101\,L\,O_2} = 1.428\,{}^{g}\!/_{L}$$

13. a. peroxide b. superoxide c. oxide d. peroxide e. oxide

14. The difference between the mass of the hydrate and the anhydrous compound is the mass of water.

$$2.879\,g\,MgSO_4 \cdot XH_2O - 1.406\,g\,MgSO_4 = 1.473\,g\,H_2O$$

We can now calculate the number of moles of water from the mass of water.

$$1.473\,g\,H_2O \times \frac{1\,mol\,H_2O}{18.0\,g\,H_2O} = 8.183 \times 10^{-2}\,mol\,H_2O$$

We now must determine the number of moles of H_2O relative to the number of moles of $MgSO_4$. (But first, we have to calculate the number of moles of $MgSO_4$.)

$$1.406\,g\,MgSO_4 \times \frac{1\,mol\,MgSO_4}{120.3\,g\,MgSO_4} = 1.169 \times 10^{-2}\,mol\,MgSO_4$$

$$\frac{8.183 \times 10^{-2}\,mol\,H_2O}{1.169 \times 10^{-2}\,mol\,MgSO_4} = 7$$

From the above equation, we learn that we have a ratio of 7 mol H_2O to 1 mol $MgSO_4$. Therefore, the formula of the hydrate is $MgSO_4 \cdot 7\ H_2O$.

15. Again, the difference between the mass of the hydrate and anhydrous compound is the mass of water. This problem is worked in exactly the same manner as the above problem. Following this procedure gives rise to the formula $Na_2CO_3 \cdot 10\ H_2O$.

CHAPTER 15

AQUEOUS EQUILIBRIA: ACIDS AND BASES

Learning Goals

1⊠ Define an acid and base according to the Arrhenius, Brønsted-Lowry, and Lewis theories.

2⊠ From Lewis structures, determine which chemical species can act as a Brønsted-Lowry acid, a Brønsted-Lowry base, or both.

3⊠ From a chemical equation for a proton-transfer reaction, identify the conjugate acid-base pairs.

4⊠ Given the extent of dissociation of an acid in water, determine whether the acid is a stronger or weaker acid than water and whether the conjugate base of the acid is a stronger or weaker base than water.

5⊠ Given a chemical equation representing a proton transfer reaction and the relative strengths of each acid and base involved in the reaction, determine whether the reaction is favored to the right or to the left.

6⊠ Calculate H_3O^+ concentration from OH^- concentration and vice versa. From these concentrations determine whether the solution is acidic, neutral, or basic.

7⊠ Interconvert pH and $[H_3O^+]$. Classify the solution as acidic, neutral, or basic.

8⊠ Given the molar concentration of a strong acid or a strong base, determine the pH of the solution.

9⊠ Given the pH of a weak-acid solution, determine the K_a of the acid.

10⊠ Given the K_a value and the initial concentration of a weak monoprotic acid, calculate the concentrations of all species at equilibrium, the pH of the solution, and the percent dissociation of the acid.

11⊠ Given the K_a values and the initial concentration of a weak diprotic acid, calculate the concentrations of all species at equilibrium and the pH of the solution.

12⊠ Given the K_b value and the initial concentration of a weak base, calculate the concentrations of all species at equilibrium and the pH of the solution.

13⊠ Interconvert K_a and K_b.

14⊠ Classify salt solutions as acidic, neutral, or basic. Calculate the pH of these solutions.

15⊠ Identify which of two substances is more acidic.

16⊠ Identify the Lewis acid and the Lewis base in a chemical reaction.

Chapter in Brief

In this chapter, you continue the study of chemical equilibrium by applying the concepts you learned in Chapter 13 to acid-base chemistry. This chapter begins with the definition of acids, bases, and conjugate acid-base pairs using the Brønsted-Lowry concept. You will learn the pH scale and how to determine the pH from the dissociation of a strong acid or strong base. You also will learn how to determine the strength of an acid or base relative to water given the extent of dissociation in water and how to determine the pH of a weak acid or weak base using either K_a or K_b. Finally, you will determine the pH of salt solutions and examine the Lewis acid-base concept.

Acid-Base Concepts: The Brønsted-Lowry Theory

1⊠
    A. Arrhenius theory of acids and bases.
        1. Acids - substances that dissociate in water to produce hydrogen ions.
        2. Bases - substances that dissociate in water to produce hydroxide ions.
        3. Limitations to theory.
            a. restricted to aqueous solutions
            b. doesn't account for the basicity of substances that don't contain OH groups

1⊠   B. Brønsted-Lowry theory of acids and bases
      1. Acid - any substance that can transfer a proton to another substance (proton donor).
      2. Base - any substance that can accept a proton (proton acceptor).
      3. Acid-base reaction - proton-transfer reactions

3⊠   C. Conjugate acid-base pair - chemical species whose formulas differ only by one proton.
      1. Conjugate base has one less proton than its acid.
        a. A^- is the conjugate base of HA
      2. Conjugate acid has one more proton than its base.
        a. BH^+ is the conjugate acid of B

    D. Acid-dissociation equilibrium.
      1. acid transfers a proton to the solvent, which acts as a base.
      2. $HA\ (aq) + H_2O\ (l) \rightleftarrows H_3O^+\ (aq) + A^-\ (aq)$

    E. Base-dissociation equilibrium.
      1. Base accepts a proton from the solvent, which acts as an acid.
      2. $NH_3\ (aq) + H_2O\ (l) \rightleftarrows OH^-\ (aq) + NH_4^+\ (aq)$

2⊠   F. All Brønsted-Lowry bases have one or more lone pairs of electrons.
      1. Unshared pair of electrons are used for bonding to the proton.

EXAMPLE:
Write the proton transfer equilibria for the following acids or bases in aqueous solution and identify the conjugate acid-base pairs in each one: $C_6H_5NH_3^+$ (anilinium ion), $H_2PO_4^-$ (acid reaction), NH_2NH_2 (hydrazine, a base), and PO_4^{3-}.

SOLUTION:
    $C_6H_5NH_3^+\ (aq) + H_2O \rightleftarrows C_6H_5NH_2 + H_3O^+\ (aq)$
      Conjugate acid-base pairs: $C_6H_5NH_3^+$ (acid)/$C_6H_5NH_2$ (base); H_2O (base)/H_3O^+ (acid)

    $H_2PO_4^-\ (aq) + H_2O \rightleftarrows HPO_4^{2-}\ (aq) + H_3O^+\ (aq)$
      Conjugate acid-base pairs: $H_2PO_4^-$ (acid)/ HPO_4^{2-} (base); H_2O (base)/H_3O^+ (acid)

    $NH_2NH_2 + H_2O \rightleftarrows NH_2NH_3^+\ (aq) + OH^-\ (aq)$
      Conjugate acid-base pairs: NH_2NH_2 (base)/$NH_2NH_3^+$ (acid); H_2O (acid)/OH^- (base)

    $PO_4^{3-}\ (aq) + H_2O \rightleftarrows HPO_4^{2-}\ (aq) + OH^-\ (aq)$
      Conjugate acid-base pairs: PO_4^{3-} (base)/HPO_4^{2-} (acid); H_2O (base)/OH^- (acid)

Acid Strength and Base Strength
    A. In an acid-dissociation equilibrium, the two bases, H_2O and A^-, are competing for protons.
5⊠   1. In every acid-base reaction, the proton is transferred to the stronger base.
4⊠   B. Strong acid - completely dissociated in aqueous solution.
      1. Have weak conjugate base.
        a. H_2O is a stronger base than A^-
4⊠   C. Weak acid - partially dissociates in aqueous solution.
      1. Have strong conjugate base.
        a. A^- is a stronger base than H_2O
    D. Inverse relationship between the strength of an acid and the strength of its conjugate base (see Table 15.1, p. 591 in text).

5⊠   *EXAMPLE:*
Determine the direction of reactions involving: acetic acid, acetate ion, hydrosulfuric acid, and HS⁻; acetic acid, acetate ion, hydrogen sulfate, and SO_4^{2-}.

SOLUTION: To determine the direction of these acid-base reactions, we need to use Table 15.1 on page 591 of your text. Keep in mind that the proton is transferred to the stronger base. Therefore, the direction of the reaction to reach equilibrium is proton transfer from the stonger acid to the stronger base to give the weaker acid and the weaker base. For acetic acid and HS⁻, the stronger acid is CH_3COOH and the stronger base is HS⁻. The reaction in the forward direction is:

$$CH_3COOH + HS^- \ (aq) \rightleftarrows CH_3CO_2^- \ (aq) + H_2S$$

For the reaction between acetic acid and SO_4^{2-}, the stronger acid is HSO_4^- and the stronger base is $CH_3CO_2^-$. Therefore, the reaction in the forward direction is:

$$HSO_4^- \ (aq) + CH_3CO_2^- \ (aq) \rightleftarrows SO_4^{2-} \ (aq) + CH_3COOH$$

Hydrated Protons and Hydronium Ions
A. Hydronium ion, H_3O^+ - ion in which H^+ is bonded to the oxygen atom of a solvent water molecule.
    1. Simplest hydrate of the proton - $[H(H_2O)]^+$.
        a. produces higher hydrates through hydrogen bonding with other water molecules
            i. general formula - $[H(H_2O)_n]^+$
    2. Symbols H^+ (aq) and H_3O^+ (aq) used to represent a proton hydrated by an unspecified number of water molecules.

Dissociation of Water
A. Water can act both as an acid and as a base.
B. Dissociation of water - one water molecule donates a proton to another water molecule.
    1. H_2O (l) + H_2O (l) $\rightleftarrows$ H_3O^+ (aq) + OH^- (aq).
    2. Ion-product constant for water, K_w.
        a. $K_w = [H_3O^+][OH^-]$
    3. Forward and reverse reactions are rapid.
    4. Position of the equilibrium lies far to the left.
    5. In an aqueous solution, $[H_3O^+][OH^-] = 1.0 \times 10^{-14}$.

6⊠   C. Relative values of H_3O^+ and OH^- concentrations can be used to determine if an aqueous solution is neutral, acidic or basic.
    1. $[H_3O^+] = \dfrac{1.0 \times 10^{-14}}{[OH^-]}$.

    2. $[OH^-] = \dfrac{1.0 \times 10^{-14}}{[H_3O^+]}$.

    3. $[H_3O^+] = [OH^-]$, neutral solution and $[H_3O^+] = 1.0 \times 10^{-7}$ M.
    4. $[H_3O^+] > [OH^-]$, acid solution and $[H_3O^+] > 1.0 \times 10^{-7}$ M.
    5. $[H_3O^+] < [OH^-]$, basic solution and $[H_3O^+] < 1.0 \times 10^{-7}$M.

EXAMPLE:
Calculate the molarity of OH⁻ in a solution with an H_3O^+ concentration of 0.35 M. Is this

solution acidic, basic or neutral? Is a solution with an OH⁻ concentration of 8.5×10^{-5} M acidic, basic or neutral?

SOLUTION: If we know either the $[H_3O^+]$ or $[OH^-]$, we can calculate the other from the relationship $K_w = [H_3O^+][OH^-]$.

$$\left[OH^-\right] = \frac{1 \times 10^{-14}}{0.35} = 2.9 \times 10^{-14} ; \text{ Since } [H_3O^+] > 1.0 \times 10^{-7} \text{ this solution is acidic.}$$

$$\left[H_3O^+\right] = \frac{1.0 \times 10^{-14}}{8.5 \times 10^{-5}} = 1.2 \times 10^{-10} ; \text{ In this case, } [H_3O^+] < 1.0 \times 10^{-7}; \text{ therefore, the solution is}$$
basic.

7⊠   **The pH Scale**
A. pH scale - logarithmic scale used to express the hydronium ion concentration.
1. $pH = -\log[H_3O^+]$.
2. $[H_3O^+] = \text{antilog}(-pH) = 10^{-pH}$.
3. Significant figures in a logarithm are the digits to the right of the decimal point.
a. number to the left of the decimal point is an exact number related to the integral power of 10 in the exponential expression for $[H_3O^+]$
B. pH decreases as $[H_3O^+]$ increases.
1. Acidic solutions, pH < 7.
2. Basic solutions, pH > 7.
3. Neutral solutions, pH = 7.

EXAMPLE:
Calculate the $[H_3O^+]$ for orange juice which has a pH of 3.80

SOLUTION:
$[H_3O^+] = \text{antilog} (-pH) = \text{antilog} (-3.80) = 1.6 \times 10^{-4}$

Measuring pH
A. Acid-base indicators (HIn) - substances that change color in a specific pH range.
1. Weak acids.
2. Have different colors in their acid (HIn) and conjugate base (In⁻) forms.
3. Change color over a range of ≈ 2 pH units.
4. Can determine the pH of a solution to within ≈ ±1 pH unit.
B. pH meter.
1. An electronic instrument that measures a pH-dependent electrical potential of the test solution.
2. Determines pH values more accurately than acid-base indicators.

8⊠   **The pH of Strong Acids and Strong Bases**
A. Strong acids - 100 % dissociated in aqueous solution ($HClO_4$, HCl, HNO_3).
1. $[H_3O^+] = [A^-] = $ initial concentration of the acid.
2. [undissociated HA] = 0.
3. Monoprotic acid, $pH = -\log[H_3O^+]$
B. Strong bases.
1. Alkali metal hydroxides, MOH.
a. water-soluble ionic solids
b. exist in aqueous solution as alkali-metal cations and hydroxide anions

      c. calculate pH from [OH⁻]
2. Alkaline-earth metal hydroxides, $M(OH)_2$ (M = Mg, Ca, Sr, Ba).
    a. less soluble than alkali hydroxides, therefore lower [OH⁻]
    b. most important strong base - lime, CaO
        i. produces $Ca(OH)_2$ when mixed with water
        ii. used in steel making, water purification, and chemical manufacture

EXAMPLE:
    Calculate the pH of a 1.25×10^{-3} M HNO_3 solution.

SOLUTION: HNO_3 is a strong acid which completely dissociates in water. Therefore, $[H_3O^+]$ = initial concentration of the undissociated acid.

$$[H_3O^+] = 1.25 \times 10^{-3} \text{ M}$$

$$pH = -\log [H_3O^+] = -\log (1.25 \times 10^{-3}) = 2.903$$

EXAMPLE:
    Calculate the pH of a 5.0×10^{-4} M KOH solution.

SOLUTION: KOH is a strong base and exists in aqueous solution as K^+ and OH⁻. Therefore, [OH⁻] = initial concentration of KOH.

$$[OH^-] = 5.0 \times 10^{-4} \text{ M}$$

The H_3O^+ concentration is calculated from the OH⁻ concentration.

$$[H_3O^+] = \frac{K_w}{[OH^-]} = \frac{1.0 \times 10^{-14}}{5.0 \times 10^{-4}} = 2.0 \times 10^{-11} \text{ M}$$

$$pH = -\log (2.0 \times 10^{-11}) = 10.70$$

Equilibria in Solutions of Weak Acids
A. Weak acids - partially dissociated; can write an acid-dissociation equilibrium equation.
    1. HA (aq) + H_2O (l) $\rightleftarrows$ H_3O^+ (aq) + A^- (aq)
B. Position of acid-dissociation equilibrium - characterized by the acid-dissociation constant, K_a.
    1. $K_a = \dfrac{[H_3O^+][A^-]}{[HA]}$.
C. The larger the value of K_a, the stronger the acid.
9⊠   D. K_a values can be determined from pH measurements.

EXAMPLE:
    The pH of a 0.200 M solution of nicotinic acid ($HC_6H_4NO_2$) is 2.78. Determine the value of K_a for this acid.

SOLUTION: To determine K_a first write the balanced equation for the dissociation equilibrium and the equilibrium equation that defines K_a.

    $HC_6H_4NO_2$ (aq) + H_2O (l) $\rightleftarrows$ $H_3O^+(aq)$ + $C_6H_4NO_2^-$ (aq)

$$K_a = \frac{[H_3O^+][C_6H_4NO_2^-]}{[HC_6H_4NO_2]}$$

To determine K_a we need to know the concentrations of the species in the equilibrium mixture. We can determine the concentration of H_3O^+ from the pH.

$[H_3O^+]$ = antilog (-pH) = antilog (-2.780) = 1.66×10^{-3} M.

Dissociation of one $HC_6H_4NO_2$ molecule produces one H_3O^+ and $C_6H_4NO_2^-$ ion; the H_3O^+ and $C_6H_4NO_2^-$ concentrations are equal.

$[H_3O^+] = [C_6H_4NO_2^-] = 1.66 \times 10^{-3}$ M

The $[HC_6H_4NO_2]$ concentration at equilibrium is equal to the initial concentration minus the amount of $HC_6H_4NO_2$ that dissociates.

$[HC_6H_4NO_2] = 0.200 - (1.66 \times 10^{-3}) = 0.198$ M

$$K_a = \frac{[H_3O^+][C_6H_4NO_2^-]}{[HC_6H_4NO]} = \frac{(1.66 \times 10^{-3})(1.66 \times 10^{-3})}{(0.198)} = 1.39 \times 10^{-5}$$

10⊠ **Calculating Equilibrium Concentrations in Weak-Acid Solutions**

A. Equilibrium concentrations and the pH of a solution of a weak-acid can be calculated from the value of K_a.

B. To solve acid-base equilibrium problems, think about the chemistry. (See Fig. 15.7, page 605 in your text).

EXAMPLE:

Calculate the concentration of all species present (H_3O^+, $C_4H_7O_2^-$, $HC_4H_7O_2$, and OH^-) and the pH in 0.250 M butyric acid ($HC_4H_7O_2$).

SOLUTION: To solve this problem, use the method found on page 605 of your text.

1. The species present initially are $HC_4H_7O_2$ (acid) and H_2O (acid or base)

2. The possible proton-transfer reactions are

$HC_4H_7O_2\ (aq) + H_2O\ (l) \rightleftarrows H_3O^+\ (aq) + C_4H_7O_2^-\ (aq)$ $\qquad K_a = 1.5 \times 10^{-5}$

$H_2O\ (l) + H_2O\ (l) \rightleftarrows H_3O^+\ (aq) + OH^-\ (aq)$ $\qquad K_w = 1.0 \times 10^{-14}$

3. Since $K_a \gg K_w$, the principal reaction is dissociation of $HC_4H_7O_2$.

4. Make a table.

Principal reaction	$HC_4H_7O_2\ (aq)$	$\rightleftarrows$	H_3O^+	+	$C_4H_7O_2^-$
Initial concentration	0.250		0		0
Change	-x		+x		+x
Equilibrium concentration	0.250 - x		+x		+x

5. Substitute the equilibrium concentrations into the equilibrium expression.

$$K_a = 1.5 \times 10^{-5} = \frac{[H_3O^+][C_4H_7O_2^-]}{[HC_4H_7O_2]} = \frac{(x)(x)}{(0.250 - x)}$$

Assume that x is negligible compared with the initial concentration of the acid; therefore $0.250 - x \approx 0.250$. Using this value in the denominator, solve for x.

$$x^2 \approx \left(1.5 \times 10^{-5}\right)\left(0.250\right)$$

$$x \approx 1.9 \times 10^{-3} \text{ *}$$

6. The big equilibrium concentrations are $[HC_4H_7O_2]$ = 0.250 -0.0019 = 0.248 M; $[H_3O^+]$ = $[C_4H_7O_2^-]$ = 1.9 x 10^{-3} M

7. The concentration of OH⁻ is obtained from the dissociation of water.

$$[OH^-] = \frac{K_w}{[H_3O^+]} = \frac{1.0 \times 10^{-14}}{1.9 \times 10^{-3}} = 5.3 \times 10^{-12}$$

8. pH = -log (1.9 x 10^{-3}) = 2.72

10▷ Percent Dissociation in Weak-Acid Solutions

A. Percent dissociation $= \dfrac{[HA] \text{ dissociated}}{[HA] \text{ undissociated}} \times 100$

1. Value depends on the acid.

2. Increases with increasing value of K_a.

EXAMPLE:

Calculate the % dissociation of a 0.25 M benzoic acid (C_6H_5COOH) solution. (K_a = 6.5 × 10^{-5})

SOLUTION: To determine the % dissociation we need to first determine the concentration of the dissociated HA. This requires that we determine the equilibrium concentration of H_3O^+ and $C_6H_5COO^-$. We will use the procedure found on page 605 of your text.

1. The species present initially are C_6H_5COOH and H_2O (acid or base).

2. The possible proton-transfer reactions are

C_6H_5COOH (aq) + H_2O (l) $\rightleftarrows$ H_3O^+ (aq) + $C_6H_5COO^-$ (aq) K_a = 6.5 × 10^{-5}

H_2O (l) + H_2O (l) $\rightleftarrows$ H_3O^+ (aq) + OH^- (aq) K_w = 1.0 × 10^{-14}

3. Since $K_a \gg K_w$, the principal reaction is dissociation of C_6H_5COOH.

4. Make a table.

---

* The initial concentration of $HC_4H_7O_2$ is known to the third decimal place (0.250). Usually x is considered negligible compared to the value of the initial concentration only if $x <$ 0.001. In this case, $x > 0.001$. However, solving for x using the quadratic equation gives x = 1.9 x 10^{-3}. Both values for x give an equilibrium concentration of 0.248 M for butyric acid.

Principal reaction	C_6H_5COOH (aq)	$\rightleftharpoons$	H_3O^+	+	$C_6H_5COO^-$
Initial concentration	0.25		0		0
Change	-x		+x		+x
Equilibrium concentration	0.25 - x		+x		+x

5. Substitute the equilibrium concentrations into the equilibrium expression.

$$K_a = 6.5 \times 10^{-5} = \frac{[H_3O^+][C_6H_5COO^-]}{[C_6H_5COOH]} = \frac{(x)(x)}{(0.25 - x)}$$

Given the value of K_a assume that x is negligible compared with the initial concentration of the acid; therefore $0.25 - x \approx 0.25$. Using this value in the denominator solve for x.

$$x^2 \approx (6.5 \times 10^{-5})(0.25)$$

$$x \approx 4.0 \times 10^{-3*}$$

6. The equilibrium concentrations of H_3O^+ and $C_6H_5COO^-$ are $[H_3O^+] = [C_6H_5COO^-] = 0.004$ M. This concentration also represents the amount of C_6H_5COOH that dissociated.

7. The % dissociation can now be calculated:

$$\frac{0.004}{0.25} \times 100 = 1.6\%$$

11⊠ **Polyprotic Acids**
   A. Polyprotic acid - acids that contain more than one dissociable proton.
   B. Dissociate in a stepwise manner.
      1. Each dissociation step has its own K_a.
   C. Stepwise dissociation constants decrease in the order $K_{a1} > K_{a2} > K_{a3}$.
      1. More difficult to remove a positively charged proton from a negative ion.
   D. Diprotic acid solution contains a mixture of acids, H_2A, HA^-, and H_2O.
      1. Strongest acid - H_2A.
         a. principal reaction - dissociation of H_2A
         b. all of H_3O^+ comes from first dissociation step

EXAMPLE:
Write the stepwise dissociation for arsenic acid (H_3AsO_4) and determine the pH of a 2.500×10^{-4} M solution. ($K_{a1} = 5.62 \times 10^{-3}$; $K_{a2} = 1.70 \times 10^{-7}$; $K_{a3} = 3.95 \times 10^{-12}$)

SOLUTION: The stepwise dissociation for arsenic acid is:
H_3AsO_4 (aq) + H_2O (l) $\rightleftharpoons$ H_3O^+ (aq) + $H_2AsO_4^-$ (aq)
$H_2AsO_4^-$ (aq) + H_2O (l) $\rightleftharpoons$ H_3O^+ (aq) + $HAsO_4^{2-}$ (aq)
$HAsO_4^{2-}$ (aq) + H_2O (l) $\rightleftharpoons$ H_3O^+ (aq) + AsO_4^{3-} (aq)

---

* The initial concentration of C_6H_5COOH is known to the second decimal place (0.25). Usually x is considered negligible compared to the value of the initial concentration only if $x < 0.01$. In this case, $x < 0.01$ and therefore, considered negligible. However, solving for x using the quadratic equation gives $x = 4.0 \times 10^{-3}$. Both values for x give an equilibrium concentration of 0.246 M for benzoic acid.

1-3. Since $K_{a1} > K_w$, we know that the principal reaction is the first dissociation step and that all of the H_3O^+ present is produced from this step. Therefore, we only need to consider the first dissociation step when calculating the pH of the solution.

4. Make a table showing the concentrations of the reactant and products.

Principal reaction	$H_3AsO_4\,(aq)$	$\rightleftarrows$	H_3O^+	$+$	$H_2AsO_4^-$
Initial concentration	2.500×10^{-4}		0		0
Change	$-x$		$+x$		$+x$
Equilibrium concentration	$(2.500 \times 10^{-4}) - x$		$+x$		$+x$

5. Assume that x is negligible; therefore, $(2.500 \times 10^{-4}) - x \approx 2.500 \times 10^{-4}$. Substitute the equilibrium concentrations into the equilibrium expression.

$$5.62 \times 10^{-3} = \frac{x^2}{2.500 \times 10^{-4}}; \qquad x = 1.185 \times 10^{-3}$$

From the value of x, we know that our assumption is not valid.[*] Therefore, we must use the quadratic equation to solve for the equilibrium concentration of H_3O^+. Using $(2.500 \times 10^{-4}) - x$ and rearranging the equilibrium expression gives:

$$5.62 \times 10^{-3}\left[\left(2.500 \times 10^{-4}\right) - x\right] = x^2; \qquad x^2 + \left(5.62 \times 10^{-3}\right)x - \left(1.405 \times 10^{-6}\right)$$

$$x = \frac{-\left(5.62 \times 10^{-3}\right) \pm \sqrt{\left(5.62 \times 10^{-3}\right)^2 - 4\left(-1.405 \times 10^{-6}\right)}}{2};$$

$x = -5.5860 \times 10^{-3}$ or $x = 2.398 \times 10^{-4}$; The answer that makes chemical sense is 2.398×10^{-4} M. This value represents the equilibrium concentration of H_3O^+.

6. Calculate the pH of the solution.

pH = -log $(2.398 \times 10^{-4}) = 3.62$

12⊠ **Equilibria in Solutions of Weak Bases**
   A. Weak bases - partially dissociated; can write a base-dissociation equilibrium equation
      1. B (aq) + $H_2O\,(l)$ $\rightleftarrows$ $BH^+\,(aq)$ + $OH^-\,(aq)$
   B. Position of base-dissociation equilibrium - characterized by the base-dissociation constant, K_b.
      1. $K_b = \dfrac{[BH^+][OH^-]}{[B]}$.
   C. Amines - organic compounds that are weak bases.
      1. Derivatives of ammonia.
      2. One or more hydrogen atoms are replaced by another group.
2⊠   3. Lone pair of electrons on nitrogen atom.
         a. can be used for bonding to a proton

_____

[*] A good rule of thumb is that x is negligible only if $K_a < 1.00 \times 10^{-3}$. (Remember, from Chapter 13, that the position of equilibrium lies to the left when $K_a < 1.00 \times 10^{-3}$).

D. To solve for equilibrium concentrations and solution pH, use the same procedure for solving weak acid problems.

EXAMPLE:

Determine the pH of a 0.075 M trimethylamine, $(CH_3)_3N$, solution. $K_b = 6.5 \times 10^{-5}$.

SOLUTION: We will use the same procedure found on page 605 of your text.

1-3. $K_b > K_w$; therefore, the principal reaction is

$$(CH_3)_3N \, (aq) + H_2O \, (l) \rightleftharpoons (CH_3)_3NH^+ \, (aq) + OH^- \, (aq)$$

4. Construct a table with concentrations of the reactant and products.

Principal reaction	$(CH_3)_3N \, (aq)$	$\rightleftharpoons$	$(CH_3)_3N^+$	+	OH^-
Initial concentration	0.075		0		0
Change	-x		+x		+x
Equilibrium concentration	0.075 - x		+x		+x

5. Substitute the equilibrium concentrations into the equilibrium expression.

$$K_a = 6.5 \times 10^{-5} = \frac{\left[(CH_3)_3 NH^+\right]\left[OH^-\right]}{\left[(CH_3)_3 N\right]} = \frac{(x)(x)}{0.075 - x}$$

Assume that x is negligible compared with the initial concentration of the base; therefore, $0.075 - x \approx x$. Using this value in the denominator, we can now solve for x. **Remember:** x represents the equilibrium concentration of the OH⁻ ion.

$$6.5 \times 10^{-5} = \frac{x^2}{0.075}; \qquad x = 0.0022$$

6. The equilibrium concentrations are: $[(CH_3)_3N] = 0.075 - 0.0022 = 0.073$ M; $[OH^-] = [(CH_3)_3NH^+] = 0.0022$ M. Knowing the equilibrium concentration of OH⁻, we can calculate the H_3O^+ concentration from the K_w expression.

$$\left[H_3O^+\right] = \frac{1.0 \times 10^{-14}}{0.0022} = 4.5 \times 10^{-12}$$

$$pH = -\log (4.5 \times 10^{-12}) = 11.35$$

13⊠ **Relation Between K_a and K_b**

A. For a conjugate acid-base pair, can calculate either K_a or K_b from the other.
B. Sum of an acid-dissociation reaction and a base-dissociation reaction is the dissociation of water.
   1. $K_a \times K_b = K_w$.
   2. For $K_a \times K_b$ to remain constant, the strength of a conjugate base must decrease as the strength of the acid increases.

14⊠   Acid-Base Properties of Salts

A. pH of a salt solution is determined by the acid-base properties of the constituent cations and anions.
   1. In an acid-base reaction, the influence of the stronger partner is dominant.
       a. strong acid + strong base → neutral solution
       b. strong acid + weak base → acidic solution
       c. weak acid + strong base → basic solution

B. Salts that yield neutral solutions.
   1. Derived from a strong base and a strong acid.
   2. Neither the cation nor the anion reacts with water to produce H_3O^+ or OH^- ions.

C. Salts that yield acidic solutions.
   1. Derived from a weak base and a strong acid.
   2. Anion is inert.
   3. Cation is a weak acid.
       a. reacts with water (hydrolysis reaction) to produce H_3O^+
       b. small, highly charged cations are also acidic

D. Salts that yield basic solutions.
   1. Derived from a strong base and a weak acid.
   2. Cation is inert.
   3. Anion is a weak base.
       a. reacts with water to produce OH^-

E. Salts that contain acidic cations and basic anions.
   1. Derived from a weak acid and a weak base.
   2. Both the cation and the anion can undergo proton-transfer reactions.
   3. pH depends on the relative acid strength of the cation and the base strength of the anion.
       a. K_a (for the cation) > K_b (for the anion); acidic solution
       b. K_b (for the anion) > K_a (for the cation); basic solution
       c. K_a (for the cation) ≈ K_b (for the anion); neutral solution

EXAMPLE:

Determine whether aqueous solutions of the following salts are acidic, neutral or basic. Write the hydrolysis reaction for those solutions which are either acidic or basic. NH_4Br, K_3PO_4, CsI, $[(CH_3)_3NH]F$.

SOLUTION: To determine the acidity of an aqueous solution of a salt, we must determine if the salt is derived from 1) a strong acid/strong base reaction, 2) weak acid/strong base reaction, 3) strong acid/weak base reaction, or 4) weak acid/weak base reaction.

NH_4Br: derived from the weak base NH_3 (*aq*) and the strong acid HBr (*aq*). Acidic (NH_4^+ is the weak conjugate acid of NH_3).

$$NH_4^+ \,(aq) + H_2O \rightleftarrows NH_3 \,(aq) + H_3O^+ \,(aq)$$

K_3PO_4: derived from the strong base KOH (*aq*) and the weak acid HPO_4^{2-} (*aq*). Basic (PO_4^{3-} is the weak conjugate base of HPO_4^{2-}.)

$$PO_4^{3-} \,(aq) + H_2O \rightleftarrows HPO_4^{2-} \,(aq) + OH^-$$

CsI: derived from the strong base CsOH (*aq*) and the strong acid HI (*aq*). Neutral.

[(CH$_3$)$_3$NH]F: derived from the weak base (CH$_3$)$_3$N and the weak acid HF (*aq*). To determine the acidity of this solution we must calculate the K_a of the cation and the K_b of the anion of the salt and compare the two.

$$K_a = \frac{1.0 \times 10^{-14}}{K_b\left[(CH_3)_3N\right]} = \frac{1.0 \times 10^{-14}}{6.5 \times 10^{-5}} = 1.5 \times 10^{-10}$$

$$K_b = \frac{1.0 \times 10^{-14}}{K_a(HF)} = \frac{1.0 \times 10^{-14}}{3.5 \times 10^{-4}} = 2.9 \times 10^{-11}$$

$K_a > K_b$; therefore, the solution is acidic. The hydrolysis reaction is:

$$(CH_3)_3NH^+ (aq) + H_2O (l) \rightleftarrows (CH_3)_3N (aq) + H_3O^+ (aq)$$

EXAMPLE:
Calculate the pH of a 0.250 M KOCl solution. (K_a for HOCl = 3.5 × 10^{-8})

SOLUTION: We must first write the hydrolysis reaction for this salt. K$^+$ is an inert cation and will not react. However, OCl$^-$ is the conjugate base of a weak acid and will react with water.

$$OCl^- (aq) + H_2O (l) \rightleftarrows HOCl (aq) + OH^- (aq)$$

We can write an equilibrium expression for this reaction:

$$K_b = \frac{[HOCl][OH^-]}{[OCl^-]} ; \qquad K_b = \frac{1.0 \times 10^{-14}}{K_a(HOCl)} = \frac{1.0 \times 10^{-14}}{3.5 \times 10^{-8}} = 2.9 \times 10^{-7}$$

1-3. $K_b > K_w$; therefore, the principle reaction is hydrolysis of OCl$^-$.

4. Construct at table.

Principal reaction	OCl$^-$ (aq)	$\rightleftarrows$	HOCl (aq)	+	OH$^-$
Initial concentration	0.250		0		0
Change	-x		+x		+x
Equilibrium concentration	0.250 - x		+x		+x

5. Assume that x is negligible; therefore, $0.250 - x \approx 0.250$.

$$2.9 \times 10^{-7} = \frac{(x)(x)}{0.250}; \qquad x = 2.7 \times 10^{-4}$$

6. The equilibrium concentration of OH$^-$ is 2.7 × 10^{-4}. We need to determine the H$_3$O$^+$ concentration to calculate pH.

$$[H_3O^+] = \frac{1.0 \times 10^{-14}}{2.7 \times 10^{-4}} = 3.7 \times 10^{-11}$$

7. pH = -log (3.7 × 10^{-11}) = 10.43

15⊠   Factors that Affect Acid Strength

    A. The extent of dissociation of an acid HA is often determined by the strength and polarity of the H-A bond.
       1. The weaker the H-A bond, the stronger the acid.
       2. The more polar the H-A bond, the stronger the acid.
    B. For binary acids.
       1. H-A bond strength decreases down a group in the periodic table.
          a. increase in the size of A down a group produces poorer orbital overlap and a weaker bond
       2. H-A bond strength is the most important factor for determining acid strength for binary acids of elements in the same column.
       3. H-A bond polarity is the most important factor for determining acid strength for binary acids of elements in the same row.
          a. increase in H-A bond polarity as the electronegativity of A increases
          b. in general, electronegativities increase from left to right across the periodic table
    C. Oxoacids, H_nYO_m (Y = a nonmetallic atom, n and m are integers).
       1. Y is always bonded to one or more hydroxyl (OH) groups.
          a. can also be bonded to one or more oxygen atoms
       2. Dissociation of oxoacid involves breaking an O-H bond.
       3. Strength of the acid increases as the O-H bond is weakened or becomes more polar.
          a. For oxoacids that contain the same number of OH groups and the same number of O atoms, acid strength increases with increasing electronegativity of Y.
          b. For oxoacids that contain the same atom Y but different numbers of oxygen atoms, acid strength increases with increasing oxidation number of Y, which increases, in turn, with an increasing number of oxygen atoms.

EXAMPLE:

Order the following by increasing acid strength: a) HF, NH_3, H_2O b) NH_3, AsH_3, PH_3 c) HIO_3, HIO, HIO_2, HIO_4 d) H_2SeO_4, H_2TeO_4, H_2SO_4.

SOLUTION: a) Across a period, acid strength is determined from the electronegativity of the nonmetal. An increase in electronegativity gives rise to an increase in acid strength. Therefore, the order of increasing acidity is $NH_3 < H_2O < HF$. b) Down a group, acid strength is determined from the size of the nonmetal atom. An increase in size gives rise to an increase in acid strength. Therefore, the order of increasing acidity is $NH_3 < PH_3 < AsH_3$. c) For oxoacids that contain the same atom Y but different numbers of oxygen atoms, acid strength increases with an increasing number of oxygen atoms (increasing oxidation number of Y). Therefore, the order of increasing acidity is $HIO < HIO_2 < HIO_3 < HIO_4$. d) For oxoacids that contain the same number of OH groups and the same number of O atoms, acid strength increases with increasing electronegativity of Y. Therefore, the order of increasing acidity is $H_2TeO_4 < H_2SeO_4 < H_2SO_4$.

16⊠   Lewis Acids and Bases

    A. Lewis base - electron-pair donor.
       1. All Lewis bases are Brønsted-Lowry bases.
       2. All Brønsted-Lowry bases are Lewis bases.
    B. Lewis acid - electron-pair acceptor.
       1. More general definition than Brønsted-Lowry.
       2. Include cations and neutral molecules having vacant valence orbitals that can accept a share in a pair of electrons from a Lewis base.
       3. Cationic Lewis acids - metal ions.
       4. Neutral Lewis acids - halides of group 3A elements and oxides of nonmetals.

Self-Test

This section is intended to test your knowledge of the material covered in this chapter. Think through these problems and make certain you understand what is going on. Ask yourself if your answer makes sense. Many of these questions are linked to the chapter learning goals. Therefore, successful completion of these problems indicates you have mastered the learning goals for this chapter. You will receive the greatest benefit from this section if you use it as a mock exam. You will then discover which topics you have mastered and which topics you need to study in more detail.

True/False

1. The Arrhenius theory of a base accounts for the basicity of substances that do not contain OH^- groups.

2. An acid-base reaction is a proton-transfer reaction according to the Bronsted-Lowry theory of acids and bases.

3. A conjugate base has one more proton than its acid.

4. In every acid-base reaction, the proton is transferred to the weaker base.

5. The strength of an acid and the strength of its conjugate base are directly related (the stronger the acid, the stronger its conjugate base).

6. The value for the percent dissociation of an acid depends on the acid.

7. For a diprotic acid, H_2A, the principal reaction is the dissociation of HA^-.

8. The pH of a salt solution prepared by reacting a strong acid with a weak base is greater than 7.

9. The strength of a binary acid decreases down a group in the periodic table.

10. For oxoacids with the same number of OH groups and the same number of O atoms, acid strength decreases with increasing electronegativity of Y.

Multiple Choice

1. For polyprotic acids
   a. the pH is determined from the last dissociation step.
   b. more than one dissociable proton is present.
   c. all of the acidic hydrogens are lost in one step.
   d. the stepwise dissociation constants increase as each H^+ is lost.

2. The pH of a salt solution
   a. = 7 for all salts.
   b. > 7 for salts dervied from a strong acid and a strong base.
   c. < 7 for salts derived from a strong acid and a weak base.
   d. = 7 for salts derived from a weak acid and a weak base.

3. The strength of a binary acid
   a. decreases across a period.
   b. increases as the strength of the H-A bond increases.
   c. increases as the electronegativity of A decreases
   d. increases down a group.

4. For oxoacids, H_nYO_m,
   a. the H is bonded to Y
   b. the strength increases as the H-O bond becomes stronger or more polar.
   c. the strength increases as the H-O bond becomes weaker or less polar.
   d. increases as the H-O bond becomes weaker or more polar.

5. Lewis acids
   a. donate electron pairs.
   b. have a more limited definition than Brønsted acids
   c. include cations and neutral molecules having filled valence orbitals.
   d. include metal ions which can act as cationic Lewis acids.

6. For the reaction HF (aq) + NH_2^- (aq) $\rightleftarrows$ F^- (aq) + NH_3 (aq), the stronger base is
   a. HF
   b. NH_2^-
   c. F^-
   d. NH_3

7. Acid-base indicators
   a. are strong acids.
   b. are substances that change color in a specific pH range.
   c. have the same colors in their acid (HIn) and conjugate base (In⁻) forms.
   d. can be used to determine the exact pH of a solution.

8. The conjugate base
   a. has one less proton than its acid.
   b. transfers a proton to the solvent in an acid-dissociation equilibrium.
   c. is comparable in strength to its conjugate acid.
   d. is always weaker than its conjugate acid.

9. Weak acids
   a. completely dissociate in water.
   b. have large values of K_a.
   c. have dissociation reactions in which the position of equilibrium lies far to the right.
   d. have dissociation reactions in which the position of equilibrium is characterized by K_a.

10. Alkaline earth hydroxides, $M(OH)_2$ (aq),
    a. are stronger bases than the alkali metal hydroxides.
    b. have a higher pH than the alkali metal hydroxides.
    c. are more soluble than the alkali metal hydroxides.
    d. have a lower pH than the alkali metal hydroxides.

Matching

Ion-product for water	a. contains more than one dissociable proton and dissociate in a stepwise manner
Conjugate acid-base pair	b. H^+ acceptor
Base-dissociation equilibrium	c. ion in which H^+ is bonded to the oxygen atom of a solvent water molecule
Polyprotic acids	d. given by $K_w = [H_3O^+][OH^-]$
Hydronium ion	e. chemical species whose formulas differ by only one proton
pH scale	f. a reaction in which the base accepts a proton from the solvent, which acts as an acid
Acid-base indicator	g. logarithmic scale used to express the hydronium ion concentration
Lewis acid	h. a substance that changes color in a specific pH range
Brønsted-Lowry base	i. substance that accepts a pair of electrons

Fill-in-the-Blank

1. The position of equilibrium for the dissociation of water lies _____.

2. The position of an acid-dissociation equilibrium is characterized by the _____

   _____.

3. The sum of an acid-dissociation reaction and a base-dissociation reaction is the _____

   _____.

4. The pH of a salt solution is determined by the _____ of the

   constituent cations and anions.

5. For salts that contain acidic cations and basic anions, the pH depends on the relative _____

   of the cation and the relative _____ of the anion.

6. The extent of dissociation of an acid is often determined by the _____

   of the H-A bond.

7. The H-A bond _____ is the most important factor for determining acid strength for

   binary acids of elements in the same row.

8. For oxoacids that contain the same atom Y but different number of oxygen atoms, acid strength

   increases with _____ of Y.

9. Lewis acids include cations and neutral molecules having _____

    that can _____ from a Lewis base.

10. _____ Lewis acids include the halides of group 3A elements and oxides of nonmetals.

Problems

1. What is the [OH⁻] for a solution of Ca(OH)$_2$ whose pH = 9.87?

2. Calculate the pH of a solution prepared by dissolving 0.583 g of Ba(OH)$_2$ in 125 mL of water.

3. Write chemical equations so that they proceed from left to right for reactions involving:
    HSO$_4^-$, SO$_4^{2-}$, CH$_3$COOH and CH$_3$CO$_2^-$; NH$_4^+$,NH$_3$, HNO$_2$, and NO$_2^-$

4. The percent dissociation for a 1.50×10^{-3} M diethylbarbituric acid (veronal) solution is 0.51%.
    Calculate the pH of this solution and determine the K_a for the acid.

5. Determine the K_a of histidine, a weak organic acid, if the pH of a 2.50×10^{-3} M solution is 5.89.

6. What is the pH of a 0.150 M solution of pyridine, a weak base whose formula is C$_6$H$_5$N? K_b = 1.8 x 10^{-9}

7. Calculate the concentration of all species present and the pH of a 0.025 M H$_2$C$_2$O$_4$ solution.

8. Calculate the pH of a 0.750 M solution of saccharin. ($K_a = 2.10 \times 10^{-12}$)

9. Calculate the pH of a 0.150 M solution of benzylamine. ($K_b = 2.14 \times 10^{-5}$)

10. Determine a) K_b for the conjugate base of formic, ascorbic, and hypochlorous acid and b) the K_a
     for the conjugate acid of hydrazine, aniline, and methylamine.

11. Determine if a solution containing the OCl⁻ and NH$_4^+$ ions is acidic, neutral, or basic.

12. 2.75 g of KCN is dissolved in 250 mL of water. What is the concentration of HCN (*aq*) at
     equilibrium? What is the pH of the solution?

13. Determine which acid is stronger and explain why.
     a. HBr (*aq*) or HI (*aq*)
     b. H$_2$Se (*aq*) or HBr (*aq*)
     c. H$_3$AsO$_4$ or H$_3$PO$_4$
     d. H$_3$PO$_4$ or H$_3$PO$_3$

14. Identify the Lewis acid and the Lewis base in the following reaction.

     Co^{3+} (*aq*) + 6 CN⁻ (*aq*) → Co(CN)$_6^{3-}$ (*aq*)

Solutions

True/False
1. F. One of the limitations of the Arrhenius theory is that it does not account for the basicity of substances that do not contain OH⁻.
2. T
3. F. A conjugate base has one less proton than its acid.
4. F. In every acid-base reaction, the proton is transferred to the stronger base.
5. F. The strength of an acid and the strength of its conjugate base are inversely related (the stronger the acid, the weaker its conjugate base).
6. T
7. F. The principal reaction for a diprotic acid is the dissociation of H_2A.
8. F. A salt solution prepared by reacting a strong acid with a weak base is acidic; therefore, the pH is less than 7.
9. F. The strength of a binary acid increases down a group in the periodic table.
10. F. Acid strength increases with increasing electronegativity of Y.

Multiple Choice
1. b
2. c
3. d
4. d
5. d
6. b
7. b
8. a
9. d
10. d

Matching
    Ion-product for water – d
    Conjugate acid-base pair – e
    Base-dissociation equilibrium – f
    Polyprotic acids – a
    Hydronium ion – c
    pH scale – g
    Acid-base indicator – h
    Lewis acid – i
    Brønsted-Lowry base – b

Fill-in-the Blank
1. far to the left
2. acid dissociation constant, K_a
3. dissociation of water
4. acid-base properties
5. acid strength; base strength
6. strength and polarity
7. polarity
8. increasing oxidation number
9. vacant valence orbitals; accept a share in a pair of electrons
10. Neutral

Problems

1. We first need to solve for the concentration of the H_3O^+ ion.

$$[H_3O^+] = \text{antilog } (-9.87) = 1.3 \times 10^{-10} \text{ M}$$

$$[OH^-] = \frac{1.0 \times 10^{-14}}{1.3 \times 10^{-10}} = 7.7 \times 10^{-5} \text{ M}$$

2. $Ba(OH)_2$ is a strong base and completely dissociates in water according to the reaction:

$$Ba(OH)_2 \ (aq) \rightarrow Ba^{2+} \ (aq) + 2 \ OH^- \ (aq).$$

We know that pH = $\log[H_3O^+]$. To determine $[H_3O^+]$, we need to determine $[OH^-]$. We can determine the $[OH^-]$ by first calculating the number of moles of $Ba(OH)_2$ present. Once we know the moles of $Ba(OH)_2$ present, we can calculate the moles of OH^- and the molarity of OH^-.

$$0.583 \text{ g Ba(OH)}_2 \times \frac{1 \text{ mol Ba(OH)}_2}{171.3 \text{ g Ba(OH)}_2} = 3.40 \times 10^{-3} \text{ mol Ba(OH)}_2$$

$$3.40 \times 10^{-3} \text{ mol Ba(OH)}_2 \times \frac{2 \text{ mol OH}^-}{1 \text{ mol Ba(OH)}_2} = 6.81 \times 10^{-3} \text{ mol OH}^- \ ;$$

$$\frac{6.81 \times 10^{-3} \text{ mol OH}^-}{0.125 \text{ L}} = 5.45 \times 10^{-2} \text{ M}$$

$$\left[H_3O^+ \right] = \frac{1.0 \times 10^{-14}}{5.45 \times 10^{-2}} = 1.83 \times 10^{-13} \ ; \qquad pH = -\log(1.83 \times 10^{-13}) = 12.74$$

3. $HSO_4^- \ (aq) + CH_3CO_2^- \ (aq) \rightleftarrows SO_4^{2-} \ (aq) + CH_3COOH$

$HNO_2 \ (aq) + NH_3 \ (aq) \rightleftarrows NO_2^- \ (aq) + NH_4^+ \ (aq)$

4. $0.51\% = \dfrac{[\text{acid}] \text{ dissociated}}{1.50 \times 10^{-3}} \times 100 \ ; \qquad [\text{acid}] \text{ dissociated} = 7.65 \times 10^{-6}$

The [acid] dissociated represents the $[H_3O^+]$ and $[A^-]$ at equilibrium. Therefore, we can calculate the pH using this concentration.

$$pH = -\log(7.65 \times 10^{-6}) = 5.11$$

The equilibrium concentration of veronal is 1.49×10^{-3} (calculated from the initial and dissociated concentrations). We can now calculate K_a.

$$K_a = \frac{\left(7.65 \times 10^{-6}\right)\left(7.65 \times 10^{-6}\right)}{\left(1.49 \times 10^{-3}\right)} = 3.93 \times 10^{-8}$$

5. Knowing the pH, we can calculate the $[H_3O^+]$ at equilibrium.

$$[H_3O^+] = \text{antilog } (-5.89) = 1.29 \times 10^{-6}.$$

255

This concentration also represents the amount of histidine that dissociated and can be used to calculate the equilibrium concentration of histidine.

$$(2.50 \times 10^{-3}) - (1.29 \times 10^{-6}) = 2.50 \times 10^{-3}$$

Since $[H_3O^+] = [A^-]$ at equilibrium, we can now calculate K_a.

$$K_a = \frac{\left(1.29 \times 10^{-6}\right)\left(1.29 \times 10^{-6}\right)}{\left(2.50 \times 10^{-3}\right)} = 6.65 \times 10^{-10}$$

6. To solve this problem, use the procedure found on page 605 in your text.

steps 1-5. $K_b > K_w$

Principal reaction	C_6H_5N (aq) + H_2O (l) $\rightleftharpoons$	$C_6H_5NH^+$ (aq) +	OH^- (aq)
Initial concentration	0.150	0	0
Change	- x	+ x	+ x
Equilibrium concentration	0.150 - x	+ x	+ x

Substituting these values into the equilibrium expression gives:

$$K_a = 1.8 \times 10^{-9} = \frac{[C_6H_5NH^+][OH^-]}{[C_6H_5N]} = \frac{(x)(x)}{(0.150 - x)}$$

Assume that $0.150 - x \approx 0.150$. The above expression then simplifies to

$$x^2 = (1.8 \times 10^{-9})(0.150); \quad x = 1.64 \times 10^{-5} \ M$$

step 6. $[OH^-] = x = 1.64 \times 10^{-5}$

step 7. $[H_3O^+] = \dfrac{1.0 \times 10^{-14}}{1.64 \times 10^{-5}} = 6.1 \times 10^{-10} \ M$

step 8. pH = -log $(6.1 \times 10^{-10}) = 9.21$

7. To solve this problem, use the procedure found on page 605 in your text.

steps 1-5: $K_{a1} \gg K_w$

Principal reaction	$H_2C_2O_4$ (aq) + H_2O (l) $\rightleftharpoons$	H_3O^+ (aq) +	$HC_2O_4^-$ (aq)
Initial concentration	0.025	0	0
Change	- x	+ x	+ x
Equilibrium concentration	0.025 - x	+ x	+ x

Substituting these values into the equilibrium expression gives:

$$K_{a1} = 5.9 \times 10^{-2} = \frac{[H_3O^+][HC_2O_4^-]}{[H_2C_2O_4]} = \frac{(x)(x)}{(0.025 - x)}$$

Due to the value of K_{a1}, we cannot assume that the value of x is negligible. Therefore, we must use the quadratic equation to solve for x. Rearranging the above expression gives:

$$(1.48 \times 10^{-3}) - (5.9 \times 10^{-2})x = x^2$$

$$x^2 + (5.9 \times 10^{-2})x - (1.48 \times 10^{-3}) = 0$$

$$x = \frac{-(5.9 \times 10^{-2}) \pm \sqrt{(5.9 \times 10^{-2})^2 - 4(-1.48 \times 10^{-3})}}{2}$$

$$x = 1.9 \times 10^{-2} \quad \text{or} \quad x = -7.8 \times 10^{-2}$$

The solution that makes chemical sense is $x = 1.9 \times 10^{-2}$

step 6: $[H_2C_2O_4] = 0.0250 - 0.019 = 0.006$ M; $[H_3O^+] = [HC_2O_4^-] = x = 0.019$

step 7: To calculate the $[C_2O_4^{2-}]$, we use the second dissociation step of $H_2C_2O_4$

$$HC_2O_4^- \ (aq) + H_2O \ (l) \rightleftarrows H_3O^+ \ (aq) + C_2O_4^{2-} \ (aq)$$

$$K_{a2} = 6.4 \times 10^{-5} = \frac{[H_3O^+][C_2O_4^{2-}]}{[HC_2O_4^-]}$$

Because $K_{a1} \gg K_{a2}$, the $[H_3O^+]$ and $[HC_2O_4^-]$ at equilibrium in the second dissociation step are equal to 0.019. (x, the $[C_2O_4^{2-}]$ in this dissociation step, is negligible.)

$$6.5 \times 10^{-5} = \frac{0.019x}{0.019}$$

$$x = 6.5 \times 10^{-5}$$

$$[C_2O_4^{2-}] = x = 6.5 \times 10^{-5} \text{ M}$$

The [OH⁻] is calculated using the concentration of H_3O^+ calculated in the first dissociation step.

$$[OH^-] = \frac{1.0 \times 10^{-14}}{0.019} = 5.26 \times 10^{-13} \text{ M}$$

step 8: pH = -log (0.019) = 1.72

8. Let the formula HA represent the formula for saccharin. To solve this problem use the proocedure outlined in this chapter and found on page 605 in your text.

steps 1-5: $K_a > K_w$

Principal reaction	HA (aq) + H₂O (l) $\rightleftarrows$	H₃O⁺ (aq) +	A⁻ (aq)
Initial concentration	0.750	0	0
Change	- x	+ x	+ x
Equilibrium concentration	0.750 - x	+ x	+ x

Substituting these values into the equilibrium expression gives:

$$K_a = 2.1 \times 10^{-12} = \frac{\left[H_3O^+\right]\left[A^-\right]}{[HA]} = \frac{(x)(x)}{(0.750 - x)}$$

Assume that $0.750 - x \approx 0.750$. The above expression then simplifies to:

$$x^2 = (2.1 \times 10^{-12})(0.750); \quad x = 1.25 \times 10^{-6}$$

step 6: $[H_3O^+] = 1.25 \times 10^{-6}$

step 7: pH = -log(1.25 $\times$ 10^{-6}) = 5.90

9. Let the formula B represent the formula for benzylamine. To solve this problem, use the procedure found on page 605 in your text.

steps 1-5: $K_b > K_w$

Principal reaction	B (*aq*) + H$_2$O (*l*) $\rightleftharpoons$	BH$^+$ (*aq*) +	OH$^-$ (*aq*)
Initial concentration	0.150	0	0
Change	- x	+ x	+ x
Equilibrium concentration	0.150 - x	+ x	+ x

Substituting these values into the equilibrium expression gives:

$$K_b = 2.14 \times 10^{-5} = \frac{\left[BH^+\right]\left[OH^-\right]}{[B]} = \frac{(x)(x)}{(0.150 - x)}$$

Assume that $0.150 - x \approx 0.150$. The above expression then simplifies to:

$$x^2 = (2.14 \times 10^{-5})(0.150) \qquad x = 1.79 \times 10^{-3}$$

step 6: [OH$^-$] = 1.79 $\times$ 10^{-3}

step 7: $\left[H_3O^+\right] = \dfrac{1.0 \times 10^{-14}}{1.79 \times 10^{-3}} = 5.59 \times 10^{-12}$

step 8: pH = -log(5.59 $\times$ 10^{-12}) = 11.3

10. a. Formic acid: $K_b = \dfrac{1.0 \times 10^{-14}}{1.8 \times 10^{-4}} = 5.6 \times 10^{-11}$

Ascorbic acid: $K_b = \dfrac{1.0 \times 10^{-14}}{8.0 \times 10^{-5}} = 1.2 \times 10^{-10}$

Hypochlorous acid: $K_b = \dfrac{1.0 \times 10^{-14}}{3.5 \times 10^{-8}} = 2.8 \times 10^{-7}$

b. Hydrazine: $K_a = \dfrac{1.0 \times 10^{-14}}{8.9 \times 10^{-7}} = 1.1 \times 10^{-8}$

   Aniline: $K_a = \dfrac{1.0 \times 10^{-14}}{4.3 \times 10^{-10}} = 2.3 \times 10^{-5}$

   Methylamine: $K_a = \dfrac{1.0 \times 10^{-14}}{3.7 \times 10^{-4}} = 2.7 \times 10^{-11}$

11. K_b for OCl⁻ was calculated in problem 10 and is 2.8×10^{-7}. K_a for NH_4^+ is

$\dfrac{1.0 \times 10^{-14}}{1.8 \times 10^{-5}} = 5.6 \times 10^{-10}$.

K_b (OCl⁻) $> K_a$ (NH_4^+). Therefore, the solution is basic.

12. KCN is a salt derived from a weak acid (HCN) and a strong base (KOH). Therefore, we can predict that this solution will be basic. The hydrolysis reaction for this salt is:

CN⁻ (aq) + H_2O (l) $\rightleftarrows$ HCN (aq) + OH⁻ (aq)

The K_b for this reaction can be calculated from K_w and K_a for HCN.

$K_b = \dfrac{1.0 \times 10^{-14}}{4.9 \times 10^{-10}} = 2.04 \times 10^{-5}$

To determine the HCN equilibrium concentration and the pH of the solution, we use the procedure found on page 605 in your text. However, we need to first determine the initial concentration of CN⁻.

$2.75 \text{ g KCN} \times \dfrac{1 \text{ mol KCN}}{65.1 \text{ g KCN}} \times \dfrac{1 \text{ mol CN}^-}{1 \text{ mol KCN}} = 0.0422 \text{ mol CN}^-$

$\dfrac{0.0422 \text{ mol CN}^-}{0.250 \text{ L}} = 0.169 \text{ M}$

steps 1-5: $K_b > K_w$

Principal reaction	CN⁻ (aq) + H_2O (l)	$\rightleftarrows$	HCN (aq) +	OH⁻ (aq)
Initial concentration	0.169		0	0
Change	-x		+x	+x
Equilibrium concentration	0.169 - x		+x	+x

Substituting these values into the equilibrium expression gives:

$K_b = 2.04 \times 10^{-5} = \dfrac{[\text{HCN}][\text{OH}^-]}{[\text{CN}^-]} = \dfrac{(x)(x)}{(0.169 - x)}$

Assume that $0.169 - x \approx 0.169$. The above expression then simplifies to:

$x^2 = (2.04 \times 10^{-5})(0.169);$ $x = 1.86 \times 10^{-3}$

259

step 6: $x = [HCN] = [OH^-] = 1.86 \times 10^{-3}$

step 7: $\left[H_3O^+\right] = \dfrac{1.0 \times 10^{-14}}{1.86 \times 10^{-3}} = 5.38 \times 10^{-12}$

step 8: $pH = -\log(5.38 \times 10^{-12}) = 11.27$

13. a. stronger acid: HI – For binary acids, acid strength increases down a group due to the increase in the size of the atom.

   b. Stronger acid: HBr – For binary acids, acid strength increases across a period due to the increase in the electronegativity of the nonmetal atom.

   c. Stronger acid: H_3PO_4 – For oxoacids, acid strength increases with increasing electronegativity on Y.

   d. Stronger acid: H_3PO_4 – For oxoacids, acid strength increases with increasing oxidation number (more oxygen atoms) on Y.

14. Lewis acid: Co^{3+} Lewis base: CN^-

CHAPTER 16

APPLICATIONS OF AQUEOUS EQUILIBRIA

Chapter Learning Goals

1⊠ From the relative strengths of acid and base in a neutralization reaction, predict whether the pH will be equal to, greater than, or less than 7.00 at the equivalence point.

2⊠ Write balanced net ionic equations for the four types of neutralization reactions.

3⊠ Describe the effect on pH when the conjugate base of a weak acid is added to a solution of the weak acid, and the conjugate acid of a weak base is added to a solution of the weak base. Calculate the concentrations of all species present at equilibrium.

4⊠ Given the initial concentration of weak acid (or weak base) and its conjugate base (or weak acid), determine the equilibrium concentrations of all species and the pH of a buffer solution.

5⊠ Calculate the pH of a buffer after the addition of OH^- or H_3O^+.

6⊠ Use the Henderson-Hasselbalch equation to calculate the pH of a buffer.

7⊠ From a table of weak acids and their K_a values, select the weak acid/conjugate base pair that would make the best buffer at a given pH.

8⊠ Calculate pH values for a strong acid-strong base titration.

9⊠ Calculate pH values for a weak acid-strong base titration.

10⊠ Given a titration curve, select which indicator(s) could be used to detect the equivalence point.

11⊠ Calculate pH values for a weak base-strong acid titration.

12⊠ Calculate pH values for a diprotic acid-strong base titration.

13⊠ Write the solubility product expression for a given ionic compound.

14⊠ Given the K_{sp} of an ionic compound, calculate its solubility and vice versa.

15⊠ Given the K_{sp} of an ionic compound, calculate its solubility in the presence of a common ion.

16⊠ Identify ionic compounds that have an enhanced solubility at low pH. Write chemical equations showing why the solubility increases as $[H_3O^+]$ increases.

17⊠ Given K_f, calculate the concentrations of the species present in a complex-ion equilibrium. Given K_{sp} and K_f, calculate the solubility of a slightly soluble ionic compound in an excess of the complexing agent.

18⊠ Write chemical equations showing how a given oxide or hydroxide exhibits amphoteric behavior.

19⊠ From K_{sp} values, determine whether a precipitate will form on mixing solutions of ionic compounds.

20⊠ Determine which metal sulfides will precipitate from a solution of metal ions on addition of H_2S at a specified pH.

Chapter in Brief

In the previous chapter, you applied the concept of equilibrium to acid-base chemistry. In this chapter, you will use those concepts to calculate the pH of mixtures of acids and bases. You will also apply the concepts of equilibria to the dissolution and precipitation of slightly soluble salts as well as the formation and dissociation of complex ions. This chapter begins with an examination of the chemistry that occurs when solutions of acids and bases of varying strength are mixed together. You will learn about the effect of a common ion in a solution of either an acid or base and how this effect can be used to prepare buffer solutions. You will also learn how to determine the pH of a buffer solution after a small amount of acid or base has been added. You will discover how to apply the principles of aqueous solution equilibria to calculate pH titration curves, determine the K_{sp} of a slightly soluble salt, calculate the solubility of a salt, and determine if a salt will precipitate from solution. Finally, you will learn how the principles of aqueous solution equilibria can be used to separate ions by selective precipitation, which is employed in the qualitative analysis of solutions of unknown ions.

1,2⊠ **Neutralization Reactions**
   A. Strong Acid-Strong Base.
      1. Net ionic equation for the neutralization reaction: $H_3O^+ (aq) + OH^- (aq) \rightarrow 2 H_2O (l)$.
      2. When equal numbers of moles of acid and base are mixed together, $[H_3O^+]$ and $[OH^-] = 1 \times 10^{-7}$ M.
      3. Reaction proceeds far to the right.
         a. equilibrium constant, K_n, for the reaction is the reciprocal of the ion-product constant for water
         b. $K_n = \dfrac{1}{[H_3O^+][OH^-]} = K_w$
      4. pH = 7.00.
   B. Weak Acid-Strong Base.
      1. Net ionic equation for the neutralization reaction involves proton transfer from HA to the strong base, OH^-.
         a. $HA (aq) + OH^- (aq) \rightarrow H_2O + A^- (aq)$
      2. $K_n = K_a\left(\dfrac{1}{K_w}\right)$.
         a. K_n equals the product of the equilibrium constants for the reactions added
         b. $HA (aq) + H_2O (l) \rightleftarrows H_3O^+ (aq) + A^- (aq)$ K_a
            $H_3O^+ (aq) + OH^- (aq) \rightleftarrows 2 H_2O (l)$ $1/K_w$
      3. Neutralization of any weak acid by a strong base goes 100% to completion.
         a. OH^- has a great affinity for protons
      4. pH > 7.00.
         a. anion of a weak acid is a weak base
   C. Strong Acid-Weak Base.
      1. Net ionic equation for the neutralization reaction involves proton transfer from the strong acid, H_3O^+ to the weak base, B.
         a. $H_3O^+ (aq) + B (aq) \rightarrow H_2O (l) + BH^+ (aq)$
      2. Equilibrium constant, (K_n), obtained from multiplying the equilibrium constants for the reactions that add to give the net ionic equation.
         a. $K_n = (K_b)\left(\dfrac{1}{K_w}\right)$
         b. $B (aq) + H_2O (l) \rightleftarrows BH^+ (aq) + OH^- (aq)$ K_b
            $H_3O^+ (aq) + OH^- (aq) \rightleftarrows 2 H_2O (l)$ $1/K_w$
      3. Neutralization of any weak base with a strong acid goes 100% to completion.
         a. H_3O^+ is a powerful proton donor
      4. pH < 7.00.
         a. cation of a weak base is a weak acid
   D. Weak Acid-Weak Base.
      1. Neutralization reaction involves proton transfer from the weak acid to the weak base.
         a. $HA (aq) + B (aq) \rightleftarrows BH^+ (aq) + A^- (aq)$
      2. Obtain K_n from K_a for the weak acid, K_b for the weak base, and $1/K_w$.
         a. $HA (aq) + H_2O (l) \rightleftarrows H_3O^+ (aq) + A^- (aq)$ K_a
            $B (aq) + H_2O (l) \rightleftarrows BH^+ (aq) + OH^- (aq)$ K_b
            $H_3O^+ (aq) + OH^- (aq) \rightleftarrows 2 H_2O (l)$ $1/K_w$
         b. $K_n = (K_a)(K_b)\left(\dfrac{1}{K_w}\right)$
      3. Less tendency to proceed to completion than neutralizations involving strong acids or strong bases.

EXAMPLE:
   Write the net ionic equation and predict the pH for the following reactions:

   HF (*aq*) + KOH $HClO_3$ (*aq*) + CH_3NH_2 (a weak base)

SOLUTION:
   The first equation involves the reaction of a weak acid with a strong base. The net ionic equation is:

   HF (*aq*) + OH⁻ (*aq*) ⇌ H_2O (*l*) + F⁻ (*aq*)

   The pH of this solution is expected to be greater than 7 (basic).

   The second equation involves the reaction of a strong acid with a weak base. The net ionic equation is:

   H_3O^+ (*aq*) + CH_3NH_2 ⇌ H_2O (*l*) + $CH_3NH_3^+$ (*aq*)

   The pH of this solution is expected to be less than 7 (acidic).

3⊠   **The Common-Ion Effect**
   A. Common-ion effect - the shift in the position of an equilibrium on addition of a substance that provides an ion in common with one of the ions already involved in the equilibrium.
      1. Example of Le Châtelier's principle.
   B. To determine the pH of a solution prepared from a weak acid and a salt of its conjugate base:
      1. Identify the acid-base properties of the various species in solution.
      2. Consider the possible proton-transfer reaction these species can undergo.
      3. Principal reaction - dissociation of the weak acid.
      4. A⁻ comes from 2 sources.
         a. salt of the conjugate base which provides the initial concentration of A⁻
         b. dissociation of weak acid which determines the change in the concentration of A⁻

EXAMPLE:
   Calculate the concentration of all species present, the pH, and the percent dissociation of nitrous acid in a solution that is 0.100 M in HNO_2 and 0.050 M in $NaNO_2$.

SOLUTION: Because the salt, $NaNO_2$ is 100% dissociated, the species present initially are HNO_2, NO_2^-, Na^+, and H_2O. Na^+ is inert; HNO_2 is a weak acid (K_a =4.5 × 10⁻⁴); NO_2^- is the conjugate base of a weak acid; and H_2O can be either an acid or base. Because HNO_2 is the strongest acid present ($K_a > K_w$), the principal reaction is transfer of a proton from HNO_2 to H_2O. We can set up the following table:

Principal reaction	HNO_2 (*aq*) + H_2O (*l*) ⇌	H_3O^+ (*aq*)	NO_2^- (*aq*)
Initial concentration (M)	0.100	0	0.050
Change (M)	-x	+x	+x
Eq. concentration (M)	0.100 - x	+ x	0.050 + x

The common ion in this problem is NO_2^-. The equilibrium equation for the principal reaction is

$$K_a = 4.5 \times 10^{-4} = \frac{[H_3O^+][NO_2^-]}{[HNO_2]} = \frac{(x)(0.050+x)}{(0.100-x)} \approx \frac{(x)(0.050)}{(0.100)}$$

x is assumed to be negligible because K_a is so small and the equilibrium is shifted to the left due to the common ion effect.

$$x = [H_3O^+] = \frac{(4.5 \times 10^{-4})(0.100)}{(0.050)} = 9.0 \times 10^{-4}$$

Note that the assumption concerning the size of x is justified.

$$pH = -\log(9.0 \times 10^{-4}) = 3.05$$

The percent dissociation of HNO_2 is

$$\text{Percent dissociation} = \frac{[HNO_2]_{dissociated}}{[HNO_2]_{initial}} \times 100 = \frac{9.0 \times 10^{-4}}{0.100} \times 100 = 0.90\%$$

4,5⊠ Buffer Solutions

A. Buffer solutions - resist drastic changes in pH.
  1. Contain a weak acid and its conjugate base or a weak base and its conjugate acid.
B. Add a small amount of base to a buffer solution.
  1. Acid component of the solution neutralized the added base.
C. Add a small amount of acid to a buffer solution.
  1. Base component of the solution neutralizes the added acid.
D. Important in biological systems.
  1. pH of human blood (pH = 7.4) controlled by conjugate acid-base pairs (H_2CO_3/HCO_3^-).
E. For a buffer prepared from a weak acid, HA, and its conjugate base, B.
  1. $[H_3O^+] = K_a \dfrac{[HA]}{[B]}$.
  2. Can use initial concentrations in the calculations.
    a. K_a is small for commonly used buffer solutions
    b. initial concentrations are relatively large
    c. change in concentration, x, is generally negligible
  3. Add acid or base to the buffer solution.
    a. must take account of neutralization before calculating $[H_3O^+]$
      i. alters the numbers of moles of acid and conjugate base
    b. concentration ratio, $[HA]/[B]$, only changes by a small amount
      i. $[H_3O^+]$ only changes by a small amount
F. Addition of acid or base to a buffer solution.
  1. Addition of OH^-.
    a. $HA\ (aq) + OH^-\ (aq)\ H_2O\ (l) + A^-\ (aq)$
  2. Addition of H_3O^+.
    a. $A^-\ (aq) + H_3O^+\ (aq) \rightleftharpoons H_2O\ (l) + HA\ (aq)$
  3. Neutralization alters the number of moles of acid and conjugate base.
G. Buffer Capacity - a measure of the amount of acid or base that a buffer solution can absorb without a significant change in pH.

1. Depends on how much weak acid and conjugate base is present.
2. The more concentrated the solution, the greater the buffer capacity.

EXAMPLE:
    Calculate the pH of a 1.0 L buffer solution containing 0.45 M HCOOH and 0.55 M HCO$_2$Na.
    Determine the pH of this solution after the addition of 0.10 mol HCl (assume no volume change).

SOLUTION: The principal reaction and equilibrium concentrations for this solution are:

Principal reaction	HCOOH (aq) + H$_2$O (l)	$\rightleftharpoons$	H$_3$O$^+$ (aq) +	HCO$_2^-$ (aq)
Initial conc. (M)	0.45		0	0.55
Change (M)	-x		+x	+x
Equilibrium conc. (M)	0.45 -x		+x	0.55 + x

If we solve the equilibrium equation for [H$_3$O$^+$], we obtain:

$$K_a = \frac{\left[H_3O^+\right]\left[HCO_2^-\right]}{[HCOOH]}$$

$$\left[H_3O^+\right] = K_a \frac{[HCOOH]}{\left[HCO_2^-\right]}$$

K_a for formic acid equals 1.8×10^{-4}. Substituting the given information into the equilibrium expression gives:

$$\left[H_3O^+\right] = \left(1.8 \times 10^{-4}\right)\frac{(0.45)}{(0.55)} = 1.5 \times 10^{-4}$$

pH = -log (1.5 x 10^{-4}) = 3.82

To determine the pH of the solution after addition of HCl, we must take into account the neutralization reaction that takes place:

Neutralization reaction	HCO$_2^-$ (aq) +	H$_3$O$^+$ (l)	$\rightleftharpoons$	HCOOH (aq) +	H$_2$O (l)
Before reaction (mol)	0.55	0.10		0.45	
Change (mol)	-0.10	-0.10		+0.10	
After reaction (mol)	0.45	0		0.55	

Assuming that the volume of the solution does not change, the concentrations of the buffer components after neutralization are:

$$\left[HCO_2^-\right] = \frac{0.45 \text{ mol}}{1.0 \text{ L}} = 0.45 \text{ M}$$

$$[HCOOH] = \frac{0.55 \text{ mol}}{1.0 \text{ L}} = 0.55 \text{ M}$$

Substituting these values into the expression for [H$_3$O$^+$], we can then calculate the pH.

$$\left[H_3O^+\right] = \left(1.8 \times 10^{-4}\right)\frac{0.55}{0.45} = 2.2 \times 10^{-4}$$

$$pH = -\log\left(2.2 \times 10^{-4}\right) = 3.66$$

6⊠ **The Henderson-Hasselbalch Equation**

    A. $pH = pK_a + \log\dfrac{[\text{base}]}{[\text{acid}]}$.

    B. Can determine how the pH affects the percent dissociation of a weak acid.

        1. $\log\dfrac{[\text{base}]}{[\text{acid}]} = pH - pK_a$.

7⊠   C. Useful for determining how to prepare a buffer solution.

        1. Select a weak acid whose pK_a is close to the desired pH of the buffer solution.

           a. pK_a of the weak acid should be within ±1 pH units of the desired pH

        2. Adjust the [base]/[acid] ratio.

    D. pH of a buffer solution does not depend on the volume of the solution.

        1. Depends only on pKa and the relative molar amounts of weak acid and conjugate base.

pH Titration Curves

    A. Acid-base titration - a titrant, a solution containing a known concentration of base (or acid), is slowly added from a buret to a solution containing an unknown concentration of acid (or base).

    B. Equivalence point - the point at which stoichiometrically equivalent quantities of acid and base have been mixed together.

    C. pH titration curve - a plot of the pH of the solution vs. the volume of added titrant.

        1. Use to identify the equivalence point in a titration.

        2. Useful in selecting a suitable indicator to signal the equivalence point.

        3. Calculate pH titration curves from the principals of aqueous solution equilibria.

8⊠ **Strong Acid-Strong Base Titrations**

    A. Before addition of any strong base.

        1. initial $[H_3O^+]$ = concentration of strong acid.

    B. Addition of strong base before the equivalence point.

        1. Decrease in $[H_3O^+]$.

           a. added base will neutralize some of the H_3O^+ present

        2. mmol H_3O^+ after neutralization = mmol H_3O^+ initial - mmol OH^- added.

           a. $[H_3O^+]$ after neutralization = $\dfrac{\text{mmol } H_3O^+ \text{ after neutralization}}{\text{total volume of acid and base}}$

    C. Addition of strong base at the equivalence point.

        1. Added just enough base to neutralize all the acid initially present.

        2. pH = 7.00.

           a. solution contains water and a salt derived from a strong base and a strong acid

    D. Addition of strong base after the equivalence point.

        1. Excess of OH^- present.

           a. mmol of excess OH^- = mmol of OH^- added - mmol of acid initially present

        2. $[OH^-]$ after neutralization = $\dfrac{\text{mmol of excess } OH^-}{\text{total volume of acid and base}}$.

        3. $[H_3O^+] = \dfrac{K_w}{[OH^-]}$.

E. pH titration curve.
   1. Plot pH vs. milliliters of strong base added.
   2. Sharp increase in pH in the region near the equivalent point.
      a. characteristic of the titration curve for any strong acid-strong base titration
      b. can use to identify the equivalence point when the concentration of the acid is unknown

9⊠   **Weak Acid-Strong Base Titrations**
    A. Before addition of strong base.
      1. pH calculated in the same manner as for a solution of a weak acid.
    B. Addition of strong base before the equivalence point.
      1. mmol of base, A^-, after neutralization = mL of strong base x [strong base].

      2. $[A^-] = \dfrac{\text{mmol of } A^-}{\text{total mL of acid and base}}$.

      3. mmol of acid, HA, after neutralization = mmol of HA initially - mmol of A^-.

      4. $[HA]$ after neutralization $= \dfrac{\text{mmol HA after neutralization}}{\text{total volume of acid \& base}}$.

      5. $pH = pK_a + \log\dfrac{[A^-]}{[HA]}$.

    C. Addition of strong base at the equivalence point.
      1. Basic salt solution.
      2. mmol A^- after neutralization = mL of strong base x [strong base].

      3. $[A^-] = \dfrac{\text{mmol of } A^- \text{ after neutralization}}{\text{total mL of acid and base}}$.

      4. Calculate pH from the method outlined in example 15.13 in your text.
        a. pH > 7.00
        b. anion of the weak acid is a base
    D. Addition of strong base after the equivalence point.
      1. mmol OH^- added = mL of strong base x [strong base].
      2. mmol A^- present = mmol A^- at the equivalence point.
      3. mmol OH^- present = mmol OH^- added - mmol A^- present .

      4. $[OH^-]$ present $= \dfrac{\text{mmol } OH^- \text{ present}}{\text{total mL of acid and base}}$.

      5. $[H_3O^+] = \dfrac{K_w}{[OH^-]}$.

10⊠   E. Can use pH range at the equivalence point to select a suitable indicator.

EXAMPLE:
    50.0 mL of 0.200 M HF is titrated with 0.100 M NaOH. How many mL of base are required to reach the equivalence point? Calculate the pH at each of the following points: a) after addition of 10.0 mL of base, b) halfway to the equivalence point, c) at the equivalence point, and d) after addition of 130.0 mL of base.

SOLUTION: To determine the number of mL of base needed to reach the equivalence point, we need to first calculate the number of mmol of HF present.

$$\text{mmol HF} = 50.0 \text{ mL} \times \frac{0.200 \text{ mmol HF}}{1 \text{ mL}} = 10.0 \text{ mmol}$$

We need 10.0 mmol of NaOH to reach the equivalence point, which means we need 100 mL of 0.100 M NaOH.

To calculate the pH at the different points in the titration, we need to use the steps which are outlined above.

After addition of 10.0 mL of NaOH:

mmol F^- = 10.0 mL × 0.100 mmol/mL = 1.00 mmol

$[F^-]$ = 1.00 mmol/60.0 mL = 1.67 × 10^{-2} M

mmol HF = 10.0 mmol - 1.00 mmol = 9.0 mmol

[HF] = 9.0 mmol/60.0 mL = 0.15 M

$$pH = pK_a + \log\frac{\left[F^-\right]}{[HF]}; \quad K_a = 3.5 \times 10^{-4}; \quad pK_a = 3.46$$

$$pH = 3.46 + \log\frac{\left(1.67 \times 10^{-2}\right)}{(0.15)} = 2.51$$

Halfway to the equivalence point, we have added 50.0 mL of NaOH (since 100 mL is required to reach the equivalence point).

mmol F^- = 50.0 mL × 0.100 mmol/mL = 5.00 mmol

$[F^-]$ = 5.00 mmol/100 mL = 0.500 M

mmol HF = 10.0 mmol - 5.00 mmol = 5.00 mmol

[HF] = 5.00 mmol/100 mL = 0.0500 mmol

$$pH = 3.46 + \log\frac{(0.0500)}{(0.500)} = 3.46$$

At the equivalence point, all the HF has been neutralized, and the pH is determined by the concentration of F^-.

mmol F^- = 100 mL × 0.100 mmol/mL = 10.0 mmol

$[F^-]$ = 10.0 mmol/150 mL = 6.67 x 10^{-2} M

F^- is the anion of a weak acid and therefore gives a basic solution.

$$K_b = \frac{K_w}{K_a} = \frac{1.0 \times 10^{-14}}{3.5 \times 10^{-4}} = 2.9 \times 10^{-11}$$

Principal reaction: $F^- (aq) + H_2O (aq) \rightleftarrows HF (aq) + OH^- (aq)$

$$K_b = 2.9 \times 10^{-11} = \frac{[HF][OH^-]}{[F^-]} = \frac{(x)(x)}{(6.67 \times 10^{-2})}$$

$x = 1.4 \times 10^{-6} = [OH^-]$

$$\left[H_3O^+\right] = \frac{1.0 \times 10^{-14}}{1.4 \times 10^{-6}} = 7.1 \times 10^{-9}$$

pH = -log (7.1 × 10^{-9}) = 8.15

After addition of 130.0 mL of NaOH

mmol OH⁻ added = 130.0 mL × 0.100 mmol/.mL = 13.0 mmol

mmol F⁻ = 10.0 mmol

mmol OH⁻ present = 13.0 mmol - 10.0 mmol = 3.0 mmol

[OH⁻] = 3.0 mmol/180.0 mL = 1.67×10^{-2} M

$$\left[H_3O^+\right] = \frac{1.0 \times 10^{-14}}{1.67 \times 10^{-2}} = 6.0 \times 10^{-13}$$

pH = -log (6.0×10^{-13}) = 12.2

11⊠   **Weak Base-Strong Acid Titrations**
   A. Before addition of strong acid.
      1. Calculate pH in the same manner as a weak base.
   B. Addition of strong acid before the equivalence point.
      1. Have a B/BH⁺ buffer solution.
         a. leveling of the titration curve in the buffer region between the start of the titration and the equivalence point
      2. Calculate pH from the Henderson Hasselbalch equation.
   C. Addition of strong acid at the equivalence point.
      1. Acidic salt solution of BH⁺.
      2. Calculate pH using the method outlined in section 15.14 of text.
      3. pH < 7.00.
   D. Addition of strong acid after the equivalence point.
      1. Excess H_3O^+ present.
      2. Calculate pH from [H_3O^+] excess.

12⊠   **Polyprotic Acid-Strong Base Titrations**
   A. Diprotic acid - has two dissociable protons and reacts with two molar amounts of OH⁻.
   B. Amino acids - both acidic and basic and can be protonated by strong acids.
      1. H_2A^+ - protonated form; neutralized to HA.
         a. H_2A^+ (aq) + H_2O (l) ⇌ H_3O^+ (aq) + HA (aq) K_{a1}
      2. HA - neutralized to A⁻.
         a. HA (aq) + H_2O (l) ⇌ H_3O^+ (aq) + A⁻ (aq) K_{a2}
   C. Before addition of strong base.
      1. Calculate the pH of a diprotic acid.
      2. Principal reaction - dissociation of H_2A^+.
   D. Addition of strong base before the first equivalence point.
      1. H_2A^+ is converted to HA.
      2. H_2A^+/HA buffer solution.
      3. Calculate [H_2A^+] and [HA] of the amino acid in the same way [HA] and [A⁻] are calculated in a weak acid-strong base titration.
      4. Use the Henderson-Hasselbalch equation to calculate pH.
   E. Addition of strong base at the first equivalence point.
      1. All of the H_2A^+ is converted to HA.
      2. Principal reaction - proton transfer between HA molecules.
         a. 2 HA (aq) ⇌ H_2A^+ (aq) + A⁻ (aq) $K = K_{a2}/K_{a1}$
      3. pH at the first equivalence point = average of pK_{a1} and pK_{a2}
         a. $pH = \dfrac{pK_{a1} + pK_{a2}}{2}$

      b. isoelectric point - the [HA] is at a maximum and the $[H_2A^+]$ and $[A^-]$ are very small and equal
        i. useful for separating mixtures of amino acids
  F. Addition of strong base between the first and second equivalence points.
    1. HA is converted to A^-.
    2. HA/A^- buffer solution.
    3. Calculate [HA] and $[A^-]$ of the amino acid in the same way [HA] and $[A^-]$ are calculated in a weak acid-strong base titration.
    4. Use the Henderson-Hasselbalch equation to calculate pH.
  G. Addition of strong base at the second equivalence point.
    1. All of the HA is converted to A^-.
    2. $[A^-] = [H_2A^+]$ initially.
    3. Solution of a basic salt.
      a. principal reaction - A^- (aq) + H_2O (l) $\rightleftarrows$ HA (aq) + OH^- (aq) $K_b = K_w/K_{a2}$
      b. calculate $[OH^-]$ from equilibrium equation; calculate $[H_3O^+]$ and pH from $[OH^-]$
  H. Addition of strong base beyond the second equivalence point.
    1. Calculate pH from the excess $[OH^-]$.

Solubility Equilibria

  A. Dissolution or precipitation of a sparingly soluble ionic compound.
    1. Principles of solubility equilibria - examines quantitative aspects of solubility and precipitation phenomena.
  B. Solubility product constant = K_{sp}.
    1. M_mX_s (s) $\rightleftarrows$ $m\ M^{n+}$ (aq) + $x\ X^{y-}$ (aq).
    2. $K_{sp} = \left[M^{n+}\right]^m \left[X^{y-}\right]^x$.

Measuring K_{sp} and Calculating Solubility From K_{sp}

  A. K_{sp} - measured by experiment.
    1. Temperature dependent.
    2. Use to calculate the solubility of a compound.
      a. solubility - amount of compound that dissolved per unit volume of saturated solution
    3. Can be calculated from the molar solubility of a compound.

EXAMPLE:
  The solubility of Ag_2CrO_4 is 6.50×10^{-5} M. Calculate K_{sp} for Ag_2CrO_4.

SOLUTION: The solubility equilibrium for Ag_2CrO_4 is

$$Ag_2CrO_4\ (s) \rightleftarrows 2\ Ag^+\ (aq) + CrO_4^{2-}\ (aq)$$

If 6.50×10^{-5} mol is the amount of Ag_2CrO_4 that dissolves in 1.00 L of solution, then the $[Ag^+] = 2(6.50 \times 10^{-5}) = 1.30 \times 10^{-4}$ M and $[CrO_4^{2-}] = 6.50 \times 10^{-5}$ M.

Substituting these values into the equilibrium expression gives:

$$K_{sp} = \left[Ag^+\right]^2\left[CrO_4^{2-}\right] = \left(1.30 \times 10^{-4}\right)^2 \left(6.50 \times 10^{-5}\right) = 1.10 \times 10^{-12}$$

EXAMPLE:
Calculate the K_{sp} for a solution of $Cd(OH)_2$ prepared in pure water given that the $[Cd^{2+}] = 1.10 \times 10^{-5}$ M.

SOLUTION: The solubility equilibrium for $Cd(OH)_2$ is

$$Cd(OH)_2 \ (s) \rightleftarrows Cd^{2+} \ (aq) + 2 \ OH^- \ (aq)$$

and the K_{sp} expression for this reaction is:

$$K_{sp} = \left[Cd^{2+}\right]\left[OH^-\right]^2$$

If $[Cd^{2+}] = 1.10 \times 10^{-5}$ then the $[OH^-]$ is 2 times that or 2.20×10^{-5}. Substituting these values into the equilibrium expression gives:

$$K_{sp} = \left(1.10 \times 10^{-5}\right)\left(2.20 \times 10^{-5}\right)^2 = 5.32 \times 10^{-15}$$

EXAMPLE:
The K_{sp} of $Fe(OH)_3$ is 2.6×10^{-39}. Calculate the molar solubility of $Fe(OH)_3$.

SOLUTION: The solubility equilibrium for $Fe(OH)_3$ is

$$Fe(OH)_3 \ (s) \rightleftarrows Fe^{3+} \ (aq) + 3 \ OH^- \ (aq)$$

Let x be the number of mol/L of $Fe(OH)_3$ that dissolves. The saturated solution then contains x mol/L of Fe^{3+} and $3x$ mol/L of OH^-. Substituting this into the equilibrium expression gives:

$$K_{sp} = 2.6 \times 10^{-39} = \left[Fe^{3+}\right]\left[OH^-\right]^3 = (x)(3x)^3$$

Solving for x gives

$$2.6 \times 10^{-39} = 27x^4; \quad x = 9.9 \times 10^{-11} \ M$$

Factors that Affect Solubility

15⊠   A. Common-ion effect.
      1. Solubility of a slightly soluble ionic compound is decreased by the presence of a common ion in the solution.

16⊠   B. The pH of the solution.
      1. Solubility of an ionic compound increases with decreasing pH of the solution if the compound contains a basic anion.

17⊠   C. Formation of complex ions.
      1. Solubility of an ionic compound increases dramatically if the solution contains a substance that can bond to the metal cation.
      2. Formation constant, K_f - equilibrium constant for formation of the complex ion from the hydrated metal cation.
        a. measures the stability of a complex ion

18⊠   D. Amphoterism.
      1. Amphoteric oxides - soluble both in strongly acidic and in strongly basic solutions.

EXAMPLE:

Determine the molar solubility of $Cu(C_2O_4)$ ($K_{sp} = 2.87 \times 10^{-8}$ at $25°$ C) in a 0.75 M $CuCl_2$ solution.

SOLUTION: The solubility equilibrium expression is:

$$Cu(C_2O_4) \rightleftarrows Cu^{2+} (aq) + C_2O_4^{2-} (aq)$$

The equilibrium expression for this reaction is:

$$K_{sp} = [Cu^{2+}][C_2O_4^{2-}]$$

Let x be the number of mol/L of $Cu(C_2O_4)$ that dissolves. We can now construct a table showing the equilibrium concentrations:

Solubility equilibrium	$Cu(C_2O_4)_2$ (s) $\rightleftarrows$	Cu^{2+} (aq) +	$C_2O_4^{2-}$ (aq)
Initial concentration		0.75	0
Equilibrium concentration (M)		0.75 + x	+ x

Given the value of K_{sp}, we can assume that x is negligible. Therefore, the equilibrium concentration of Cu^{2+} is approximately 0.75 M. Substituting these values into the equilibrium expression gives:

$$2.87 \times 10^{-8} = (0.75)(x); \quad x = 3.8 \times 10^{-8}$$

EXAMPLE:

The initial pH of a solution containing $Sn(OH)_2$ was 9.35 after addition of excess NH_3. Determine the molar solubility of $Sn(OH)_2$ in this solution. (For $Sn(OH)_2$, $Ksp = 5.4 \times 10^{-27}$.)

SOLUTION: This problem also is a common ion problem. We can determine the [OH⁻] of the solution from the pH. We can then determine the molar solubility of $Sn(OH)_2$ in the same manner used in the above example. The solubility expression for this compound is:

$$Sn(OH)_2 \rightleftarrows Sn^{2+} (aq) + 2 OH^- (aq)$$

The equilibrium expression is:

$$K_{sp} = [Sn^{2+}][OH^-]^2$$

From the pH of the solution, we can calculate the [OH⁻] from the K_w and [H_3O^+].

$$[H_3O^+] = \text{antilog} (-9.35) = 4.47 \times 10^{-10}$$

$$[OH^-] = \frac{1 \times 10^{-14}}{4.47 \times 10^{-10}} = 2.24 \times 10^{-5}$$

Let x be the number of mol/L of $Sn(OH)_2$ that dissolves. We can now construct a table showing the equilibrium concentrations.

Solubility equilibrium	$Sn(OH)_2$ (s)	$\rightleftharpoons$	Sn^{2+} (aq)	+	OH^- (aq)
Initial concentration			0		2.24×10^{-5}
Equilibrium concentration (M)			$+ x$		$2.24 \times 10^{-5} + x$

Given the very small value of K_{sp}, we can assume that x is negligible. Substituting the equilibrium concentrations into the K_{sp} expression gives:

$$5.4 \times 10^{-27} = (x)(2.24 \times 10^{-5}); \quad x = 2.41 \times 10^{-22}$$

EXAMPLE:
Write a balanced, net ionic equation for the dissolution reaction between AgBr and $Na_2S_2O_3$ and calculate the equilibrium constant give K_{sp} (AgBr) = 5.4×10^{-13} and K_f {$Ag(S_2O_3)_2^{3-}$} = 2.0×10^{13}.

SOLUTION:
The net ionic equation for the dissolution of AgBr in $Na_2S_2O_3$ is obtained by combining the solubility expression for AgBr and the formation expression for $Ag(S_2O_3)_2^{3-}$.

$$AgBr \ (s) \rightleftharpoons Ag^+ \ (aq) + Br^- \ (aq) \qquad K_{sp} = 5.4 \times 10^{-13}$$

$$Ag^+ \ (aq) + 2 \ S_2O_3^{2-} \rightleftharpoons Ag(S_2O_3)_2^{3-} \qquad K_f = 2.0 \times 10^{13}$$

Adding the two equations together and canceling out what is common on both the reactant and product sides gives:

$$\cancel{AgBr \ (s)} \rightleftharpoons \cancel{Ag^+ \ (aq)} + Br^- \ (aq) \qquad K_{sp} = 5.4 \times 10^{-13}$$

$$\cancel{Ag^+ \ (aq)} + 2 \ S_2O_3^{2-} \ (aq) \rightleftharpoons Ag(S_2O_3)_2^{3-} \ (aq) \qquad K_f = 2.0 \times 10^{13}$$

$$\overline{AgBr \ (s) + 2 \ S_2O_3^{2-} \ (aq) \rightleftharpoons Ag(S_2O_3)_2^{3-} \ (aq) + Br^- \ (aq) \qquad K = K_{sp} \times K_f = 10.8}$$

19⊠ **Precipitation of Ionic Compounds**
   A. Ion product (IP) - determines if a precipitate will form when solutions containing the constituent ions are mixed.
       1. For the salt, $M_m X_s$; IP = $[M^{n+}]^m [X^{y-}]^s$.
       2. Defined in the same way as K_{sp}.
       3. Concentrations are initial concentrations, not equilibrium concentrations.
   B. IP > K_{sp}.
       1. Supersaturated solution.
       2. Precipitation occurs.
   C. IP = K_{sp}.
       1. Saturated solution.
       2. Equilibrium exists.
   D. IP < K_{sp}.
       1. Unsaturated solution.
       2. Precipitation will not occur.

EXAMPLE:
Will a precipitate form when 150 mL of 0.75 M $Zn(NO_3)_2$ is mixed with 250 mL of 1.50 M Na_2CO_3?

SOLUTION: To determine if a precipitate will form we must ask ourselves two questions: 1) What precipitate could form by mixing these two solutions and 2) Is the IP > K_{sp} for this precipitate? We can answer the first question by writing a metathesis reaction for the reaction between $Zn(NO_3)_2$ and Na_2CO_3.

$$Zn(NO_3)_2 \ (aq) \ + \ Na_2CO_3 \ (aq) \ \rightarrow \ 2 \ NaNO_3 \ (aq) \ + \ ZnCO_3 \ (s)$$

(We know that $ZnCO_3$ is insoluble based upon the solubility rules in Chapter 4.) If the concentrations of Zn^{2+} and CO_3^{2-} are high enough, the $ZnCO_3$ should precipitate out of solution. The solubility expression for $ZnCO_3$ is:

$$ZnCO_3 \ (s) \ \rightleftarrows \ Zn^{2+} \ (aq) \ + \ CO_3^{2-} \ (aq) \qquad K_{sp} = 1.2 \times 10^{-10}$$

The ion product expression is:

$$IP = [Zn^{2+}][CO_3^{2-}]$$

We can calculate the IP from the concentrations of Zn^{2+} and CO_3^{2-}. However, don't forget you mixed two solutions to give a total volume of 400 mL. Therefore, the concentrations of Zn^{2+} and CO_3^{2-} must be calculated based upon the dilution that occurs upon mixing the two solutions. We will use the equation $M_iV_i = M_fV_f$.

$$\left[Zn^{2+} \right] = \frac{(0.75 \ M)(150 \ mL)}{400 \ mL} = 0.28 \ M$$

$$\left[CO_3^{2-} \right] = \frac{(1.50 \ M)(250 \ mL)}{400 \ mL} = 0.938 \ M$$

IP = (0.28)(0.938) = 0.26;   IP > K_{sp}; Therefore, a precipitate will form.

Separation of Ions by Selective Precipitation

A. To separate a mixture of ions in solution, add a reagent that will precipitate some of the ions but not others.

20▷ B. Can separate insoluble metal sulfides from the more soluble metal sulfides

1. carry separation out in acidic solution

2. $MS \ (s) \ + \ 2 \ H_3O^+ \ (aq) \ \rightleftharpoons \ M^{2+} \ (aq) \ + \ H_2S \ (aq) \ + \ 2 \ H_2O \ (l)$

3. K_{spa}, solubility product in acid

   a. $K_{spa} = \dfrac{[M^{2+}][H_2S]}{[H_3O^+]^2}$

4. $Q_c > K_{spa}$; metal sulfide precipitates out.

5. Adjust $[H_3O^+]$ so that $Q_c > K_{spa}$ for the more insoluble metal sulfide.

   a. The more insoluble metal sulfide will precipitate out from the more soluble metal sulfide.

274

Qualitative Analysis
    A. Procedure for identifying the ions present in an unknown solution
    B. Traditional scheme of analysis for metal cations involves the separation of 20 cations into five groups by selective precipitation. (see Fig. 16.16, page 674 in your text)
        1. determine the presence or absence of the ions in each group with further separations and tests.
    C. Excellent vehicle for developing laboratory skills and learning about acid-base, solubility, and complex-ion equilibria.

Self-Test

This section is intended to test your knowledge of the material covered in this chapter. Think through these problems and make certain you understand what is going on. Ask yourself if your answer makes sense. Many of these questions are linked to the chapter learning goals. Therefore, successful completion of these problems indicates you have mastered the learning goals for this chapter. You will receive the greatest benefit from this section if you use it as a mock exam. You will then discover which topics you have mastered and which topics you need to study in more detail.

True/False

1. When HNO_3 reacts with an equimolar amount of NaOH, the pH of the solution will be equal to 7.00.

2. When HClO reacts with an equimolar amount of NaOH, the pH of the solution will be less than 7.00.

3. When the conjugate base of a weak acid is added to a solution of the weak acid, the equilibrium shifts to the right.

4. The best acid/base conjugate pair to use for preparing a buffer with a pH = 3.35 is HNO_2/NO_2^-.

5. The best indicator to use in the titration of a weak acid with $K_a \approx 1 \times 10^{-6}$ is methyl red.

6. The solubility of FeS increases with increasing pH.

7. The solubility of $PbCl_2$ decreases in a solution that contains both $PbCl_2$ and NaCl.

8. $PbSO_4$ will precipitate from a solution containing 1.5×10^{-3} M $Pb(NO_3)_2$ and 0.05 M Na_2SO_4.

9. The solubility of an ionic compound increases dramatically if the solution contains a substance that can bond to the metal cation.

Fill-in-the-Blank

1. In the reaction between a strong acid and weak base, the net ionic equation for the neutralization

   reaction involves _____ from the _____ to the

   _____ .

2. Neutralization reactions between a _____ and _____ have less of a

   tendency to proceed to completion than neutralization reactions involving

   _____ or _____ .

3. The shift in the position of an equilibrium on addition of a substance that provides an ion in common with one of the ions already involved in the equilibrium is referred to as the _____ _____.

4. A buffer solution contains a _____ or a _____ _____ and is able to resist drastic changes in _____.

5. A measure of the amount of acid or base that a buffer solution can absorb without a significant change in pH is referred to as the _____.

6. In preparing a buffer solution with a specific pH, you should select a weak acid whose _____ value is close to the desired pH of the buffer solution.

7. The pH range at the _____ of a titration curve can be used to select a suitable indicator for the titration.

8. The amount of compound that dissolves per unit volume of a saturated solution is the

   _____.

9. The _____ measures the stability of a complex ion.

10. The _____ determines if a precipitate will form when solutions containing the constituent ions are mixed.

Problems

1. Write balanced, net ionic equations and predict the pH for reactions between:
   a. HNO_3 and KOH
   b. HNO_3 and NH_2NH_2
   c. benzoic acid and KOH
   d. benzoic acid and pyridine

2. Calculate the pH of a solution containing 0.15 mol of N_2H_4 and 0.10 mol of N_2H_5Cl in 1.0 L.

3. Determine the pH and concentration of all species present in a buffer solution containing 0.50 mol HClO and 0.35 mol NaClO in 1.0 L of solution. Determine the change in pH upon addition of 0.10 mol NaOH. Determine the change in pH upon addition of 0.10 mol HCl.

4. Determine the ratio of lactic acid to lactate ion required for preparing a buffer solution whose pH is 4.25.

5. Determine the change in pH that will occur upon addition of 0.150 mol HCl (assuming no volume change) to a 1 L buffer solution prepared from 0.25 M HCooh and 0.10 M HCO_2Na. How does the buffer capacity of this solution compare to the buffer capacity of the solution in the example on page 266?

6. Calculate the pH of 50.0 mL of 0.0500 M HCl after addition of the following volumes of 0.1000M NaOH: a) 0.0 mL; b) 10.0 mL; c) 25.0 mL; d) 30.0 mL.

7. Calculate the pH of 50.0 mL of 0.0500 M NaCN after addition of the following volumes of 0.1000 M HCl: a) 0.0 mL; b) 10.0 mL; c) 25.0 mL; d) 30.0 mL. (**Warning!** This titration should never be attempted in the laboratory!)

8. The molar solubility of $Mg(OH)_2$ is 1.12×10^{-4} M. Calculate the value of K_{sp}.

9. Calculate the molar solubility for $Cu_3(PO_4)_2$ given that the $K_{sp} = 1.4 \times 10^{-37}$.

10. Calculate the solubility of a solution of $BaSO_4$ which contains 0.50 M Na_2SO_4. ($K_{sp} = 1.1 \times 10^{-10}$)

11. Which of the following compounds are more soluble in acidic solution than in pure water?
    a. CuBr   b. Ag_2S c. AgCN   d. MgF_2

12. What are the concentrations of Cu^{2+} and $Cu(NH_3)_4^{2+}$ in a solution prepared by adding 0.25 mol of $Cu(NO_3)_2$ to 1.0 L of 5.0 M NH_3? ($K_f = 1.1 \times 10^{13}$)

13. Will a precipitate form when 250 mL of 1.75 M $CaCl_2$ is mixed with 250 mL of 2.50 M Na_3PO_4?

14. Determine if it is possible to separate Cu^{2+} from Fe^{2+} by bubbling H_2S through a 0.30 M HCl solution that contains 0.005 M Cu^{2+} and 0.005 M Fe^{2+}. (K_{spa} for CuS = 6×10^{-16} and K_{spa} for FeS = 6×10^2.)

15. A student is given an unknown solution that may contain metal cations found in the qualitative analysis scheme on page 674 in your text. Upon addition of HCl to this solution, a precipitate (ppt A) forms. After separating out the precipitate, the remaining solution (soln a) was treated with H_2S. No visible change occurred. When this same solution (soln a + H_2S) was treated with NH_3, a precipitate (ppt B) formed. After separating out ppt B, $(NH_4)_2CO_3$ was added to the remaining solution (soln b). A white precipitate formed. Flame tests performed on soln b gave a fleeting violet color.

    a. What groups of ions are present? From the information given, what specific ions can be identified?

    b. Ppt A was treated with NH_3. No visible change occurred. The precipitate was separated from the solution (soln c). After treating the precipitate with hot water, the precipitate dissolved. The solution containing the dissolved precipitate was treated with K_2CrO_4 and a yellow precipitate formed. Soln c was treated with excess HNO_3 and a precipitate formed. What ions are present?

Solutions

True/False
1. T
2. F. When a weak acid (HClO) reacts with a strong base (NaOH) the pH of the solution is greater than 7.00.
3. F. Adding the conjugate base of a weak acid to a solution of a weak acid increases the concentration of a product in an equilibrium reaction. Therefore, the reaction shifts to the left.
4. T. ($pK_a = 3.35$ for HNO_2)
5. F. The best indicator to use is phenolphthalein.(see Fig. 16.6 on page 649 in your text)
6. F. The solubility of FeS increases with decreasing pH since FeS contains the basic anion S^{2-}. Increasing the amount of H_3O^+ in a solution removes the S^{2-} from solution, shifting the solubility equilibrium to the right. This is illustrated by the following equations:
    FeS (s) $\rightleftarrows$ Fe^{2+} (aq) + S^{2-} (aq); S^{2-} (aq) + H_3O^+ (aq) $\rightleftarrows$ HS^- (aq) + OH^- (aq)
7. T
8. T. (I.P. > K_{sp})
9. T

Fill-in-the-Blank
1. proton transfer; strong acid; weak base
2. weak acid; weak base; strong acids; strong bases
3. common ion effect
4. weak acid and its conjugate base; weak base and its conjugate acid; pH
5. buffer capacity
6. pK_a
7. equivalence point
8. solubility
9. K_f
10. ion product (I.P.)

Problems
1. a. HNO_3 and KOH: This is a reaction between a strong acid and a strong base. The net ionic equation is: H_3O^+ (aq) + OH^- (aq) → 2 H_2O (l); pH = 7

   b. HNO_3 and NH_2NH_2: NH_2NH_2 is a weak base. The net ionic equation is:

   $$H_3O^+ (aq) + NH_2NH_2 \rightarrow H_2O (l) + NH_2NH_3^+ (aq); \quad pH < 7.00$$

   c. benzoic acid (C_6H_5COOH) and KOH: Benzoic acid is a weak acid. The net ionic equation is:

   $$C_6H_5COOH + OH^- (aq) \rightarrow C_6H_5CO_2^- (aq) + H_2O (l); \quad pH > 7.00$$

   d. benzoic acid and pyridine (C_5H_5N): Benzoic acid is a weak acid and pyridine is a weak base. The net ionic equation is:

   $$C_6H_5COOH (aq) + C_5H_5N (aq) \rightleftarrows C_6H_5CO_2^- (aq) + C_5H_5NH^+ (aq);$$

   pH is determined by comparing the K_a of benzoic acid to the K_b of pyridine. $K_a > K_b$; therefore, the solution is acidic.

2. Because the salt N_2H_5Cl is 100% dissociated, the species present initially are N_2H_4, $N_2H_5^+$, Cl^-, and H_2O. N_2H_4 is the strongest base present ($K_b \gg K_w$). This gives the following table:

Principal reaction	N_2H_4 (aq) + H_2O (l) $\rightleftarrows$	$N_2H_5^+$ (aq) +	OH^- (aq)
Initial concentration (M)	0.15	0.10	0
Change (M)	- x	+ x	+ x
Eq. concentration (M)	0.15 – x	0.10 + x	+ x

The common ion in this problem is $N_2H_5^+$. The equilibrium equation for the principal reaction is:

$$K_b = 8.9 \times 10^{-7} = \frac{[N_2H_5^+][OH^-]}{[N_2H_4]} = \frac{(0.10+x)(x)}{(0.15-x)} \approx \frac{(0.10)(x)}{(0.15)}$$

x is assumed to be negligible because K_b is so small and the equilibrium is shifted to the left due to the common ion effect.

$$x = [OH^-] = \frac{(8.9 \times 10^{-7})(0.15)}{(0.10)} = 1.3 \times 10^{-6}$$

Note that the assumption concerning the size of x is justified. Since we now know the [OH⁻], we can calculate the [H₃O⁺].

$$\left[H_3O^+\right] = \frac{1.0 \times 10^{-14}}{1.3 \times 10^{-6}} = 7.7 \times 10^{-9}$$

pH = -log (7.7×10^{-9}) = 8.11

3. The principal reaction and equilibrium concentrations for this solution are:

Principal reaction	HClO (aq) + H₂O (l)	⇌	H₃O⁺ (aq) +	ClO⁻ (aq)
Initial concentration (M)	0.50		0	0.35
Change (M)	- x		+ x	+ x
Equilibrium concentration (M)	0.50 - x		+ x	0.35 + x

If we solve the equilibrium equation for H₃O⁺, we obtain

$$K_a = \frac{\left[H_3O^+\right]\left[ClO^-\right]}{[HClO]}$$

$$\left[H_3O^+\right] = K_a \frac{[HClO]}{\left[ClO^-\right]}$$

K_a for HClO is 3.5×10^{-8}. Substituting the given information into the equilibrium expression gives:

$$\left[H_3O^+\right] = \left(3.5 \times 10^{-8}\right)\frac{(0.50)}{(0.35)} = 5.0 \times 10^{-8}$$

pH = -log (5.0×10^{-8}) = 7.30

To determine the pH of the solution after addition of NaOH, we must take into account the neutralization reaction that takes place:

Principal reaction	HClO (aq) + OH⁻ (aq)	⇌	H₂O (l) +	ClO⁻ (aq)
Before reaction (mol)	0.50	0.10		0.35
Change (mol)	- 0.10	-0.10		+ 0.10
After reaction (mol)	0.40	0		0.45

Substituting these values into the expression for [H₃O⁺], we can then calculate the pH.

$$\left[H_3O^+\right] = \left(3.5 \times 10^{-8}\right)\frac{(0.40)}{(0.45)} = 3.1 \times 10^{-8}$$

pH = -log(3.1×10^{-8}) = 7.51

To determine the pH of the solution after addition of NaOH, we must take into account the neutralization reaction that takes place:

Principal reaction	ClO^- (aq) +	H_3O^+ (aq)	$\rightleftharpoons$	H_2O (l)	+	$HClO$ (aq)
Before reaction (mol)	0.35	0.10				0.50
Change (mol)	-0.10	-0.10				+0.10
After reaction (mol)	0.25	0				0.60

Substituting these values into the expression for $[H_3O^+]$, we can then calculate the pH.

$$\left[H_3O^+\right] = \left(3.5 \times 10^{-8}\right)\frac{(0.60)}{(0.25)} = 8.4 \times 10^{-8}$$

$$pH = -\log(8.4 \times 10^{-8}) = 7.08$$

4. We can solve this problem by rearranging the Henderson-Hasselbalch equation:

$$\log\frac{[base]}{[acid]} = pH - pK_a$$

The pK_a for lactic acid is $-\log(1.4 \times 10^{-4}) = 3.85$

Substituting the pH and pK_a values gives:

$$\log\frac{[lactate]}{[lactic\ acid]} = 4.25 - 3.85 = 0.40$$

Taking the antilog of both sides gives:

$$\frac{[lactate]}{[lactic\ acid]} = 2.51$$

Notice that this gives the ratio of the concentration of the lactate ion to lactic acid. To determine the ratio of the concentration of lactic acid to the lactate ion, we simply take the reciprocal of 2.51.

$$\frac{[lactic\ acid]}{[lactate]} = 0.398$$

5. Before we can determine the change in pH that occurs, we need to calculate the pH of the buffer solution. Using the Henderson-Hasselbalch equation we get:

$$pH = pK_a + \log\frac{\left[HCO_2^-\right]}{[HCOOH]} \qquad pH = 3.74 + \log\frac{0.10}{0.25} = 3.34$$

To determine the pH of the solution after addition of HCl, we must take into account the neutralization reaction that takes place:

Neutralization reaction	HCO_2^- (aq) + H_3O^+ (aq) $\rightarrow$		$HCOOH$ (aq) + H_2O (l)
Before reaction (mol)	0.10	0.150	0.25
Change (mol)	-0.10	-0.10	-0.10
After reaction (mol)	0	0.050	0.35

Note that the limiting reactant in this problem is the HCO_2^- ion. This buffer is swamped out by the HCl present, and the pH is calculated from the $[H_3O^+]$ after the neutralization reaction.

pH = -log(0.050) = 1.30

The buffer capacity of this solution is much less than the buffer capacity of the 0.45 M HCOOH/0.55 M HCO_2^- which experienced only a 0.16 change in pH.

6. a. 0.0 mL of NaOH: $[H_3O^+]$ = 0.0500 M; pH = 1.30

    b. 10.0 mL of NaOH: mmol H_3O^+ = mmol H_3O^+ initial - mmol OH⁻ added
    mmol H_3O^+ = (50.0 mL)(0.0500 mmol/mL) - (10.0 mL)(0.1000 mmol/mL) = 1.50 mmol

    $$\left[H_3O^+\right]\text{after neutralization} = \frac{(1.50\text{ mmol})}{(60.0\text{ mL})} = 0.0250 ; \quad pH = -\log(0.0250) = 1.60$$

    c. 25.0 mL of NaOH: At this point, the number of mmol of H_3O^+ = the number of mmol of OH⁻, which is the equivalence point. Since this is a strong acid/strong base titration, the pH = 7.00.

    d. 30.0 mL of NaOH: mmol of excess OH⁻ = mmol of OH⁻ added - mmol of HCl initial
    mmol of excess OH⁻ = (30.0 mL)(0.1000 mmol/mL) - (50.0 mL)(0.0500 mmol/mL) = 0.5000 mmol

    $$[OH^-]\text{ after neutralization} = \frac{0.50\text{ mmol}}{80\text{ mL}} = 6.25 \times 10^{-3}$$

    $$\left[H_3O^+\right] = \frac{1.0 \times 10^{-14}}{6.25 \times 10^{-3}} = 1.6 \times 10^{-12}; \quad pH = -\log(1.6 \times 10^{-12}) = 11.80$$

7. a. 0.0 mL HCl: pH is calculated in the same manner as a weak base.

Principal reaction	CN^- (aq) + H_2O (l) $\rightleftharpoons$	HCN (aq) +	OH^- (aq)
Initial concentration (M)	0.0500	0	0
Change (M)	-x	+x	+x
Equilibrium concentration (M)	0.0500 - x	+ x	+ x

$$K_b = \frac{K_w}{K_a} = \frac{1.0 \times 10^{-14}}{4.9 \times 10^{-10}} = 2.04 \times 10^{-5} = \frac{[HCN][OH^-]}{[CN^-]} = \frac{(x)(x)}{(0.0500 - x)} \approx \frac{(x)^2}{(0.0500)}$$

$$x = [OH^-] = 1.01 \times 10^{-3}\text{ M}; \quad \left[H_3O^+\right] = \frac{1.0 \times 10^{-14}}{1.01 \times 10^{-3}} = 9.9 \times 10^{-12};$$

pH = -log (9.9 x 10⁻¹²) = 11.00

b. 10.0 mL HCl: mmol HCN after neutralization = mmol of HCl added
    mmol HCN after neutralization = (10.0 mL)(0.1000 mmol/ mL) = 1.00 mmol HCN

$$[HCN] = \frac{1.00 \text{ mmol}}{60.0 \text{ mL}} = 1.67 \times 10^{-2}$$

mmol CN^- after neutralization = mmol CN^- initial - mmol HCN
mmol CN^- after neutralization = (50.0 mL)(0.0500 mmol/mL) - 1.00 mmol = 1.50 mmol CN^-

$$[CN^-] = \frac{1.50 \text{ mmol}}{60.0 \text{ mL}} = 2.50 \times 10^{-2}$$

We can calculate the pH from the Henderson-Hasselbalch equation, using the pK_a for HCN.

$$pH = pK_a + \log\frac{[CN^-]}{[HCN]}; \qquad pH = 9.31 + \log\frac{(2.50 \times 10^{-2})}{(1.67 \times 10^{-2})} = 9.49$$

c. 25.0 mL HCl: mmol HCN after neutralization = (25.0 mL)(0.1000 mmol/mL) = 2.50 mmol

$$[HCN] = \frac{2.50 \text{ mmol}}{75.0 \text{ mL}} = 3.33 \times 10^{-2}$$

mmol CN^- after neutralization = mmol CN^- initial - mmol HCN
mmol CN^- after neutralization = 2.50 - 2.50 = 0

The pH is determined from the equilibrium expression for HCN.

$$K_a = 4.9 \times 10^{-10} = \frac{(x)^2}{3.33 \times 10^{-2}}; \quad x = [H_3O^+] = 4.04 \times 10^{-6}; pH = -\log(4.04 \times 10^{-6}) = 5.39$$

d. 30.0 mL HCl: mmol H_3O^+ added = (30.0 mL)(0.1000 mmol/mL) = 3.00 mmol
    mmol HCN present = 2.50 mmol
    mmol H_3O^+ present = mmol H_3O^+ added - mmol CN^- present = 3.00 mmol - 2.50 mmol = 0.50 mmol

$$[H_3O^+] = \frac{0.50 \text{ mmol}}{80.0 \text{ mL}} = 6.25 \times 10^{-3}; \qquad pH = -\log(6.25 \times 10^{-3}) = 2.20$$

8. The solubility equilibrium for $Mg(OH)_2$ is:

$$Mg(OH)_2 (s) \rightleftarrows Mg^{2+} (aq) + 2 OH^- (aq)$$

If 1.12×10^{-4} mol is the amount of $Mg(OH)_2$ that dissolves in 1.0 L of solution then the $[Mg^{2+}] = 1.12 \times 10^{-4}$ M and the $[OH^-] = 2(1.12 \times 10^{-4}) = 2.24 \times 10^{-4}$ M. Substituting these values into the equilibrium expression gives:

$$K_{sp} = [Mg^{2+}][OH^-]^2 = (1.12 \times 10^{-4})(2.24 \times 10^{-4})^2 = 5.62 \times 10^{-12}$$

9.

Solubility equilibrium	$Cu_3(PO_4)_2$ (s)	$\rightleftharpoons$	3 Cu^{2+} (aq)	+	2 PO_4^{3-} (aq)
Equilibrium concentration (M)			+3 x		+2 x

$$K_{sp} = 1.4 \times 10^{-37} = \left[Cu^{2+}\right]^3\left[PO_4^{3-}\right]^2 = (3x)^3(2x)^2 = 108x^5$$
$$x = 1.7 \times 10^{-8}$$

10.

Solubility equilibrium	$BaSO_4$ (s)	$\rightleftharpoons$	Ba^{2+} (aq)	+	SO_4^{2-} (aq)
Equilibrium concentration (M)			+ x		0.50 + x

$$K_{sp} = 1.1 \times 10^{-10} = \left[Ba^{2+}\right]\left[SO_4^{2-}\right] = (x)(0.50 + x) \approx (x)(0.50)$$
$$x = 2.2 \times 10^{-10}$$

11. a. CuBr: Not more soluble. Br^- is the conjugate base of a very strong acid and is very unreactive.
   b. Ag_2S: More soluble. S^{2-} is the conjugate base of a weak acid.
   c. AgCN: More soluble. CN^- is the conjugate base of a very weak acid.
   d. MgF_2: More soluble. F^- is the conjugate base of a weak acid.

12. Because K_f for $Cu(NH_3)_4^{2+}$ is large (1.1×10^{13}), nearly all the Cu^{2+} from $Cu(NO_3)_2$ will be converted to $Cu(NH_3)_4^{2+}$.

$$Cu^{2+} (aq) + 4 NH_3 (aq) \rightleftharpoons Cu(NH_3)_4^{2+}$$

We will use the procedure found in example 16.11 (page 668 in your text) to calculate $[Cu^{2+}]$ and $[Cu(NH_3)_4^{2+}]$. Conversion of 0.25 mol/L of Cu^{2+} to $Cu(NH_3)_4^{2+}$ consumes 1.00 mol/L of NH_3 (due to a 1:4 mole to mole ratio of Cu^{2+} to NH_3). Assuming 100% conversion to $Cu(NH_3)_4^{2+}$ gives the following concentrations:

$[Cu^{2+}] = 0$ M

$[Cu(NH_3)_4^{2+}] = 0.25$ M

$[NH_3] = 5.0 - 1.00 = 4.0$ M

After conversion of Cu^{2+} to $Cu(NH_3)_4^{2+}$, assume that a small amount of back reaction occurs, producing Cu^{2+}.

$$Cu(NH_3)_4^{2+} \rightleftharpoons Cu^{2+} (aq) + 4 NH_3 (aq)$$

Dissociation of x mol/L of $Cu(NH_3)_4^{2+}$ produces x mol/L of Cu^{2+} and $4x$ mol/L of NH_3. The equilibrium concentrations are:

$[Cu(NH_3)_4^{2+}] = 0.25 - x$

$[Cu^{2+}] = x$

$[NH_3] = 4.0 + 4x$

The table below summarizes this reasoning under the balanced equation:

	Cu^{2+} (aq) +	4 NH_3 (aq) $\rightleftharpoons$	$Cu(NH_3)_4^{2+}$ (aq)
Initial concentration (M)	0.25	5.0	0
After 100% reaction (M)	0	4.0	0.25
Equilibrium concentration (M)	x	$4.0 + 4x$	$0.25 - x$

Substituting the equilibrium concentrations into the expression for K_f and making the approximation that x is negligible compared with 0.25 gives:

$$K_f = 1.1 \times 10^{13} = \frac{[Cu(NH_3)_4^{2+}]}{[Cu^{2+}][NH_3]^4} = \frac{(0.25 - x)}{(x)(4.0 + x)^4} \approx \frac{(0.25)}{(x)(4.0)^4}$$

$$[Cu^{2+}] = x = \frac{(0.25)}{(1.1 \times 10^{13})(4.0)^4} = 8.9 \times 10^{-17} \text{ M}$$

$$[Cu(NH_3)_4] = 0.25 - (8.9 \times 10^{-17} \text{ M} = 0.25 \text{ M}$$

13. The equation for this reaction is:

$$3 \text{ } CaCl_2 \text{ } (aq) + 2 \text{ } Na_3PO_4 \text{ } (aq) \rightarrow Ca_3(PO_4)_2 \text{ } (s) + 6 \text{ } NaCl$$

To determine if a precipitate will form, we must calculate the ion product and compare its value to K_{sp}. (Remember, the concentrations we use for calculating ion product must be determined using $M_iV_i = M_fV_f$. Upon mixing, both solutions are diluted by ½.) The ion product expression is:

$$\text{I.P.} = [Ca^{2+}]^3[PO_4^{3-}]^2$$

$$[Ca^{2+}] = 0.875 \text{ M} \qquad [PO_4^{3-}] = 1.25 \text{ M}$$

$$\text{I.P.} = (0.875)^3(1.25)^2 = 1.05$$

K_{sp} for $Ca_3(PO_4)_2 = 2.1 \times 10^{-33}$; I.P. $>> K_{sp}$; therefore, a precipitate will form.

14. K_{spa} (CuS) = 6 x 10^{-16}; K_{spa} (FeS) = 6 x 10^2

$$Q_c = \frac{[Cu^{2+}][H_2S]^2}{[H_3O^+]^2} = \frac{(0.005)(0.10)^2}{(0.30)^2} = 5.6 \times 10^{-4}$$

Since K_{spa} (CuS) $< Q_c < K_{spa}$ (FeS); Cu^{2+} will selectively precipitate out of solution.

15. a. The presence of ppt. A indicates that group I ions are present. Ppt B contains group III cations since a precipitate formed upon addition of H_2S and NH_3. The white precipitate formed after addition of $(NH_4)_2CO_3$ to soln. b indicates that cations from group IV are present. The fleeting violet color in the flame test of soln. b indicates that K^+ is present.

b. Since ppt. A dissolved after addition of hot water, it appears that Pb^{2+} is present. This is confirmed by addition of K_2CrO_4. The resulting yellow precipitate is $PbCrO_4$. However, we cannot rule out the possibility of the presence of Ag^+. Upon addition of NH_3, no visible change occurred. However, it is possible that AgCl dissolved. We can test for the presence of Ag^+ ion by the addition of HNO_3. HNO_3 will react with NH_3 according to the following neutralization equilibrium:

$$H_3O^+ (aq) + NH_3 (aq) \rightleftarrows NH_4^+ (aq) + H_2O (l) \qquad\qquad K_n = 1.8 \times 10^9$$

Because K_n is greater than K_f for $Ag(NH_3)_2^+$, the neutralization reaction will predominate, removing NH_3 from the formation equilibrium. This causes Ag^+ to be available for the solubility equilibrium and AgCl to precipitate out.

CHAPTER 17

ENTROPY, FREE ENERGY, AND EQUILIBRIUM

Chapter Learning Goals

1⊠ Qualitatively determine whether simple chemical or physical changes are spontaneous.
2⊠ Qualitatively predict whether the sign of ΔS is positive or negative for a chemical or physical change.
3⊠ On the basis of probability, determine which of two states has the higher entropy.
4⊠ Calculate the standard entropy of reaction from the standard molar entropies of products and reactants.
5⊠ Determine whether a reaction is spontaneous by determining the sign of ΔS_{total}.
6⊠ Use the equation $\Delta G = \Delta H - T\Delta S$ to calculate the free energy of reaction and to determine the temperature at which a nonspontaneous reaction becomes spontaneous.
7⊠ Calculate the standard free energy of reaction from standard free energies of formation.
8⊠ Calculate the free energy of reaction for a system having nonstandard pressures and concentrations.
9⊠ From the standard free energy of reaction, calculate the value of the equilibrium constant.

Chapter in Brief

In Chapter 8, you were introduced to the concepts of entropy and free energy and how the values of these thermodynamic functions can determine the spontaneity of a chemical reaction. In this chapter, you will explore these concepts in greater detail. Your study begins with a look at spontaneous processes as well as a brief review of enthalpy, entropy and their relationship to spontaneous processes. You will also examine the relationship among entropy, probability, and temperature, and learn how to calculate the standard entropies of reaction. You will examine the second law of thermodynamics and how the thermodynamic properties of enthalpy, entropy, and free energy are interwoven to determine the spontaneity of a reaction. You will learn how to calculate free energy changes for reactions which take place at both standard and nonstandard state conditions. Your study of thermodynamic ends by examining the relationship between free energy and chemical equilibrium.

Spontaneous Processes

1⊠
    A. Spontaneous process - one that proceeds on its own without any external influence.
    B. Spontaneous reaction always moves a system toward equilibrium.
       1. Spontaneity of forward or reverse reaction depends on:
         a. temperature
         b. pressure
         c. composition of the reaction mixture
       2. $Q < K$; reaction proceeds in the forward direction.
       3. $Q > K$; reaction proceeds in the reverse direction.
    C. Spontaneity of a reaction is no indication of the speed of the reaction.

1⊠
    EXAMPLE:
    Determine which of the following processes are spontaneous and which are nonspontaneous.
    a. The odor of a dead skunk fills your car as you drive by.
    b. The handle of a metal spoon in a pot of hot soup is hot.
    c. The reaction of N_2 and H_2 to produce NH_3 at 300 K and partial pressures of 1.25 atm, 3.00 atm, and 2.50 atm, respectively. $K_p = 4.4 \times 10^5$ at 300 K.

SOLUTION: Both a and b are spontaneous processes. To determine if c is spontaneous or not, we must determine Q for the reaction. If $Q < K$, then the reaction will proceed spontaneously in the forward direction. If $Q < K$, then the reaction will proceed spontaneously in the reverse direction. The equilibrium expression for this reaction is:

$$N_2 (g) + 3 H_2 (g) \rightleftarrows 2 NH_3 (g)$$

$$Q = \frac{[NH_3]^2}{[N_2][H_2]^3} = \frac{(2.50)^2}{(1.25)(3.00)^3} = 0.185$$

$Q < K$; The reaction is spontaneous in the forward direction.

Enthalpy, Entropy, and Spontaneous Processes: A Brief Review

A. Spontaneous reactions.
   1. Some are accompanied by conversion of potential energy to heat.
      a. reactions are exothermic
   2. Some spontaneous reactions are endothermic.
      a. system moves spontaneously to a state of higher potential energy by absorbing heat from the surroundings
   3. Enthalpy alone does not account for the direction of spontaneous change.
   4. Second thermodynamic driving force for spontaneous change.
      a. nature's tendency to move to a condition of maximum randomness or disorder
B. Molecular systems tend to move spontaneously to a state of maximum randomness or disorder.
C. Entropy, S - molecular randomness or disorder.
   1. State function.
   2. $\Delta S = S_{final} - S_{initial}$.
   3. ΔS is positive, increase in randomness or disorder of a system.
      a. solid $\rightarrow$ liquid
      b. liquid $\rightarrow$ gas
      c. a reaction results in an increase in the number of gaseous molecules
      d. dissolution of certain solutes
   4. ΔS is negative, decrease in randomness or disorder of a system.

EXAMPLE:
Predict the sign of ΔS in the system for
a. $I_2 (s) \rightarrow I_2 (g)$
b. $H_2O (g) \rightarrow H_2O (l)$
c. $PbS (s) + 2 HNO_3 (aq) \rightarrow H_2S (g) + Pb(NO_3)_2 (aq)$

SOLUTION:
   a. ΔS is positive; There is an increase in randomness when the number of moles of gas is increased.
   b. ΔS is negative; There is a decrease in randomness when gaseous molecules are condensed into a liquid.
   c. ΔS is positive; There is an increase in randomness when the number of moles of gas is increased.

Entropy and Probability

    A. Systems tend to a state of maximum randomness because the random state, which is more probable, can be achieved in more ways.

    B. Probabilities of ordered and random states are proportional to the number of ways that the state can be achieved.

3⊠   C. Boltzmann - the entropy of a particular state is related to the number of ways that the state can be achieved.

        1. $S = k \ln W$.

           a. $\ln W =$ the number of ways that the state can be achieved

           b. $k = 1.38 \times 10^{-23}$ J/K

    D. Gas expands spontaneously because the state of greater volume is more probable.

        1. For an ideal gas.

           a. $\Delta S = R \ln \dfrac{V_{final}}{V_{initial}}$

           b. $P = nRT/V$

           c. $\Delta S = R \ln \dfrac{P_{initial}}{P_{final}}$

              i. the entropy of a gas increases when its pressure decreases at constant temperature

              ii. the entropy of a gas decreases when the pressure increases at constant temperature

Entropy and Temperature

    A. Entropy - associated with molecular motion.

    B. Random molecular motion increases as the temperature of a substance increases.

        1. Increase in the average kinetic energy of the molecules.

           a. total energy is distributed among the individual molecules in a number of ways

           b. Boltzmann - the more ways (W) that the energy can be distributed the greater the randomness of the state and the higher the entropy

        2. At absolute zero, every substance is a solid where particles are rigidly fixed in a crystalline structure.

    C. Third law of thermodynamics - The entropy of a perfectly ordered crystalline substance at 0 K is zero.

4⊠   **Standard Molar Entropies and Standard Entropies of Reaction**

    A. Standard molar entropy, S^o - the entropy of 1 mol of the pure substance at 1 atm pressure and a specified temperature, usually 25^oC.

        1. Absolute entropies - measured with respect to an absolute reference point.

    B. Standard entropy of reaction, ΔS^o.

        1. $\Delta S^o = S^o$(products) - S^o(reactants).

           a. multiply the ΔS^o value for each substance by the stoichiometric coefficient of that substance in the balanced equation.

        2. Increase in entropy whenever a molecule breaks into two or more pieces.

EXAMPLE:

Calculate the ΔS^o_{rxn} for

$$BaCO_3 \, (s) + H_2SO_4 \, (l) \rightarrow BaSO_4 \, (s) + H_2O \, (l) + CO_2 \, (g)$$

SOLUTION: The ΔS^o_{rxn} can be calculated using information found in appendix B in your text.

    ΔS^o ($BaSO_4$) = 132 J/mol·K ΔS^o (H_2O) = 69.9 J/mol·K

    ΔS^o (CO_2) = 213.6 J/mol·K ΔS^o ($BaCO_3$) = 112 J/mol·K

    ΔS^o (H_2SO_4) = 156.9 J/mol·K

$$\Delta S^{\circ}_{rxn} = [(132 \text{ J/mol·K}) + (69.9 \text{ J/mol·K}) + (213.6 \text{ J/mol·K})] - [(112 \text{ J/mol·K}) + (156.9 \text{ J/mol·K})]$$
$$= 146.6 \text{ J/mol·K}$$

Entropy and the Second Law of Thermodynamics

A. Value of ΔG - criterion for spontaneity.
   1. $\Delta G = \Delta H - T\Delta S$.
   2. $\Delta G > 0$; nonspontaneous reaction.
   3. $\Delta G < 0$; spontaneous reaction.
   4. $\Delta G = 0$; reaction is at equilibrium.
B. First law of thermodynamics - In any process, spontaneous or nonspontaneous, the total energy of a system and its surroundings is constant.
   1. Helps keep track of energy flow between system and the surroundings.
   2. Does not indicate the spontaneity of the process.
C. Second law of thermodynamics - In any spontaneous process, the total entropy of a system and its surroundings always increases.
   1. Provides a clear-cut criterion of spontaneity.
   2. Direction of spontaneous change is always determined by the sign of the total entropy change.
   3. $\Delta S_{total} = \Delta S_{system} + \Delta S_{surroundings}$.
      a. $\Delta S_{total} > 0$; spontaneous reaction
      b. $\Delta S_{total} < 0$; nonspontaneous reaction
      c. $\Delta S_{total} = 0$; reaction at equilibrium
D. All reactions proceed spontaneously in the direction that increases the entropy of the system plus surroundings.
E. To determine ΔS_{total}, need to know ΔS_{system} and $\Delta S_{surroundings}$.
   1. ΔS_{system} - the entropy of reaction.
      a. calculate from standard molar entropies
   2. at constant pressure, $\Delta S_{surr} = \dfrac{-\Delta H_{rxn}}{T}$.
   3. $\Delta S_{total} = \Delta S^{\circ}_{rxn} - \dfrac{\Delta H_{rxn}}{T}$.

EXAMPLE:

Calculate ΔS_{total} for the following reaction at 298 K and determine if the reaction is spontaneous under standard state conditions.

$$AgNO_3 \ (aq) + NaCl \ (aq) \rightarrow AgCl \ (s) + NaNO_3 \ (aq)$$

SOLUTION:

$$\Delta S_{total} = \Delta S_{system} - \frac{\Delta H_{rxn}}{T};$$

You may find it helpful to write the net ionic equation for this reaction:

$$Ag^+ \ (aq) + Cl^- \ (aq) \rightarrow AgCl \ (s)$$

$$\Delta S_{system} = [S^{\circ}(AgCl)] - [S^{\circ}(Ag^+) + S^{\circ}(Cl^-)]$$

$$\Delta S_{system} = [96.2] - [72.7 + 56.5] = -33.0 \text{ J/K}$$
$$\Delta H^{\circ}_{rxn} = [\Delta H^{\circ}_f(AgCl)] - [\Delta H^{\circ}_f(Ag^+) + \Delta H^{\circ}_f(Cl^-)]$$

$$\Delta H^{\circ}_{rxn} = [-127.1] - [105.6 + (-167.2)] = -65.5$$

When substituting the above values into the equation for ΔS_{total}, **watch your units!**

$$\Delta S_{total} = -33.0 \text{ J / K} - \frac{-6.55 \times 10^4 \text{ J}}{298 \text{ K}} = 187 \text{ J / K}.$$

6⊠   **Free Energy**
   A. Restate the second law in terms of the thermodynamic properties of the system.
      1. Chemists are more interested in the system.
   B. Free energy: $G = H - TS$.
      1. TS = part of the system's energy that is already disordered.
      2. $H - TS$ = part of the system's energy that is still ordered and therefore free to cause spontaneous change by becoming disordered.
      3. State function.
         a. $\Delta G = \Delta H - T\Delta S$
   C. Relationship between free energy and spontaneity.
      1. $\Delta S_{total} = \Delta S - \dfrac{\Delta H}{T}$; $-T\Delta S_{total} = \Delta H - T\Delta S$.
      2. $-T\Delta S_{total} = \Delta G$.
         a. ΔG and ΔS_{total} have opposite signs
      3. In any spontaneous process at constant temperature and pressure, the free energy of the system decreases.
      4. Temperature acts as a weighting factor that determines the relative importance of the enthalpy and entropy changes (see Table 17.2, page 704 in text).
         a. can estimate the temperature at which ΔG changes from positive to negative with
            $$\Delta G = \Delta H - T\Delta S$$
      5. To estimate the temperature at which ΔG° changes from a positive to a negative value, set $\Delta G^{\circ} = \Delta H^{\circ} - T\Delta S^{\circ} = 0$.
         a. $T = \dfrac{\Delta H^{\circ}}{\Delta S^{\circ}}$

Standard Free-Energy Changes for Reactions
   A. Standard free-energy change, ΔG° - the change in free energy that occurs when reactants in their standard states are converted to products in their standard states.
      1. ΔG° - an extensive property and refers to the number of moles indicated in the chemical equation.
      2. Calculate ΔG° from the standard enthalpy change, ΔH°, and the standard entropy change, ΔS°.
         a. $\Delta G^{\circ} = \Delta H^{\circ} - T\Delta S^{\circ}$

EXAMPLE:
Determine ΔG° for the reaction:

$$2 \text{ NaCl } (s) + \text{H}_2\text{SO}_4 \, (l) \rightarrow \text{Na}_2\text{SO}_4 \, (s) + 2 \text{ HCl } (g)$$

Given the following information:

	ΔH° (kJ/mol)	S° (J/mol·K)
NaCl (s)	-411.2	72.1
H$_2$SO$_4$ (l)	-814.0	156.9
Na$_2$SO$_4$ (s)	-1387.1	149.6
HCl (g)	-92.3	186.8

SOLUTION:
$\Delta H^\circ_{rxn} = [(-1387.1) + (2 \times -92.3)] - [(2 \times -411.2) + (-814.0)] = 64.7$ kJ

$\Delta S^\circ_{rxn} = [(149.6) + (2 \times 186.8)] - [(2 \times 72.1) + (156.9)] = 222.1$ J/K

When calculating ΔG°, **watch your units!** (It is a common mistake for students to mix the kJ of ΔH° with the J of ΔS°.) Since we are using standard enthalpies and entropies, $T = 298$ K.

$\Delta G^\circ = 64.7$ kJ $- (298)(0.2221$ kJ/K$) = -1.49$ kJ (The reaction is spontaneous.)

Standard Free Energies of Formation
A. Standard free energy of formation, ΔG°_f, the free-energy change for formation of 1 mol of the substance in its standard state from the most stable form of the constituent elements in their standard states.

   1. For an element in its most stable form at 25°C: $\Delta G^0_f = 0$.

B. ΔG^0_f measures a substance's thermodynamic stability with respect to its constituent elements.

   1. ΔG^0_f is negative - substance is stable and does not decompose to its constituent elements.

   2. ΔG^0_f is positive - substance is thermodynamically unstable.

      a. may not decompose to its constituent elements if the rate of decomposition is slow

6⊠   C. Can use ΔG^0_f to calculate standard free-energy changes for reactions.

   1. $\Delta G^\circ = \Delta G^\circ_f (\text{products}) - \Delta G^\circ_f (\text{reactants})$

EXAMPLE:
Calculate ΔG°_{rxn} for the following reaction:

C_2H_4 (g) + HBr (g) → C_2H_5Br (g)

Given the following information.

ΔG°_f (kJ / mol)

C_2H_4 (g)	68.4
HBr (g)	-53.4
C_2H_5Br	-61.9

SOLUTION:
$\Delta G^\circ_{rxn} = [-61.9] - [(68.4) + (-53.4)] = -76.9$ kJ

7⊠   **Free-Energy Changes and Composition of the Reaction Mixture**
A. To calculate the free-energy change for a reaction when the reactants and products are present at non-standard-state pressures and concentrations.

   1. $\Delta G = \Delta G^0 + RT \ln Q$.

      a. $Q =$ the reaction quotient - an expression having the same form as the equilibrium constant expression (concentrations are not necessarily at equilibrium)

EXAMPLE:
Calculate ΔG for the reaction in the above example if $P(C_2H_4) = 1.57$ atm, $P(HBr) = 2.01$ atm, $P(C_2H_5Br) = 0.83$ atm.

SOLUTION:
Substituting the value for ΔG°_{rxn} obtained in the above example, we have

$$\Delta G = -76,900 \frac{J}{mol} + 8.314 \frac{J}{mol \cdot K}(298 \text{ K}) \ln \frac{(0.83)}{(1.57)(2.01)} = -80,209 \frac{J}{mol} = -80.2 \frac{kJ}{mol}$$

8⊠   **Free Energy and Chemical Equilibrium**
   A. The total free energy of a reaction mixture changes as the reaction progresses toward equilibrium.
   B. $\Delta G = \Delta G^\circ + RT \ln Q$
      1. When the reaction mixture is mostly reactants:
         a. $Q \ll 1$
         b. $RT \ln Q \ll 0$
         c. $\Delta G < 0$
         d. total free energy decreases as the reaction proceeds spontaneously in the forward direction
      2. When the reaction mixture is mostly products:
         a. $Q \gg 1$
         b. $RT \ln Q > 0$
         c. $\Delta G > 0$
         d. total free energy decreases as the reaction proceeds spontaneously in the reverse direction
   C. Free energy curve - shows how the total free energy of a reaction mixture changes as the reaction progresses. (see Fig. 17.10, page 713 in text)
      1. Goes through a minimum somewhere between pure reactants and pure products.
         a. the system is at equilibrium
   D. Relationship between free energy and the equilibrium constant.
      1. $\Delta G^\circ = -RT \ln K$.
   E. Relationship between ΔG° and K.
      1. $\Delta G^\circ < 0$, $\ln K > 0$ - equilibrium mixture is mainly products.
      2. $\Delta G^\circ > 0$, $\ln K < 0$ - equilibrium mixture is mainly reactants.
      3. $\Delta G^\circ = 0$, $\ln K = 0$ - equilibrium mixture contains comparable amounts of reactions and products.
   F. Properties of nature that determine the direction and extent of a chemical reaction.
      1. ΔG°_{rxn}
         a. ΔH°_f and ΔS°_{rxn}

EXAMPLE:
Calculate K for the reaction:

$$C_2H_4 (g) + HBr (g) \rightarrow C_2H_5Br (g)$$

SOLUTION:
Using the ΔG°_{rxn} calculated earlier and rearranging the above equation, we have

$$\ln K = \frac{-76,900 \frac{J}{mol}}{-\left[8.314 \frac{J}{mol \cdot K}(298 \text{ K})\right]} = 31.0$$

Taking the antilog of both sides gives $K = 3. \times 10^{13}$.

Self Test

This section is intended to test your knowledge of the material covered in this chapter. Think through these problems and make certain you understand what is going on. Ask yourself if your answer makes sense. Many of these questions are linked to the chapter learning goals. Therefore, successful completion of these problems indicates you have mastered the learning goals for this chapter. You will receive the greatest benefit from this section if you use it as a mock exam. You will then discover which topics you have mastered and which topics you need to study in more detail.

True/False

1. The spontaneity of a reaction also gives an indication of the speed of a reaction.

2. All spontaneous reactions are exothermic reactions.

3. The spontaneity of a reaction is determined by the enthalpy of the reaction.

4. Molecular systems tend to move spontaneously to a state of maximum randomness or disorder.

5. The entropy of a particular state is related to the number of ways that the state can be achieved.

6. Whenever a molecule breaks into two or more pieces, there is a decrease in entropy.

7. The first law of thermodynamics helps keep track of the energy flow between the system and surroundings and also indicates the spontaneity of the process.

8. The direction of spontaneous change is always determined by the sign of the total entropy change.

9. In determining the spontaneity of a reaction, the temperature is the weighting factor that determines the relative importance of the enthalpy and entropy changes.

10. When a reaction mixture is mostly reactants, the total free energy increases as the reaction proceeds spontaneously in the forward direction.

Matching

Spontaneous process

    a. in any process, spontaneous or nonspontaneous, the total energy of a system and its surroundings is constant.

Entropy

    b. the free-energy change for formation of 1 mol of the substance in its standard state from the most stable form of the constituent elements in their standard states.

Third law of thermodynamics

    c. the entropy of 1 mol of the pure substance at 1 atm pressure and a specified temperature, usually 25°C.

Standard molar entropy

d. a process that proceeds on its own without any external influence

First law of thermodynamics

e. in any spontaneous process, the total entropy of a system and its surroundings always increases.

Second law of thermodynamics

f. molecular randomness or disorder

Standard free energy of formation

g. the entropy of a perfectly ordered crystalline substance at 0 K is zero.

Fill-in-the-Blank

1. A spontaneous reaction always moves a system toward _____.

2. Molecular systems tend to move spontaneously to a state of maximum _____.

3. The probabilities of ordered and random states are proportional to _____.

4. Random molecular motion increases as the _____ of a substance increases.

5. The entropy of a gas _____ when its pressure decreases at constant temperature.

6. The direction of spontaneous change is always determined by the _____ of the total entropy change.

7. All reactions proceed spontaneously in the direction that increases the _____ _____.

8. In any spontaneous process at constant temperature and pressure, the free energy of the system _____.

9. _____ measures a substance's thermodynamic stability with respect to its constituent elements.

10. An equilibrium mixture is mainly products when ΔG^0 is _____ and ln K _____.

Problems

1. Explain, in terms of probability, which state has the higher entropy.
   a. A file cabinet neatly organized in alphabetical order or a three year-old's toy box.
   b. A perfectly ordered crystal of salt or frozen slush (after the road was salted).
   c. 1 mol of CO_2 gas at STP or 1 mol of CO_2 gas at 273 K in a volume of 15.5 L.

2. Determine the sign of ΔS for the following processes or reactions:
   a. an increase in the volume of a gas at constant temperature
   b. formation of gaseous products from solid reactants
   c. $CaO\ (s)\ +\ 2\ NH_4Cl\ (s)\ \rightarrow\ 2\ NH_3\ (g)\ +\ CaCl_2\ (s)$
   d. $4\ NH_3\ (g)\ +\ 3\ O_2\ (g)\ \rightarrow\ 2\ N_2\ (g)\ +\ 6\ H_2O\ (g)$
   e. $^{235}_{92}U$ is separated from a mixture of $^{235}_{92}U$ and $^{238}_{92}U$.

3. Calculate the standard entropy of reaction for:

    a. Na_2CO_3 (s) + 2 HCl (aq) → 2 NaCl (aq) + CO_2 (g) + H_2O (g)

    b. 4 NH_3 (g) + 3 O_2 (g) → 2 N_2 (g) + 6 H_2O (g)

4. Given ΔS°_{total} =1.814 × 10^4 J/K and ΔS°_{rxn} = 310.8 J/K, calculate ΔH°_{rxn} for the following reaction:

    2 C_4H_{10} (g) + 13 O_2 (g) → 8 CO_2 (g) + 10 H_2O (g)

5. Calculate ΔS if 1 mol of methane gas at STP expands to a volume of 30.5 L at 273 K and 1 atm of pressure.

6. Determine ΔG° for the following reaction using the equation $\Delta G^{\circ} = \Delta H^{\circ} - T\Delta S^{\circ}$.

    Pb (s) + PbO_2 (s) + 2 H_2SO_4 (l) → 2 $PbSO_4$ (s) + 2 H_2O (l)

    Determine the temperature at which the reverse reaction becomes spontaneous.

7. Calculate the standard free energy of reaction from the standard free energy of formation for the following reaction:

    2 C_2H_2 (g) + 5 O_2 (g) → 4 CO_2 (g) + 2 H_2O (l)

8. Calculate the free energy change for the following reaction if the partial pressures are 3.0 atm for SO_2 and SO_3 and 1.5 atm for O_2.

    2 SO_2 (g) + O_2 (g) ⇌ 2 SO_3 (g)

9. Determine the equilibrium constant for the following reaction using the information found in Appendix B.

    Na_2CO_3 (s) + 2 HCl (aq) ⇌ 2 NaCl (aq) + CO_2 (g) + H_2O (l)

10. For the reaction:

    2 KCl (s) ⇌ Cl_2 (g) + 2 K (s)

    a. Calculate the ΔG°_{rxn} from ΔG°_f.
    b. Calculate the temperature at which the reaction becomes spontaneous.
    c. Calculate the equilibrium constant for this reaction.

Solutions

True/False
1. F. The spontaneity of a reaction is determined by thermodynamics; the speed of the reaction is determined by kinetics. A reaction can be spontaneous and very slow.
2. F. Some spontaneous reactions move to a state of higher potential energy by absorbing heat from the surroundings.
3. F. Enthalpy alone does not account for the direction of spontaneous change. ΔS_{total} determines the spontaneity of a reaction.
4. T

5. T
6. F. When a molecule breaks in two, there is an increase in the disorder of the system, which results in an increase of the entropy of the system.
7. F. The first law of thermodynamics does not indicate the spontaneity of a process. It only keeps track of the energy flow between the system and surroundings.
8. T
9. T
10. F. If the reaction proceeds spontaneously in the forward direction, the free energy decreases since a spontaneous reaction has a negative ΔG.

Matching
    Spontaneous process - d
    Entropy - f
    Third law of thermodynamics - g
    Standard molar entropy - c
    First law of thermodynamics - a
    Second law of thermodynamics - e
    Standard free energy of formation - b

Fill-in-the-Blank
1. equilibrium
2. disorder or randomness
3. the number of ways the state can be achieved
4. temperature
5. increases
6. sign
7. entropy of the system plus surroundings
8. decreases
9. ΔG_f^o
10. $< 0; > 0$

Problems
1. Keep in mind that the probabilities of ordered and random states are proportional to the number of ways that the state can be achieved.
   a. The toy box has the higher entropy. A file cabinet organized in alphabetical order can only be achieved in one way. Anyone who has ever spent any amount of time with a three year-old knows that there are an infinite number of ways the toys in his toy box can be arranged.
   b. The salt on the road has the higher entropy. Again, a perfectly ordered crystal of salt can only be arranged in one way. However, the crystals of salt found on the road after a winter storm are intermingled with snow and ice and therefore, arranged in many different ways.
   c. The 1 mol of CO_2 gas at STP has the higher entropy because the state of greater volume is more probable. (Remember that 1 mol of gas at STP has a volume of 22.4 L.)

2. a. ΔS is positive; an increase in the volume of a gas at constant temperature leads to a decrease in the pressure which leads to an increase in the entropy
   b. ΔS is positive; more randomness with gaseous particles
   c. ΔS is positive; product side of the reaction has more moles of gas
   d. ΔS is positive; product side of the reaction has more moles of gas
   e. ΔS is negative; moving from more disorder to less disorder by reducing the number of ways the state can be achieved

3. Use the data found in Appendix B of your text and be sure to pay attention to the state of the compound.
   a. Before solving this problem, you may find it useful to write the net ionic equation for the reaction:

   $Na_2CO_3(s) + 2H^+(aq) \rightarrow 2Na^+(aq) + CO_2(g) + H_2O(g)$
   $\Delta S_{rxn}^0 = [2 \times S^0(Na^+) + S^0(CO_2) + S^0(H_2O)] - [S^0(Na_2CO_3) + 2S^0(H^+)]$
   $\Delta S_{rxn}^0 = [(2 \times 59.0) + 213.6 + 188.7)] - [135.0 + (2 \times 0)]$
   $\Delta S_{rxn}^0 = 385.3 \text{ J/K}$

   b. $\Delta S_{rxn}^\circ = [2S^\circ(N_2) + 6S^\circ(H_2O)] - [4S^\circ(NH_3) + 3S^\circ(O_2)]$
   $\Delta S_{rxn}^\circ = [(2 \times 191.5) + (6 \times 188.7)] - [(4 \times 192.3) + (3 \times 205.0)]$
   $\Delta S_{rxn}^0 = 131.0 \text{ J/K}$

4. Using the equation $\Delta S_{total}^\circ = \Delta S_{rxn}^0 - \dfrac{\Delta H_{rxn}^\circ}{T}$, we can very easily solve for ΔH using the information given.

   $$\Delta H_{rxn}^\circ = -T\left(\Delta S_{total}^\circ - \Delta S_{rxn}^\circ\right) = -298\left(1.814 \times 10^4 \frac{J}{mol \cdot K} - 310.8 \frac{J}{mol \cdot K}\right) = -5.313 \times 10^6 \frac{J}{mol}$$

   $$= -5.313 \times 10^3 \frac{kJ}{mol}$$

5. Using the equation $\Delta S = R \ln \dfrac{V_{final}}{V_{initial}}$, we can calculate the value of ΔS.

   $$\Delta S = 8.314 \frac{J}{mol \cdot K} \ln \frac{30.5 \text{ L}}{22.4 \text{ L}} = 2.57 \frac{J}{mol \cdot K}$$

6. $\Delta H_{rxn}^0 = [2\Delta H_f^0(PbSO_4) - 2\Delta H_f^0(H_2O)] + [\Delta H_f^0(Pb) + \Delta H_f^0(PbO_2) + 2\Delta H_f^0(H_2SO_4)]$
   $\Delta H_{rxn}^0 = [(2 \times -919.9) + (2 \times -285.8)] - [0 + (-277) + (2 \times -814.0)] = -503.4 \text{ kJ}$
   $\Delta S_{rxn}^0 = [2S^0(PbSO_4) + 2S^0(H_2O)] - [S^0(Pb) + S^0(PbO_2) + 2S^0(H_2SO_4)]$
   $\Delta S_{rxn}^0 = [(2 \times 148.6) + (2 \times 69.9)] - [64.8 + 68.6 + (2 \times 156.9)] = -10.2 \text{ J/K}$

   Watch your units!
   $\Delta G_{rxn}^0 = \Delta H_{rxn}^0 - T\Delta S_{rxn}^0 = (-5.064 \times 10^5 \text{ J}) - (298 \text{ K})(-10.2 \text{ J/K}) = -5.034 \times 10^5 \text{ J} = -503.4 \text{ kJ}$

   To determine the temperature at which the reaction becomes spontaneous use:

   $$T = \frac{\Delta H^0}{\Delta S^0} = \frac{-5.064 \times 10^5 \text{ J}}{-10.2 \text{ J/K}} = 4.96 \times 10^4 \text{ K}$$

7. $\Delta G_{rxn}^0 = [4G_f^0(CO_2) + 2G_f^0(H_2O)] - [2G_f^0(C_2H_2) + 5G_f^0(O_2)]$
   $\Delta G_{rxn}^0 = [(4 \times -394.4) + (2 \times -237.2)] - [(2 \times 209.2) + (5 \times 0)] = -2.470 \times 10^3 \text{ kJ}$

8. $\Delta G = \Delta G^0 + RT \ln Q$

$\Delta G^0 = [2G_f^0(SO_3)] - [2G_f^0(SO_2) + G_f^0(O_2)]$

$\Delta G^0 = [(2 \times -371.1)] - [(2 \times -300.2) + 0] = -141.8$ kJ

$\Delta G = -1.418 \times 10^5$ J $+ (8.314$ J$/$K$)(298$ K$) \ln \dfrac{(3.0)^2}{(3.0)^2(1.5)} = -1.43 \times 10^5$ J

9. $\Delta G^0 = -RT \ln K$

To calculate ΔG^0, write the net ionic equation.

$Na_2CO_3 (s) + 2 H^+ (aq) + 2 Cl^- (aq) \rightleftarrows 2 Na^+ (aq) + CO_2 (g) + H_2O (l)$

$\Delta G_{rxn}^0 = [2G_f^0(Na^+) + G_f^0(CO_2) + G_f^0(H_2O)] - [G_f^0(Na_2CO_3) + G_f^0(H^+)]$

$\Delta G_{rxn}^\circ = \left[(2 \times -261.9) + (-394.4) + (-237.2)\right] - \left[(-1044.5) + (2 \times 0)\right] = -110.9$ kJ

-1.109×10^5 J $= -(8.314$ J$/$K$)(298$ K$) \ln K$

$\ln K = \dfrac{-1.109 \times 10^5 \text{ J}}{-(8.314 \text{ J}/\text{K})(298 \text{ K})} = 45; \quad K = 2.8 \times 10^{19}$

10. To solve this problem, we need to use the data found in Appendix B in your text.

   a. $\Delta G_{rxn}^\circ = [0 + 0] - [(2 \times -409.2)] = 818.4$ kJ/mol

   b. To calculate the temperature at which the reaction becomes spontaneous, we need to first calculate the ΔS_{rxn}° and ΔH_{rxn}°.

   $\Delta S_{rxn}^\circ = [223.0 + (2 \times 64.2)] - [(2 \times 82.6)] = 186.2$ J/mol·K

   $\Delta H_{rxn}^\circ = [0 + 0] - [(2 \times -436.7)] = 873.4$ kJ/mol

   We can now solve for the temperature (**watch your units**).

   $$T = \dfrac{873.4 \dfrac{\text{kJ}}{\text{mol}}}{0.1862 \dfrac{\text{kJ}}{\text{mol} \cdot \text{K}}} = 4.691 \times 10^3 \text{ K}$$

   c. $\ln K = \dfrac{\Delta G_{rxn}^\circ}{-RT} = \dfrac{818,400 \dfrac{\text{J}}{\text{mol}}}{-\left[\left(8.314 \dfrac{\text{J}}{\text{mol} \cdot \text{K}}\right)(298 \text{ K})\right]} = -330.2; \quad K = e^{-330.2}$

CHAPTER 18

ELECTROCHEMISTRY

Chapter Learning Goals

1⊠ Sketch a galvanic cell, identifying the anode and cathode half-reactions, the sign of each electrode, and the direction of electron and ion flow.
2⊠ Write balanced, chemical equations for reactions occurring in a galvanic cell.
3⊠ Write and interpret shorthand notations for galvanic cells.
4⊠ Interconvert cell potential and free-energy change for a reaction.
5⊠ Use a table of standard reduction potentials to calculate standard cell potentials.
6⊠ Use a table of standard reduction potentials to rank substances in order of increasing oxidizing strength or reducing strength and to determine whether a reaction is spontaneous.
7⊠ Use the Nernst equation to calculate cell potentials for reactions occurring under nonstandard conditions.
8⊠ From a measured cell potential for a reaction involving hydrogen ion and a reference cell potential, calculate the pH of the solution.
9⊠ Calculate equilibrium constants from standard cell potentials and vice versa.
10⊠ Write balanced chemical equations for reactions occurring in common batteries.
11⊠ Describe the reactions that occur when iron rusts.
12⊠ Describe half-cell and overall reactions occurring in electrolytic processes.
13⊠ Perform electrolytic cell calculations interconverting current and time, charge, moles of electrons, and moles (or grams) of product.

Chapter in Brief

Electrochemistry is the area of chemistry concerned with the interconversion of chemical and electrical energy. An electrochemical cell is the device used for this interconversion. In this chapter, you will examine the principles involved in the design and operation of electrochemical cells as well as some of the important connections between electrochemistry and thermodynamics. You will learn how the table of standard reduction potentials was derived and how this table can be used to obtain an enormous amount of chemical information. You will also learn how to use the Nernst equation to calculate cell potentials under nonstandard-state conditions and how this equation is used to determine the pH of a solution. You will then apply your knowledge of galvanic cells to the study of batteries and the process of corrosion. Finally, you will examine the commercial applications and the quantitative aspects of electrolysis and electrolytic cells.

1⊠ **Galvanic Cells**
    A. Galvanic cell - a spontaneous chemical reaction generates an electric current.
    B. Electrolytic cell - an electric current drives a nonspontaneous reaction.
    C. Redox reaction.
        1. Oxidation - a loss of electrons (an increase in oxidation number).
        2. Reduction - a gain of electrons (a decrease in oxidation number).
        3. Represent oxidation and reduction aspects of the reaction with half-reactions.
        4. Oxidizing agent - species that causes oxidation to occur and is itself reduced.
        5. Reducing agent - species that causes reduction to occur and is itself oxidized.
        6. If a spontaneous reaction is carried out in a beaker:
            a. oxidizing agent and reducing agent are in direct contact
            b. electrons are directly transferred
            c. enthalpy of reaction is lost to the surroundings for an exothermic reaction

7. If spontaneous reaction is carried out in a galvanic cell:
   a. chemical energy released by the reaction is converted to electrical energy

D. For the reaction: Zn (s) + Cu^{2+} (aq) → Zn^{2+} (aq) + Cu (s).
   1. Use a Daniell cell, a type of galvanic cell, to carry out the reaction. (See Figure 18.2, page 729 in text)
      a. consists of two half-cells
         i.. a beaker with a strip of Zn in a solution of $ZnSO_4$
         ii. a beaker with a strip of Cu in a solution of $CuSO_4$
      b. electrodes - strips of zinc and copper
      c. salt bridge - a U-shaped tube that contains a gel permeated with a solution of an inert electrolyte
   2. The electrons can be transferred only through the wire.
      a. oxidation and reduction half-reactions occur at separate electrodes
      b. electric current flows through the wire
   3. Anode - the electrode at which oxidation takes place.
      a. the negative (-) electrode
      b. produces electrons
   4. Cathode - the electrode at which reduction takes place.
      a. the positive (+) electrode
      b. consumes electrons
2⊠   5. Anode and cathode half-reaction must add to give the overall cell reaction.
   6. Salt bridge maintains electrical neutrality by a flow of ions.
      a. anions flow through the salt bridge from the cathode to the anode compartment
      b. cations migrate through the salt bridge from the anode to the cathode compartment
   7. Electrons move through the external circuit from the anode to the cathode.

EXAMPLE:
Describe how you would construct a galvanic cell based on the following reaction:

$$Pb^{2+} (aq) + Zn (s) → Pb (s) + Zn^{2+} (aq)$$

SOLUTION: Let's start by taking the overall cell reaction and breaking it into two half-reactions.

$$Pb^{2+} (aq) + 2 e^- → Pb (s)$$

$$Zn (s) → Zn^{2+} (aq) + 2 e^-$$

Looking at the two half-reactions, we find that the Pb^{2+} is being reduced, and the Zn is being oxidized. Therefore, the anode compartment of our cell would consist of a strip of zinc metal immersed in a solution containing Zn^{2+} ions (such as zinc nitrate). The cathode compartment would consist of a strip of lead immersed in a solution containing Pb^{2+} ions (such as lead (II) nitrate). The two half-cells would be connected to each other with a salt bridge and an external wire. Electrons flow through the wire from the zinc anode to the lead cathode. Anions move from the cathode compartment towards the anode while cations migrate from the anode compartment toward the cathode.

3⊠   **Shorthand Notation for Galvanic Cells**
   A. Single vertical line, |, represents a phase boundary.
   B. Double vertical line, ||, represents a salt bridge.
   C. Shorthand for the anode half-cell is always written on the left of the salt-bridge symbol, followed on the right of the symbol by the shorthand for the cathode half-cell.
      1. Electrons move through the external circuit from left to right.

2. For Zn (s) + Cu^{2+} (aq) → Zn^{2+} (aq) + Cu (s):
   Zn (s) | Zn^{2+} (aq) || Cu^{2+} (aq) | Cu (s).

D. Cell involving a gas.
   1. List the gas immediately adjacent to the appropriate electrode.

E. Detailed notation includes ion concentrations and gas pressures.

EXAMPLE:

Give the shorthand notation for a galvanic cell that employs the overall reaction

$$Pb(NO_3)_2 \ (aq) + Ni \ (s) \rightarrow Pb \ (s) + Ni(NO_3)_2 \ (aq)$$

Give a brief description of the cell.

SOLUTION: The two half-reactions for this overall reaction are:

$$Pb^{2+} \ (aq) + 2 \ e^- \rightarrow Pb \ (s)$$

$$Ni \ (s) \rightarrow Ni^{2+} \ (aq) + 2 \ e^-$$

From these half-reactions, we know that lead is being reduced and nickel is being oxidized. Therefore, Ni is the anode and Pb is the cathode. The cell notation is:

$$Ni \ (s) | Ni^{2+} \ (aq) || Pb^{2+} \ (aq) | Pb \ (s)$$

This cell would consist of a strip of nickel as the anode dipping into an aqueous solution of $Ni(NO_3)_2$ and a strip of Pb as the cathode dipping into an aqueous solution of $Pb(NO_3)_2$. The two half-cells would be connected by a salt bridge and a wire.

Cell Potentials and Free-Energy Changes for Cell Reactions

A. Electromotive force (emf) - the driving force (electrical potential) that pushes the negatively charged electrons away from the anode and pulls them toward the cathode.
   1. Also called the cell potential (E) or the cell voltage.
   2. Potential of a galvanic cell is a positive quantity.

B. Coulomb (C) - the amount of charge transferred when a current of 1 ampere (A) flows for 1 s.
   1. $1 \ J = 1 \ C \times 1 \ V$

C. Cell potential - measured with a voltmeter.
   1. Gives a positive reading when the + and - terminals of the voltmeter are connected to cathode (+) and anode (-), respectively.
      a. can use voltmeter-cell connections to determine which electrode is the anode and which is the cathode

D. Two driving forces of a chemical reaction: cell potential, E and free-energy change, ΔG.
   1. Related by $\Delta G = -nFE$.
      a. n = number of moles of electrons transferred in the reaction
      b. F (faraday) - the electrical charge on 1 mol of electrons
         i. $1 \ F = 96{,}500 \ C/mol \ e^-$
      c. ΔG and E have opposite signs
         i. spontaneous reaction has a positive cell potential but negative ΔG

E. Standard cell potential, E^o - the cell potential when both reactants and products are in their standard states.
   1. Solutes at 1 M concentration.
   2. Gases at a partial pressure of 1 atm.
   3. Solids and liquids in pure form.
   4. $T = 25^o$C.
F. $\Delta G^o = -nFE^o$.

Standard Reduction Potentials

A. $E^o_{cell} = E^o_{anode} + E^o_{cathode}$
   1. Can't measure potential of a single electrode.
   2. Measure a potential difference by placing a voltmeter between two electrodes.
   3. Develop a set of standard half-cell potentials.
      a. choose an arbitrary standard half-cell as a reference point and assign an arbitrary potential
      b. express the potential of all other half-cells relative to the reference half-cell
B. Standard hydrogen electrode (S.H.E.) - reference half-cell.
   1. Corresponding half-reaction - assigned an arbitrary potential of exactly 0 V.
      a. $2 H^+$ (aq, 1 M) $+ 2 e^- \rightarrow H_2$ (g, 1 atm) E^o 0 V
   2. Shorthand notation for S. H. E.
      a. H^+ (1 M) | H_2 (1 atm) | Pt (s)
C. Determine standard potentials for half-cells by constructing a galvanic cell in which the half-cell of interest is paired up with the standard hydrogen electrode.
   1. Standard oxidation potential - the corresponding half-cell potential for an oxidation half-reaction.
   2. Standard reduction potential - the corresponding half-cell potential for a reduction half-reaction.
   3. Whenever the direction of a half-reaction is reversed, the sign of E^o must be reversed.
      a. The standard oxidation potential and the standard reduction potential always have the same magnitude, but they have opposite signs
   4. Construct a table of standard reduction potentials (Appendix D in text).
D. Conventions used in constructing a table of half-cell potentials:
   1. The half-reactions are written as reductions.
      a. oxidizing agents and electrons are on the reactant side
      b. reducing agents are on the product side
   2. The half-cell potentials are standard reduction potentials.
      a. also known as standard electrode potentials
   3. The half-reactions are listed in order of decreasing standard reduction potential.
      a. strongest oxidizing agents are located in the upper left of the table
      b. strongest reducing agents are in the lower right of the table
   4. Ordering of half-reactions correspond to ordering of the oxidation reactions in the activity series.
      a. the more active metals at the top of the activity series have the more positive oxidation potential (more negative reduction potential)

Using Standard Reduction Potentials

A. Table of standard reduction potentials - summarizes an enormous amount of chemical information.
   1. Can arrange any two or more oxidizing or reducing agents in order of increasing strength.
   2. Predict the spontaneity or nonspontaneity of thousands of redox reactions.
      a. combine half-reactions of interest and use $E^o_{cell} = E^o_{anode} + E^o_{cathode}$
      b. may need to multiply half-reactions by some factor to ensure that electrons cancel

302

      i. do not multiply values of E^o for the half-reactions by that factor
  B. E^o values are independent of the amount of reaction.
    1. $\Delta G^o = -nFE^o$
      a. ΔG^o is an extensive property because it depends on the amount of substance
      b. change the amount of substance that reacts, ΔG^o changes by the same amount as does
        n, the number of electrons transferred
      c. $E^o = -\Delta G^o/nF$ remains constant
  C. Can predict the spontaneity of a reaction by knowing the location of the oxidizing and
    reducing agent in the table.
6⊠   1. An oxidizing agent can oxidize any reducing agent that lies below it in the table.
      a. E^o for overall reaction must be positive

EXAMPLE:
 Write the balanced net ionic equation, and calculate E^o for the following galvanic cell:

  $Al\ (s)\,|\,Al^{3+}\ (aq)\,|\,|\,Cu^{2+}\ (aq)\,|\,Cu\ (s)$

SOLUTION: $Al\ (s)$ is the anode and therefore undergoes oxidation while $Cu\ (s)$ is the cathode and
  Cu^{2+} therefore, undergoes reduction. The half-reactions and their cell potentials are:

  $Al\ (s) \rightarrow Al^{3+}\ (aq)\ +\ 3\ e^{-}$ $E^o = +1.66$ V

  $Cu^{2+}\ (aq)\ +\ 2\ e^{-} \rightarrow Cu\ (s)$ $E^o = +0.34$ V
 (Notice that the sign for E^o for the Al/Al^{3+} half-reactions has been reversed.) To write the
 balanced net ionic equation, we need to make sure that the electrons cancel out on both sides.
 Therefore, we need to multiply the top reaction by 2 and the bottom reaction by 3.

  $2\ Al\ (s) \rightarrow 2\ Al^{3+}\ (aq)\ +\ 6\ e^{-}$ $E^o = +1.66$ V

  $3\ Cu^{2+}\ (aq)\ +\ 6\ e^{-} \rightarrow 3\ Cu\ (s)$ $E^o = +0.34$ V

 Notice that although we multiplied the coefficients in both half-reactions by the factor of 2 and 3
 respectively, we did not multiply the values of E^o by these factors. This is because E^o values are
 independent of the amount of reaction.

  $E^o_{cell} = E^o_{Al \rightarrow Al^{3+}} + E^o_{Cu^{2+} \rightarrow Cu} = +1.66$ V $+ 0.34$ V $= 2.00$ V

EXAMPLE:
 Using the table of standard reduction potentials and without calculating the cell potential,
 determine if the following reactions are spontaneous.

  $Ag\ (s)\ +\ Cu^{2+}\ (aq) \rightarrow 2\ Ag^{+}\ (aq)\ +\ Cu\ (s)$

  $MnO_4^{-}\ (aq)\ +\ 8H^{+}\ (aq)\ +\ 10\ Br^{-}\ (aq) \rightarrow Mn^{2+}\ (aq)\ +\ 4\ H_2O\ (l)\ +\ 5\ Br_2\ (l)$

SOLUTION: Remember to predict a spontaneous reaction the reducing agent must lie below the
 oxidizing agent in the table of standard reduction potentials. For the first reaction, the reducing
 agent, Ag, lies above the oxidizing agent, Cu^{2+}. Therefore, the first reaction is nonspontaneous.
 For the second reaction, the reducing agent, Br^{-}, lies below the oxidizing agent, MnO_4^{-}.
 Therefore, the second reaction is spontaneous.

7⊠   **Cell Potentials and Composition of the Reaction Mixture: The Nernst Equation**
   A. Cell potentials depend on temperature and on the composition of the reaction mixture.
      1. $\Delta G = \Delta G^o + RT \ln Q$.
         a. $\Delta G = -nFE$; $\Delta G^o = -nFE^o$
      2. $-nFE = -nFE^o + RT \ln Q$.
   B. Nernst equation: $E = E^o - \dfrac{0.0592}{n} \log Q$ (in volts at 25°C).
      1. Enables us to calculate cell potentials under nonstandard-state conditions.

EXAMPLE:
   Calculate E_{cell} for the following cell reaction:

   $$2 \, Cr \, (s) + 3 \, Pb^{2+} \, (aq) \rightarrow 2 \, Cr^{3+} \, (aq) + 3 \, Pb \, (s)$$

   $$[Pb^{2+}] = 0.15 \text{ M}; \; [Cr^{3+}] = 0.50 \text{ M}$$

SOLUTION: The half-reactions for this equation are

   $2 \, Cr \, (s) \rightarrow 2 \, Cr^{3+} + 6 \, e^-$ $\qquad\qquad\qquad$ $E^o = +0.74$ V

   $3 \, Pb^{2+} \, (aq) + 6 \, e^- \rightarrow 3 \, Pb \, (s)$ $\qquad\qquad$ $E^o = -0.13$ V

   $E^{\circ}_{cell} = +0.74 + (-0.13) = +0.61$ V

The Nernst equation for this reaction is:

$$E_{cell} = E^o_{cell} - \frac{0.0592}{6} \log \frac{\left[Cr^{3+} \right]^2}{\left[Pb^{2+} \right]^3}$$

Substituting the information given and solving for E_{cell} $E_{cell} = +0.61 - \dfrac{0.0592}{6} \log \dfrac{(0.5)^2}{(0.15)^3} = 0.59$

8⊠   **Electrochemical Determination of pH**
   A. Important application of Nernst equation - electrochemical determination of pH using a pH meter.
   B. Consider a cell with a hydrogen electrode as the anode and a second reference electrode as the cathode.
      1. Pt $(s) | H_2$ (1 atm) $| H^+$ (? M) $| \, |$ reference cathode.
      2. $E_{cell} = 0.0592 \text{pH} + E_{ref}$.
      3. pH is a linear function of the cell potential.
         a. $\text{pH} = \dfrac{E_{cell} - E_{ref}}{0.0592}$
         b. can measure the pH of a solution by measuring E_{cell}
   C. Actual pH measurements use a glass electrode with a calomel electrode as the reference.

EXAMPLE:

The following cell has a potential of 0.49 V. Calculate the pH of the solution in the anode compartment.

$$\text{Pt } (s) \, | H_2 \, (g) \, (1 \text{ atm}) \, | H^+ \, (pH = ?) \, || Cl^- \, (aq) \, (1M) \, | Hg_2Cl_2 \, (s) \, | Hg \, (l)$$

SOLUTION: The cell reaction is

$$Hg_2Cl_2 \, (s) \, + \, H_2 \, (g) \, \rightarrow \, 2 \, Hg \, (l) \, + \, 2 \, Cl^- \, (aq) \, + \, 2 \, H^+ \, (aq)$$

and the standard cell potential can be calculated from the data in Appendix D.

$$E^\circ = E^\circ_{H_2 \rightarrow H^+} + E^\circ_{Hg_2Cl_2 \rightarrow Hg, Cl^-} = 0.00 \text{ V} + 0.28 \text{ V} = 0.28 \text{ V}$$

The pH can be calculated using the equation:

$$pH = \frac{E_{cell} - E_{ref}}{0.0592}$$

$$pH = \frac{0.49 \text{ V} - 0.28 \text{ V}}{0.0592} = 3.55$$

Standard Cell Potentials and Equilibrium Constants

A. Standard free-energy change for a reaction is related to both the standard cell potential and the equilibrium constant.
   1. $\Delta G^\circ = -nFE^\circ$.
   2. $\Delta G^\circ = -RT \ln K$.
B. Can combine the two equations.

9⊠
   1. $E^\circ = \dfrac{0.0592}{n} \log K$.
   2. Most common use - calculating equilibrium constants from standard cell potentials.
C. Equilibrium constants for redox reactions tend to be either very large or very small in comparison with equilibrium constants for acid-base reactions.
   1. Positive value of E° corresponds to $K > 1$.
   2. Negative value of E° corresponds to $K < 1$.
D. Three different ways to determine the value of an equilibrium constant K:

   1. K from concentration data: $K = \dfrac{[C]^c [D]^d}{[A]^a [B]^b}$.

   2. K from thermochemical data: $\ln K = \dfrac{-\Delta G^\circ}{RT}$.

   3. K from electrochemical data: $\ln K = \dfrac{nFE^\circ}{RT}$.

EXAMPLE:

Calculate the equilibrium constant for the following reaction at 25°C.

$$5 \, S_2O_8^{2-} \, (aq) \, + \, I_2 \, (s) \, + \, 6 \, H_2O \, (l) \rightarrow 10 \, SO_4^{2-} \, (aq) \, + \, 2 \, IO_3^- \, (aq) \, + \, 12 \, H^+ \, (aq)$$

SOLUTION: The half-reactions for this reaction are:

$$S_2O_8^{2-} (aq) + 2 e^- \rightarrow 2 SO_4^{2-} (aq) \qquad\qquad E^0 = +2.01 \text{ V}$$

$$I_2 (s) + 6 H_2O (l) \rightarrow 2 IO_3^- (aq) + 12 H^+ (aq) + 10 e^- \qquad E^0 = -1.20 \text{ V}$$

$$E^{\circ}_{cell} = 2.01 + (-1.20) = +0.81 \text{ V}$$

The value of n for this reaction is 10. We can now solve for K.

$$\log K = \frac{(10)(0.81)}{(0.0592)} = 137; \quad K = 10^{137}$$

10⊠ **Batteries**
   A. Most important practical application of galvanic cells is their use as batteries.
   B. Features required in a battery depend on the application.
   C. General features.
      1. Compact and lightweight.
      2. Physically rugged and inexpensive.
      3. Provide a stable source of power for relatively long periods of time.
   D. Lead storage battery.
      1. Used as a reliable source of power for starting automobiles for more than three-quarters of a century.
      2. 12 V battery - six 2 V cells connected in series.
      3. Anode - a series of lead grids packed with spongy lead.
      4. Cathode - a series of grids packed with lead dioxide, dipped into an aqueous solution of H_2SO_4 (38% w/w).
      5. Electrode half-reactions and the overall cell reaction.

Anode: $Pb (s) + HSO_4^- (aq) \rightarrow PbSO_4 (s) + H^+ (aq) + 2 e^-$ $\qquad E^0 = 0.296 \text{ V}$

Cathode: $PbO_2 (s) + 3 H^+ (aq) + HSO_4^- (aq) + 2 e^- \rightarrow 2 PbSO_4 (s) + 2 H_2O (l)$ $\quad E^0 = 1.628 \text{ V}$

Overall: $Pb (s) + PbO_2 (s) + 2 H^+ (aq) + 2 HSO_4^- (aq) \rightarrow 2 PbSO_4 (s) + 2 H_2O (l)$ $\quad E^0 = 1.924 \text{ V}$

      6. $PbSO_4$ adheres to the surface of the electrodes.
         a. recharge by using an external source of direct current to drive the cell reaction in the reverse, nonspontaneous direction
   E. Dry-Cell Batteries (Leclanché cell) - common household batteries.
      1. Anode - Zn metal can.
      2. Cathode - inert graphite rod surrounded by a paste of solid MnO_2 and carbon black.
      3. Electrolyte - a moist paste of NH_4Cl and $ZnCl_2$ in starch.
         a. surrounds the MnO_2 containing paste
         b. acidic - causes corrosion of the Zn anode ($Zn \rightarrow Zn^{2+}$)
      4. Alkaline dry cell - modified version of Leclanché.
         a. replace acidic NH_4Cl (acidic) with NaOH or KOH
         b. electrode reactions - oxidation of zinc and reduction of manganese dioxide
            i. produces ZnO due to basic conditions
            ii. zinc corrodes more slowly
            iii. battery has a longer life
         c. produces higher power and more stable current and voltage
            i. more efficient ion transport in the alkaline electrolyte
      5. Mercury battery - used in watches, heart pacemakers, and other devices.
         a. small size

    b. anode - Zn (same as dry cell)
    c. cathode - steel in contact with HgO in an alkaline medium of KOH and $Zn(OH)_2$
F. Nickel-Cadmium batteries - used in calculators and portable power tools.
    1. Rechargeable.
    2. Anode - cadmium metal.
    3. Cathode – NiO(OH) supported on nickel metal
    4. Solid products of electrode reaction adhere to the surface of the electrodes.
        a. allows battery to be recharged
G. Lithium batteries – light weight, high voltage, rechargeable battery
    1. Anode – lithium metal.
        a. highest standard oxidizing potential
    2. Cathode – metal oxide or sulfide that can incorporate Li^+.
    3. Electrolyte – lithium salt in an organic solvent.
H. Fuel cells - a galvanic cell in which one of the reactants is a traditional fuel.
    1. Reactants are not self-contained within the cell.
        a. supplied from an external reservoir
    2. Best-known - hydrogen-oxygen fuel cell.
        a. used in space vehicles as a source of electric power

Corrosion

A. Corrosion - the oxidative deterioration of a metal.
B. Well-known example of corrosion - conversion of iron to rust.
    1. Requires both oxygen and water.
    2. Involves pitting of the metal surface.
        a. rust is deposited at a location physically separated from the pits
11⊠ C. Proposed mechanism for formation of rust - an electrochemical process in which iron is oxidized in one region of the surface and oxygen is reduced in another region.
    1. Anode region: $Fe\ (s) \rightarrow Fe^{2+}\ (aq) + 2\ e^-$ $E^0 = 0.45$ V
    2. Cathode region: $O_2\ (g) + 4\ H^+\ (aq) + 4\ e^- \rightarrow 2\ H_2O\ (l)$ $E^0 = 1.23$ V
    3. Electrons flow from the anode to the cathode through the metal.
    4. Ions migrate through the water droplets.
        a. Fe^{2+} reacts with O_2 and is oxidized to Fe^{3+}
        b. Fe^{3+} reacts with H_2O to form $Fe_2O_3 \cdot xH_2O\ (s)$ (rust)
    5. Explains why cars rust more rapidly when road salt is used to melt snow and ice.
        a. dissolved salt in water greatly increases the conductivity of the electrolyte
    6. O_2 - able to oxidize all metals except a few.
        a. O_2/H_2O half-reaction lies above the M^{n+}/M half-reaction
D. Prevention of corrosion - shield the metal surface from oxygen and moisture.
    1. Durable surface coating - metals such as chromium, tin, or zinc.
    2. Galvanizing - coating by dipping into a bath of molten zinc.
    3. Cathodic protection - protecting a metal from corrosion by connecting it to a second metal that is more easily oxidized.

12⊠ **Electrolysis and Electrolytic Cells**

A. Electrolytic cell - an electric current is used to drive a nonspontaneous reaction.
    1. Processes occurring in galvanic and electrolytic cells are the reverse of each other.
B. Electrolysis - the process of using an electric current to bring about chemical change.
C. Electrolytic cell.
    1. Two electrodes that dip into an electrolyte and are connected to a battery or some other source of direct electric current.
        a. battery - an electron pump, pushing electrons into one electrode and pulling them out of the other electrode

2. Anode - electrode where oxidation takes place.
   a. positive sign
   b. the battery pulls electrons out of it
3. Cathode - electrode where reduction takes place.
   a. negative sign
   b. the battery pushes electrons into it

D. Electrolysis of molten NaCl.
   1. Cathode – attracts Na^+.
      a. $Na^+ + e^- \rightarrow Na\ (l)$
   2. Anode – attracts Cl^-.
      a. $2\ Cl^- \rightarrow Cl_2\ (g) + e^-$

E. Electrolysis of aqueous NaCl.
   1. Electrode reactions in an aqueous solution can differ from those for a molten salt.
   2. Cathode reaction can involve the reduction of Na^+ or the reduction of water.
      a. actual reaction – the reduction of water
      b. $2\ H_2O\ (l) + 2\ e^- \rightarrow H_2\ (g) + 2\ OH^-\ (aq)$
   3. Anode reaction can involve the oxidation of Cl^- or the oxidation of water.
      a. actual reaction – the oxidation of Cl^- due to overvoltage
      b. $2\ Cl^-\ (aq) \rightarrow Cl_2\ (g) + 2\ e^-$
   4. Overvoltage – amount of voltage needed above the calculated standard reduction (or oxidation) potential for electrolysis to occur.
      a. needed when the half-reaction has a substantial barrier for electron transfer (slow rate).
         i. surmounts barrier
         ii. reaction proceeds at satisfactory rate
      b. small overvoltage needed for solution or deposition of metals
      c. large overvoltage needed for formation of O_2 or H_2
      d. can't predict; need experimental evidence if cell potentials are similar
   5. Overall cell reaction:
      $2\ Cl^-\ (aq) + 2\ H_2O\ (l) \rightarrow Cl_2\ (g) + 2\ H_2\ (g) + 2\ OH^-\ (aq)$
      a. Na^+ is a spectator ion and reacts with the OH^- to form NaOH

F. Electrolysis of water.
   1. Electrolysis of any aqueous solution requires the presence of an electrolyte to carry the current in solution.
   2. If the ions of the electrolyte are less easily oxidized and reduced than water is, then water will react at both electrodes.
   3. Anode: $2\ H_2O\ (l) \rightarrow O_2\ (g) + 4\ H^+\ (aq) + 4\ e^-$.
   4. Cathode: $4\ H_2O\ (l) + 4\ e^- \rightarrow 2\ H_2\ (g) + 4\ OH^-\ (aq)$.
   5. Overall cell reaction:
      $2\ H_2O\ (l) \rightarrow O_2\ (g) + 2\ H_2\ (g)$

12⊠ **Commercial Applications of Electrolysis**
   A. Manufacture of sodium - produced commercially in a Downs cell by electrolysis of a molten mixture of NaCl and $CaCl_2$.
      1. Liquid Na produced at the cylindrical steel cathode is less dense than the molten salt and thus floats to the top part of the cell, where it is drawn off into a suitable container.
   B. Manufacture of chlorine and sodium hydroxide - electrolysis of aqueous NaCl.
      1. Basis of chlor-alkali industry.
      2. Anode and cathode reactions for electrolysis of aqueous NaCl carried out in membrane cell.
         a. membrane keeps Cl_2 and OH^- apart but allows a current of Na to flow
   C. Manufacture of aluminum - Hall-Heroult process.

1. Electrolysis of a molten mixture of Al_2O_3 and cryolite (Na_3AlF_6) at $1000°$ C in a cell with graphite electrodes.
   a. success of process - the use of cryolite as a solvent
2. Electrode reactions involve the formation of complex ions.
   a. ions are reduced at the cathode to produce Al (l)
   b. ions are oxidized at the anode to produce O_2 (g)
      i. O_2 (g) reacts with the graphite electrode to produce CO_2 (g)
      ii. Requires frequent replacement of the anodes
3. Largest single consumer of electricity in the U.S.
   a. 1 mol of electrons produces only 9 g of Al
   D. Electrorefining - the purification of a metal by means of electrolysis.
   E. Electroplating - the coating of one metal on the surface of another using electrolysis.
      1. Cathode - object to be plated (carefully cleaned).
      2. Electrolytic cell contains a solution of ions of the metal to be deposited.

13⊠   **Quantitative Aspects of Electrolysis**
   A. The amount of substance produced at an electrode by electrolysis depends on the quantity of charge passed through the cell.
      1. Follows directly from the stoichiometry of the reaction and the atomic weight of the product.
   B. Moles of electrons passed through a cell are determined from the electric current and the time that the current flows.

$$\text{Moles of } e^- = \text{charge(C)} \times \frac{1 \text{ mol } e^-}{96,500 \text{ C}}$$

   C. Sequence of conversion used to calculate the mass or volume of product produced by passing a known current for a fixed period of time.

| current and time | → | charge | → | moles of e^- | → | moles of product | → | grams or liters of product |

   D. Think of electrons as reactants in a balanced equation and proceed as with any other stoichiometry problem.

EXAMPLE:
How many grams of Cl_2 would be produced in the electrolysis of molten NaCl by a current of 4.25 A for 35.0 min?

SOLUTION: (Remember that a coulomb is an A·s or that an ampere is C/s.)

$$2 \, Cl^- \rightarrow Cl_2 + 2 \, e^-$$

moles of electrons = 2

$$4.25 \, \frac{C}{s} \times 35.0 \text{ min} \times \frac{60 \text{ s}}{1 \text{ min}} \times \frac{1 \text{ mol } e^-}{96,500 \text{ C}} \times \frac{1 \text{ mol } Cl_2}{2 \text{ mol } e^-} \times \frac{70.9 \text{ g } Cl_2}{1 \text{ mol } Cl^-} = 3.28 \text{ g } Cl_2$$

EXAMPLE:
The Dow process isolates Mg (s) from sea water. The final step in this process involves the electrolysis of molten $MgCl_2$ to the metal. How long would it take to produce 25 lb of Mg (s) at a current of 20 A?

SOLUTION: First write the electrolysis reaction:

$$Mg^{2+} + 2\,e^- \rightarrow Mg\,(s)$$

We can now use this equation to solve our problem.

$$25\text{ lb Mg} \times \frac{1000\text{ g}}{2.2046\text{ lb}} \times \frac{1\text{ mol Mg}}{24.3\text{ g Mg}} \times \frac{2\text{ mol e}^-}{1\text{ mol Mg}} \times \frac{96{,}500\text{ A}\cdot\text{s}}{1\text{ mol e}^-} \times \frac{1}{20\text{ A}} \times \frac{1\text{ h}}{3600\text{ s}} \times \frac{1\text{ day}}{24\text{ h}} = 52.1\text{ days}$$

Self-Test

This section is intended to test your knowledge of the material covered in this chapter. Think through these problems and make certain you understand what is going on. Ask yourself if your answer makes sense. Many of these questions are linked to the chapter learning goals. Therefore, successful completion of these problems indicates you have mastered the learning goals for this chapter. You will receive the greatest benefit from this section if you use it as a mock exam. You will then discover which topics you have mastered and which topics you need to study in more detail.

True/False

1. An electrolytic cell is one in which a spontaneous chemical reaction generates an electric current.

2. The reducing agent in a redox reaction is the species that causes reduction to occur and it itself is oxidized.

3. The cathode is the electrode where oxidation takes place.

4. The cell potential of a galvanic cell is positive.

5. Whenever the direction of a half-reaction is reversed, the sign of E^o must be reversed.

6. In constructing a table of half-cell potentials, the half-reactions are written as reduction reactions.

7. An oxidizing agent can oxidize any reducing agent that lies below it in the table of standard reduction potentials.

8. Rust is deposited in the same location where pitting of the metal surface has occurred.

9. The amount of substance produced at an electrode by electrolysis depends on the quantity of charge passed through the cell.

10. In an electrolytic cell, the anode is the electrode where reduction takes place.

Multiple Choice

1. For the following galvanic cell Ni $(s)\,|\,Ni^{2+}\,(aq)\,||\,Br^-\,(aq)\,|\,Br_2\,(l)\,|\,Pt\,(s)$
   a. the cathode is Ni (s).
   b. electrons flow from the Pt (s) electrode to the Ni (s) electrode.
   c. the Ni^{2+} ions flow to the anode.
   d. the electrons flow from the Ni (s) electrode to the Pt (s) electrode.

2. The reaction carried out in a galvanic cell
   a. is spontaneous, and therefore, has a negative $E°$ value.
   b. is nonspontaneous, and therefore, has a negative $E°$ value.
   c. is spontaneous, and therefore, has a positive $E°$ value.
   d. is nonspontaneous, and therefore, has a positive $E°$ value.

3. For the net reaction $Cr\,(s)\ +\ Fe^{3+}\,(aq)\ \rightarrow\ Cr^{3+}\,(aq)\ +\ Fe\,(s)$
   a. $Cr\,(s)$ is the oxidizing agent.
   b. $Fe\,(s)$ is the cathode.
   c. $Fe^{3+}\,(aq)$ is the reducing agent.
   d. $Cr\,(s)$ undergoes reduction.

4. Given the following reduction half-cells:

   $PbO_2\,(s)\ +\ 3\,H^+\,(aq)\ +\ HSO_4^-\,(aq)\ +\ 2\,e^-\ \rightarrow\ PbSO_4\,(s)\ +\ 2\,H_2O\,(l)$ $E° = 1.628\ V$
   $Cr_2O_7^{2-}\,(aq)\ +\ 14\,H^+\,(aq)\ +\ 6\,e^-\ \rightarrow\ 2\,Cr^{3+}\,(aq)\ +\ 7\,H_2O\,(l)$ $E° = 1.33\ V$
   $SO_4^{2-}\,(aq)\ +\ 4\,H^+\,(aq)\ +\ 4\,H^+\,(aq)\ +\ 2\,e^-\ \rightarrow\ H_2SO_3\,(aq)\ +\ H_2O\,(l)$ $E° = 0.17\ V$
   $2\,CO_2\,(g)\ +\ 2\,H^+\,(aq)\ +\ 2\,e^-\ \rightarrow\ H_2C_2O_4\,(aq)$ $E° = -0.45\ V$

   a. the strongest oxidizing agent is $PbSO_4\,(s)$.
   b. $PbSO_4\,(s)$ will spontaneously react with $CO_2\,(g)$.
   c. the weakest oxidizing agent is $PbO_2\,(s)$.
   d. $H_2C_2O_4\,(aq)$ will spontaneously react with $PbO_2\,(s)$.

5. In the electrolysis of molten BaI_2
   a. the Ba^{2+} ions migrate towards the cathode.
   b. the I^- ions migrate towards the cathode.
   c. water undergoes oxidation at the anode.
   d. the Ba^{2+} ions migrate towards the anode.

6. In the table of standard reduction potentials
   a. the strongest reducing agents are located in the bottom left of the table.
   b. the strongest oxidizing agents are located in the top left of the table.
   c. the strongest reducing agents are located in the top left of the table.
   d. the strongest oxidizing agents are located in the bottom left of the table.

7. In an electrolytic cell
   a. the cathode has a positive sign.
   b. reduction occurs at the anode.
   c. the anode has a negative sign.
   d. reduction occurs at the cathode.

8. In an electrolytic cell
   a. anions migrate towards the cathode.
   b. cations migrate towards the anode.
   c. ions migrate through a salt bridge.
   d. anions migrate towards the anode.

9. Alkaline dry cells
   a. are rechargeable nickel-cadmium batteries.
   b. have a cathode in which steel is in contact with HgO in an alkaline medium.
   c. contain an electrolyte of a moist paste of NH_4Cl and $ZnCl_2$.

311

    d. contain an electrolyte of a moist paste of NaOH and $ZnCl_2$.

10. The amount of substance produced at an electrode by electrolysis depends
    a. on the quantity of reactant present.
    b. on the quantity of charge passed through the cell.
    c. on the spontaneity of the reaction.
    d. on the size of the electrolytic cell.

Fill-in-the-Blank

1. _____ is the area of chemistry concerned with the interconversion of chemical and electrical energy.

2. It's convenient to separate overall cell reactions into _____ because oxidation and reduction occur at separate _____.

3. The standard cell potential is the sum of the _____ for the anode half-reaction and the _____ for the cathode half-reaction.

4. Standard half-cell potentials are defined relative to an arbitrary value of 0 V for the _____ _____.

5. _____ are used to arrange oxidizing and reducing agents in order of increasing strength.

6. A _____ is a convenient, portable source of electrical energy consisting of one or more galvanic cells.

7. A _____ differs from an ordinary battery in that the reactants are continuously supplied to the cell.

8. The process of _____ involves covering iron with another metal, such as zinc in order to prevent corrosion.

9. The _____ is used to produce aluminum metal from a mixture of Al_2O_3 and cryolite.

10. The _____ is the amount of voltage needed above the calculated standard reduction (or oxidation) potential for electrolysis to occur.

Matching

    Galvanic cell   a. the oxidative deterioration of a metal

    Electrolytic cell   b. a U-shaped tube that contains a gel permeated with a solution of an inert electrolyte

    Oxidizing agent   c. an equation used to calculate cell potentials under nonstandard-state conditions.

Reducing agent	d. an electrochemical cell in which a spontaneous chemical reaction generates an electric current
Cathode	e. the coating of one metal on the surface of another using electrolysis
Anode	f. the species that causes oxidation to occur and is itself reduced
Salt bridge	g. the cell potential when both reactants and products are in their standard states
Electromotive force	h. the amount of charge transferred when a current of one ampere flows for one second
Coulomb	i. the purification of a metal by means of electrolysis
Standard cell potential	j. the process of using an electric current to bring about chemical change
Nernst equation	k. the electrode at which oxidation takes place
Corrosion	l. an electrochemical cell in which an electric current drives a nonspontaneous reaction
Electrolysis	m. the species that causes reduction to occur and is itself oxidized
Electrorefining	n. the electrode at which reduction takes place
Electroplating	o. the driving force that pushes the negatively charged electrons away from the anode and pulls them toward the cathode

Problems

1. Write the cell notation for the galvanic cells formed by using the following pairs of half-reactions.

   a. $NO_3^- (aq) + 4 H^+ (aq) + 3 e^- \rightarrow NO (g) + 2 H_2O (l)$
      $MnO_4^- (aq) + 8 H^+ (aq) + 5 e^- \rightarrow Mn^{2+} (aq) + 4 H_2O (l)$

   b. $Fe^{3+} (aq) + 3 e^- \rightarrow Fe (s)$
      $Ag^+ (aq) + e^- \rightarrow Ag (s)$

2. Write the overall balanced reaction and the half-reactions for MnO_4^- (in acid) reacting with $FeCl_2$ to produce Mn^{2+} and Fe^{3+}. Describe the galvanic cell you would use to carry out this reaction and give the shorthand notation for that cell.

313

3. Calculate $E°$ for the following cell:

$$Hg\ (l)\ |\ Hg_2Cl_2\ (s)\ |\ Cl^-\ (aq)\ ||\ Hg_2^{2+}\ |\ Hg(l)$$

4. Predict if the following reactions occur spontaneously in aqueous solution.

   a. $Ca\ (s)\ +\ Cd^{2+}\ (aq)\ \rightarrow\ Ca^{2+}\ (aq)\ +\ Cd\ (s)$

   b. $2\ Ag\ (s)\ +\ Ni^{2+}aq)\ \rightarrow\ 2\ Ag^+\ (aq)\ +\ Ni\ (s)$

   c. $SO_4^{2-}\ (aq)\ +\ 4\ H^+\ (aq)\ +\ 2\ I^-\ (aq)\ \rightarrow\ H_2SO_3\ (aq)\ +\ I_2\ (s)\ +\ H_2O\ (l)$

5. Calculate $\Delta G°$ and K for the following reaction:

$$O_2\ (g)\ +\ 4\ H^+\ (aq)\ +\ 4\ Fe^{2+}\ (aq)\ \rightarrow\ 2\ H_2O\ (l)\ +\ 4\ Fe^{3+}$$

6. $E°$ for a galvanic cell in which Sn^{2+} is reduced to $Sn\ (s)$ is +1.04 V. What is the potential at the anode? What metal is oxidized at the anode? Write the half-reaction which occurs at the anode.

7. Determine the Cl^- concentration if the following cell has $E = -0.30$ V.

$$C\ (s)\ |\ Cl_2\ (g, 1\ atm)\ |\ Cl^-\ (aq)\ ||\ MnO_4^-\ (aq,\ 0.010\ M),\ H^+\ (pH = 3.87),\ Mn^{2+}\ (aq,\ 0.10\ M)\ |\ Pt\ (s)$$

8. Calculate E and ΔG for the following reaction:

$$3\ Zn\ (s)\ +\ 2\ Cr^{3+}\ (aq)\ \rightarrow\ 3\ Zn^{2+}\ (aq)\ +\ 2\ Cr\ (s)$$
$$[Cr^{3+}] = 0.050\ M;\ [Zn^{2+}] = 0.035\ M$$

9. $E°_{cell} = 2.48$ V for the following reaction:

$$NiO_2\ (s)\ +\ 4\ H^+\ (aq)\ +\ 2\ Ag\ (s)\ \rightarrow Ni^{2+}\ (aq)\ +\ 2\ H_2O\ (l)\ +2\ Ag^+\ (aq)$$

   Calculate the pH of the solution if $E_{cell} = 2.06$ V and $[Ag^+]$ and $[Ni^{2+}] = 0.010$ M.

10. Calculate the K_{sp} for $AgBr\ (s)$ using the table of standard reduction potentials.

11. Determine $\Delta G°$ and $E°$ given that $K = 2.35 \times 10^{-8}$ and $n = 2$.

12. Calculate the amount of product produced at the cathode in the electrolysis of molten NaBr if a current of 35 A is applied for 6.0 hours.

13. How many liters of O_2 are produced when 1.19×10^3 C are passed through water at a pressure of 755 mm Hg and 25°C?

14. How many grams of $Fe(OH)_2$ are produced at an iron anode when a basic solution undergoes electrolysis at a current of 5.00 A for 3 hours?

314

Solutions

True/False
1. F. In an electrolytic cell, electricity is used to drive a nonspontaneous reaction.
2. T
3. F. Reduction takes place at the cathode.
4. T
5. T
6. T
7. T
8. F. The mechanism described for rusting indicates that the rust is deposited in a spot other than where the pitting occurs.
9. T
10. F. Oxidation takes place at the anode.

Multiple Choice
1. d
2. c
3. b
4. d
5. a
6. b
7. d
8. d
9. d
10. b

Fill-in-the-Blank
1. Electrochemistry
2. half-reactions; electrodes
3. oxidation potential; reduction potential
4. standard hydrogen electrode
5. Tables of standard reduction potentials
6. battery
7. fuel cell
8. galvanizing
9. Hall-Heroult process
10. overvoltage

Matching
    Galvanic cell - d
    Electrolytic cell - l
    Oxidizing agent - f
    Reducing agent - m
    Cathode - n
    Anode - k
    Salt bridge - b
    Electromotive force - o
    Coulomb - h
    Standard cell potential - g
    Nernst equation - c
    Corrosion - a
    Electrolysis - j

Electrorefining - i
Electroplating - e

Problems

1. For the reactions to occur in a galvanic cell, they must have a positive E°_{cell}. Remember that an oxidizing agent will spontaneously react with a reducing agent that lies below it in the table of standard reduction potentials.

   a. The half-reaction involving MnO_4^- lies above the half-reaction involving NO_3^-; therefore, the oxidizing agent (species reduced; cathode) is the MnO_4^-. The cell could possibly be

   $$Pt\ (s)\,|\,NO\ (g)\,|\,NO_3^-\ (aq),\ H^+\ (aq)\,|\,|\,MnO_4^-\ (aq),\ H^+\ (aq),\ Mn^{2+}\ (aq)\,|\,Pt\ (s)$$

   b. The half-reaction involving Ag^+ lies above the half-reaction involving Fe^{3+}. Therefore, Ag^+ is the oxidizing agent (species reduced; cathode) and Fe (s) is the reducing agent (species oxidized; anode).

   $$Fe\ (s)\,|\,Fe^{3+}\ (aq)\,|\,|\,Ag^+\ (aq)\,|\,Ag\ (s)$$

2. We can approach this problem two different ways. We can either use the rules we learned in Chapter 4 on balancing redox reactions, or we can use the table of standard reduction potentials (in which the half-reactions are already balanced) to write the balanced half-reactions. Keep in mind, the reaction involving MnO_4^- is in acid; therefore, H^+, must be involved in the reaction. The two half-reactions are:

   $$MnO_4^-\ (aq)\ +\ 8\ H^+\ (aq)\ +\ 5\ e^-\ \rightarrow\ Mn^{2+}\ (aq)\ +\ 4\ H_2O\ (l)$$

   $$Fe^{2+}\ (aq)\ \rightarrow\ Fe^{3+}\ (aq)\ +\ e^-$$

   To write the overall, balanced redox reaction, we must multiply the oxidation reaction by 5 so that electrons lost will equal electrons gained. The overall, balanced reaction is

   $$MnO_4^-\ (aq)\ +\ 8\ H^+\ (aq)\ +\ 5\ Fe^{2+}\ (aq)\ \rightarrow\ Mn^{2+}\ (aq)\ +\ 4\ H_2O\ (l)\ +\ 5\ Fe^{3+}\ (aq)$$

   A possible galvanic cell for this reaction would consist of an anode compartment with a Pt (s) electrode immersed in a solution of Fe^{2+}, and a cathode compartment with a graphite electrode immersed in a solution of MnO_4^-. The shorthand notation for this cell is:

   $$Pt\ (s)\,|\,Fe^{2+}\ (aq),\ Fe^{3+}\ (aq)\,|\,|\,MnO_4^-\ (aq),\ Mn^{2+}\ (aq),\ H^+\ (aq)\,|\,C\ (s)$$

3. The two half-reactions and their potentials for this cell are:

   Anode: $2\ Hg\ (l)\ +\ 2\ Cl^-\ (aq)\ \rightarrow\ Hg_2Cl_2\ (s)\ +\ 2\ e^-$ $\qquad\qquad E^\circ = -0.28\ V$

   Cathode: $Hg_2^{2+}\ (aq)\ +\ 2\ e^-\ \rightarrow\ 2\ Hg\ (l)$ $\qquad\qquad E^\circ = 0.80\ V$

   $$E^\circ_{cell} = 0.80\ V + (-0.28\ V) = 0.52V$$

4. a. Spontaneous; Cd^{2+}, the reducing agent, lies above Ca, the oxidizing agent, in the table of standard reduction potentials.

b. Nonspontaneous; Ag, the reducing agent, lies above Ni^{2+}, the oxidizing agent, in the table of standard reduction potentials.

c. Nonspontaneous; I^-, the reducing agent, lies above SO_4^{2-}, the oxidizing agent, in the table of standard reduction potentials.

5. The half-reactions are:

   Anode: $4 Fe^{2+} (aq) \rightarrow 4 Fe^{3+} (aq) + 4 e^-$ $E^o = -0.77$ V

   Cathode: $O_2 (g) + 4 H^+ (aq) + 4 e^- \rightarrow 2 H_2O (l)$ $E^o = +1.23$ V

   $E^o_{cell} = 1.23 + (-0.77) = 0.46$ V

   $\Delta G^o = -nFE^o_{cell} = -(4)(96,500)90.46) = -1.8 \times 10^5$ J $= -180$ kJ

   $\log K = \dfrac{nE^o_{cell}}{0.0592} = \dfrac{(4)(0.46)}{0.0592} = 31; \qquad K = 10^{31}$

6. $E^o_{cell} = E^o_{anode} + E^o_{cathode}$ $+1.04$ V $= E^o_{anode} + (-0.14$ V$)$ $E^o_{anode} = 1.18$ V

   Remember that the potential at the anode represents the oxidation potential. Therefore, the reduction potential is -1.18 V. From the table of standard reduction potentials we find that Mn (s) has a reduction potential of -1.18 V. The half-reaction at the anode is:

   $Mn (s) \rightarrow Mn^{2+} (aq) + 2 e^-$

7. The half-reactions for the cell are:

   Anode: $2 Cl^- (aq) \rightarrow Cl_2 (g) + 2 e^-$ $E^o = -1.36$ V

   Cathode: $MnO_4^- (aq) + 8 H^+ (aq) + 5 e^- \rightarrow Mn^{2+} (aq) + 4 H_2O (l)$ $E^o = +1.51$ V

   The overall reaction for the cell is:

   $10 Cl^- (aq) + 2 MnO_4^- (aq) + 16 H^+ (aq) \rightarrow 5 Cl_2 (g) + 2 Mn^{2+} (aq) + 8 H_2O (l)$

   $E^o_{cell} = E^o_{anode} + E^o_{cathode} = (-1.36\text{ V}) + 1.51\text{ V} = 0.15$ V

   Knowing the overall balanced equation and the E^o_{cell} we can now calculate the concentration of Cl^- (aq).

   $E_{cell} = E^o_{cell} - \dfrac{0.0592}{n} \log \dfrac{[Mn^{2+}]^2 P^5_{Cl_2}}{[H^+]^{16}[MnO_4^-]^2[Cl^-]^{10}}; \quad n = 10$

   $-0.30\text{ V} = 0.15\text{ V} - \dfrac{0.0592}{10} \log \dfrac{(0.10)^2 (1)^5}{(1.35 \times 10^{-4})^{16}(0.010)^2[Cl^-]^{10}}$

$$76.0 = \log \frac{(0.1)^2}{(1.35 \times 10^{-4})^{16}(0.010)^2} - \log[\text{Cl}^-]^{10}$$

$$12.1 = -10 \log[\text{Cl}^-]; \quad [\text{Cl}^-] = 6.2 \times 10^{-2}$$

8. To calculate ΔG, we first need to calculate E_{cell} using the Nernst equation. The half-reactions are:

    3 Zn (s) → 3 Zn^{2+} (aq) + 6 e$^-$ $E^0 = +0.76$ V

    2 Cr^{3+} (aq) + 6 e$^-$ → 2 Cr (s) $E^0 = -0.74$ V

    $$E^{\circ}_{cell} = 0.76 + (-0.74) = 0.02 \text{ V}$$
    $$E_{cell} = 0.02 - \frac{0.0592}{6} \log \frac{(0.035)^3}{(0.050)^2} = 0.04$$
    $$\Delta G = -nFE = -(6)(96,500)(0.04) = -2 \times 10^4 \text{ J} = -20 \text{ kJ}$$

9. From the Nernst equation, we have
    $$E_{cell} = E^{\circ}_{cell} - \frac{0.0592}{n} \log \frac{[\text{Ni}^{2+}][\text{Ag}^+]^2}{[\text{H}^+]^4}$$
    Substituting the information gives:
    $$2.06 = 2.48 - \frac{0.0592}{2} \log \frac{(0.01)(0.01)^2}{[\text{H}^+]^4}$$
    $$14 = \log \frac{(1 \times 10^{-6})}{[\text{H}^+]^4} = \log(1 \times 10^{-6}) - \log[\text{H}^+]^4$$

    $$20 = 4(-\log [\text{H}^+]) = 4 \text{ pH}$$

    pH = 5

10. The overall reaction we are interested in is:

    AgBr (s) → Ag$^+$ (aq) + Br$^-$ (aq)

    Using the table of standard reduction potentials, we can break this overall reaction into the following two half-reactions:

    AgBr (s) + e$^-$ → Ag (s) + Br$^-$ (aq) $E^{\circ} = 0.07$ V

    Ag (s) → Ag$^+$ (aq) + e$^-$ $E^{\circ} = -0.80$ V

    $$E^{\circ}_{cell} = 0.07 \text{ V} + (-0.80 \text{ V}) = -0.73 \text{ V}$$

    $$-0.73 \text{ V} = \frac{0.0592}{1} \log K$$

$$K = 4.67 \times 10^{-13}$$

11. Using the equation $\Delta E^{\circ} = \dfrac{0.0592}{n} \log K$ we find

$$\Delta E^{\circ} = \frac{0.0592}{2} \log\left(2.35 \times 10^{-8}\right) = -0.226 \text{ V}$$

$$\Delta G^{\circ} = -nFE^{\circ} = -(2 \text{ mol e}^-)(96{,}500 \text{ C/mol e}^-)(-0.226 \text{ V}) = 4.36 \times 10^4 \text{ C·V} = 4.36 \times 10^4 \text{ J}$$

12. The reaction occurring at the cathode is:

$$\text{Na}^+ + \text{e}^- \rightarrow \text{Na } (l)$$

$$35 \frac{\text{C}}{\text{s}} \times 6 \text{ h} \times \frac{3600 \text{ s}}{\text{h}} \times \frac{1 \text{ mol e}^-}{96{,}500 \text{ C}} \times \frac{1 \text{ mol Na}}{1 \text{ mol e}^-} \times \frac{23.0 \text{ g Na}}{1 \text{ mol Na}} = 180 \text{ g Na}$$

13. To calculate the volume of O_2, we first need to calculate the moles of O_2 produced and then use the ideal gas law. One of the half-reactions for the electrolysis of water is:

$$2 \text{ H}_2\text{O } (l) \rightarrow \text{O}_2 (g) + 4 \text{ H}^+ (aq) + 4 \text{ e}^-$$

$$1.19 \times 10^3 \text{ C} \times \frac{1 \text{ mol e}^-}{96{,}500 \text{ C}} \times \frac{1 \text{ mol O}_2}{4 \text{ mol e}^-} = 3.08 \times 10^{-3} \text{ mol O}_2$$

$$V = \frac{nRT}{P} = \frac{(3.08 \times 10^{-3})(0.0821)(298)}{\left(755 \text{ mm Hg} \times \dfrac{1 \text{ atm}}{760 \text{ mm Hg}}\right)} = 0.0759 \text{ L}$$

14. The reaction is $\text{Fe } (s) + 2 \text{ OH}^- (aq) \rightarrow \text{Fe(OH)}_2 (s) + 2 \text{ e}^-$

$$5.00 \frac{\text{C}}{\text{s}} \times 3 \text{ h} \times \frac{3600 \text{ s}}{1 \text{ h}} \times \frac{1 \text{ mol e}^-}{96{,}500 \text{ C}} \times \frac{1 \text{ mol Fe(OH)}_2}{2 \text{ mol e}^-} \times \frac{89.8 \text{ g Fe(OH)}_2}{1 \text{ mol e}^-} = 25.1 \text{ g Fe(OH)}_2$$

CHAPTER 19

THE MAIN-GROUP ELEMENTS

Chapter Learning Goals

1⊠ Determine which of two main-group elements has: (a) the more metallic character; (b) the higher ionization energy; (c) the larger atomic radius; (d) the higher electronegativity; (e) the more acidic oxide; (f) the more ionic hydride; (g) the more ionic oxide.

2⊠ Contrast the chemical and physical properties of the second period main-group elements with the properties of the heavier members in the same groups.

3⊠ Compare the properties of the group 3A elements. Include valence electron configurations, common oxidation states, and trends in atomic radii, ionic radii, first ionization energies, and electronegativities.

4⊠ Draw the structure of and describe the bonding in diborane.

5⊠ Compare the properties of the group 4A elements. Include valence electron configurations; common oxidation states; and trends in atomic radii, first ionization energies, and electronegativities. Describe how each is found in nature and give one method of preparation and one commercial use.

6⊠ Describe the carbon allotropes.

7⊠ Briefly describe the chemistry of carbon oxides, carbonates, cyanides, and carbides, including commercial uses of these compounds.

8⊠ Given the formula of a silicate-containing mineral, determine the charge, the number of shared oxygens, and the structure of the silicate.

9⊠ Compare the properties of the group 5A elements. Include valence electron configurations, common oxidation states, and trends in atomic radii, first ionization energies, and electronegativities. Describe how each is found in nature and give one commercial use.

10⊠ Give an example of a nitrogen-containing compound for each common oxidation state exhibited by nitrogen. For each compound, sketch its electron-dot structure and describe its geometry.

11⊠ Briefly describe the chemistry of ammonia, hydrazine, and the nitrogen oxides.

12⊠ Give an example of a phosphorus-containing compound for each common oxidation state exhibited by phosphorus. For each compound, sketch its electron-dot structure and describe its geometry.

13⊠ Show how phosphoric acids can be interconverted by removing or adding water molecules.

14⊠ Compare the properties of the group 6A elements. Include valence electron configurations, common oxidation states, and trends in atomic radii, ionic radii, first ionization energies, electron affinities, electronegativities, and redox potentials for $X + 2 H^+ + 2 e^- \rightarrow H_2X$.

15⊠ Give an example of a sulfur-containing compound for each common oxidation state exhibited by sulfur. For each compound, write a balanced chemical equation for its preparation, sketch its Lewis electron-dot structure, and describe its geometry.

16⊠ Give an example of a halogen oxoacid, HXO_n, for $n = 1, 2, 3$, and 4. Name each acid, sketch its Lewis electron-dot structure, and describe its geometry.

Chapter in Brief

In this chapter, you will explore the chemistry of the Group 3A through 7A elements, paying particular attention to boron, carbon, silicon, nitrogen, phosphorus, and sulfur. You will examine how the chemistry of the elements in each group is determined by the valence electron configurations and how this configuration affects the common oxidation states, atomic radii, ionic radii, first ionization energies, and electronegativities of the elements. You will also examine how the size and electronegativities of the second row elements changes their chemistry relative to the other elements in the group. You will then see how the chemical concepts you have been studying throughout this study guide can be applied in understanding the chemistry of specific compounds of the elements.

1☒ **A Review of General Properties and Periodic Trends**
    A. Z_{eff} increases across the periodic table.
        1. Each additional valence electron does not completely shield the additional nuclear charge.
            a. atom's electrons are more strongly attracted to the nucleus
            b. ionization energy increases; atomic radius decreases
            c. electronegativity increases
            d. metallic character decreases and nonmetallic character increases across the table
    B. Atomic radius increases from the top to the bottom of a group in the periodic table.
        1. Additional shells of electrons are occupied.
        2. Valence electrons are farther from the nucleus.
            a. ionization energy and electronegativity generally decrease
            b. metallic character increases
            c. nonmetallic character decreases
    C. Metals form ionic compounds with nonmetals.
    D. Nonmetals tend to form covalent, molecular compounds with each other.

2☒ **Distinctive Properties of the Second-Row Elements**
    A. Properties of second-row elements differ markedly from those of heavier elements in the same periodic group.
    B. Second-row atoms have especially small sizes and especially high electronegativities.
        1. Accentuate nonmetallic behavior.
            a. BeO - amphoteric (other group 2A element oxides are basic)
            b. Boron forms mainly covalent, molecular compounds
            c. hydrogen bonding interactions - restricted to compounds of the highly electronegative second-row elements N, O, and F
    C. Second-row elements lack valence d orbitals.
        1. Have only four valence orbitals.
            a. form a maximum of four covalent bonds
    D. Small size of second-row atoms - allows formation of multiple bonds involving π overlap of the $2p$ orbitals.
        1. $3p$ orbitals are more diffuse.
            a. longer bond distances
            b. poor π overlap
        2. π bonds involving p orbitals are rare for elements of the third and higher rows.

3☒ **The Group 3A Elements**
    A. Gallium - the largest liquid range of any metal.
        1. Used in making GaAs, a semiconductor.
    B. Indium - also used in making semiconductor devices.
    C. Thallium - extremely toxic and has no commercial uses.
    D. Valence electron configuration : ns^2np^1.
        1. Most stable oxidation state for Ga and In: +3.
        2. Most stable oxidation state for Tl: +1.
    E. Properties - consistent with increasing metallic character down the group.
        1. Metal (except B).
        2. Boron - much higher electronegativity and much smaller atomic radius.
            a. shares valence electrons
            b. has nonmetallic character

Boron

A. Obtained from BBr_3.
   1. $2 BBr_3 (g) + 3 H_2 (g) \rightarrow 2 B (s) + 6 HBr (g)$.
B. Crystalline boron - strong, hard, high-melting substance.
   1. Chemically inert at room temperature.
   2. Desirable component in high-strength composite materials.
C. Boron halide - highly reactive, volatile, covalent compounds.
   1. BX_3 molecules.
   2. Behave as Lewis acids.
      a. uses its vacant $2p$ orbital in accepting a share in a pair of electrons from a Lewis base
D. Boron hydrides (boranes) - volatile, molecular compounds with formulas B_nH_m.
   1. Simplest - diborane (B_2H_6).
   2. $2 BH_2$ groups are connected by two bridging H atoms.
      a. geometry around B atoms - tetrahedral
      b. bridging B-H bonds are significantly longer than the terminal B-H bonds

4⊠
   3. Diborane - 12 valence electrons; electron deficient.
      a. B atoms use sp^3 hybrid orbitals to bond to four neighboring H atoms
         i. formed from overlap of a boron sp^3 hybrid orbital and hydrogen $1s$ orbital
      b. three-center, two-electron bond - joins each bridging H atom to both B atoms
      c. electron density between adjacent atoms is less than in an ordinary 2c-2e bond
      d. two electrons in B-H-B bridge are spread out over three atoms

5⊠ ## The Group 4A Elements

A. Especially important, both in industry and in living organisms.
B. Increase in metallic character down a group in the periodic table.
C. Valence electron configuration: ns^2np^2.
   1. Most common oxidation state: $+4$.
      a. covalent
   2. +2 oxidation state occurs for Sn and Pb.
      a. most stable for Pb
      b. ionic compounds
   3. No simple $M^{4+} (aq)$ ions for any of the group 4A elements.

Carbon

A Uncombined form, carbon is found as diamond and graphite.
6⊠
B. Diamond.
   1. Has a covalent network structure in which each C atom uses sp^3 hybrid orbitals to form a tetrahedral array of σ bonds.
      a. interlocking 3-dimensional network of strong bonds
      b. hardest known substance
      c. highest known melting point
   2. Electrical insulator.
      a. valence electrons are localized in the σ bonds
6⊠
C. Graphite.
   1. 2-dimensional sheetlike structure.
   2. Uses sp^2 hybrid orbitals.
      a. forms trigonal planar σ bonds to three neighboring C atoms
      b. remaining p orbital ($\perp$ plane of sheet) to form a π bond
      c. π bonds delocalized and free to move in plane of sheet
      d. electrical conductivity 10^{20} greater than diamond
   3. Useful as an electrode material.

   4. Carbon sheets held together by London dispersion forces.
      a. can easily slide over each other
      b. slippery feel
      c. used as a lubricant

6⊠ D. Fullerene - third crystalline allotrope of carbon; found in the soot formed by vaporizing graphite.
      1. Spherical C_{60} molecules with the shape of a soccer ball.
      2. Prepared by electrically heating a graphite rod in a helium atmosphere.
      3. Molecular substance.
         a. soluble in nonpolar, organic solvents
         b. forty amorphous forms that resemble graphite

7⊠ F. Oxides of carbon.
      1. CO: colorless, odorless, toxic gas.
         a. forms when C or hydrocarbon fuels are burned in a limited supply of oxygen
         b. used for the industrial synthesis of methanol
         c. toxicity - due to its ability to bond strongly to the Fe^{2+} atom of hemoglobin
            i. impairs the ability of hemoglobin to carry O_2 to the tissues
      2. CO_2 - colorless, odorless, nonpoisonous gas.
         a. formed when fuels burn in an excess of O_2
         b. byproduct of yeast-catalyzed fermentation of sugar in the manufacture of alcoholic beverages
         c. produced when metal carbonates are reacted with acids
         d. used in beverages
            i. CO_2 solutions - mildly acidic
            ii. $CO_2\,(g) + H_2O\,(l) \rightleftarrows H^+\,(aq) + HCO_3^-\,(aq)$
            iii. gives bite to carbonated beverages
         e. used in fire extinguishers
            i. nonflammable and 1.5 times more dense than air
            ii. settles over fire and cuts off source of O_2

7⊠ G. Carbonates.
      1. Carbonic acid - forms two series of salts: carbonates and hydrogen carbonates.
      2. Na_2CO_3 – soda ash.
         a. used in making glass
         b. $Na_2CO_3 \cdot 10\,H_2O$ – washing soda
            i. used in laundering textiles
            ii. CO_3^{2-} removes cations from hard water
            iii. produces OH^- which removes grease from fabrics
      3. $NaHCO_3$ - baking soda.
         a. reacts with acidic substances in food
         b. produces CO_2
            i. causes dough to rise

7⊠ H. Hydrogen cyanide and cyanides.
      1. HCN - highly toxic, volatile substance.
         a. $CN^-\,(aq) + H^+\,(aq) \rightarrow HCN\,(aq)$
      2. CN^- (pseudohalide ion) - behaves like halides.
         a. forms insoluble AgCN
      3. CN^- acts as a Lewis base.
         a. bonds through the lone pair of electrons on carbon

         b. toxicity of HCN and cyanides - due to the strong bonding of CN⁻ to Fe^{3+} in cytochrome
           oxidase (important enzyme in food metabolism)
           i. enzyme is unable to function
           ii. halts cellular energy production
           iii. rapid death
      4. CN⁻ used to extract Au and Ag from their ores.
         a. produces $M(CN)_2^-$ ion
         b. $M(CN)_2^-$ reduced by Zn to produce M

7⊠   I. Carbides - binary compounds of carbon.
      1. Carbon atom has a negative oxidation state.
         a. carbides of active metals
         b. interstitial carbides of transition metals
         c. covalent network carbides

Silicon

    A. Hard, gray, semiconducting solid that melts at $1410°$C.
      1. Diamond structure.
      2. Relatively poor overlap of π orbitals.
         a. no graphitelike allotrope
      3. Naturally found combined with oxygen (SiO_2 and various silicate minerals).
    B. Ultrapure silicon needed for making solid-state semiconductor devices.
      1. Convert to $SiCl_4$: $Si\,(s)\,+\,2\,Cl_2\,(g)\,\rightarrow\,SiCl_4\,(l)$.
         a. separate by fractional distillation
      2. Convert $SiCl_4$ back to elemental silicon by reduction with hydrogen.
         a. $SiCl_4\,(g)\,+\,2\,H_2\,(g)\,\rightarrow\,Si\,(s)\,+\,4\,HCl\,(g)$
      3. Further purification - zone refining.
    C. Silicates - 90% of earth's crust.
      1. Silicon oxoanions.
      2. Basic structural building block - SiO_4 tetrahedron.
      3. Common minerals - anions in which two or more O atoms bridge between Si atoms to give
         rings, chains, layers, and extended 3-dimensional structures.

8⊠   D. Sharing of two oxygen atoms per SiO_4 tetrahedron $\rightarrow$ cyclic anions or infinitely extended
      chain anions (repeating unit $Si_2O_6^{4-}$).
      1. $Si_6O_{18}^{12-}$ - cyclic anion.
         a. present in beryl, $Be_3Al_2Si_6O_{18}$
         b. emerald gemstone $-$ 2% of Al^{3+} in beryl is replaced with Cr
    E. Additional sharing of O atoms gives:
      1. The $(Si_4O_{11}^{6-})_n$ - double-stranded chain anions.
         a. found in asbestos minerals
           i. tremolite - $Ca_2Mg_5(Si_4O_{11})_2(OH)_2$
         b. asbestos is fibrous material
           i. bonds between chains are relatively weak and easily broken
      2. The $(Si_4O_{10}^{4-})_n$ - infinitely extended 2-dimensional layer anions.
         a. found in clay minerals, micas, and talc
         b. talc - $Mg_3(OH)_2(Si_4O_{10})$
         c. mica is sheetlike material
           i. bonds between two-dimensional layers are relatively weak and easily broken
    F. Silica - an infinitely extended three-dimensional structure in which the layer anions
      $(Si_4O_{10}^{4-})_n$ are stacked on top of one another and SiO_4 and AlO_4 tetrahedra share all four of
      their corners with neighboring tetrahedra.
      1. Quartz - crystalline form of SiO_2.

2. Aluminosilicates (feldspars) - most abundant of all minerals.
   a. partially substitute the Si^{4+} in SiO_2 with Al^{3+}
   b. orthoclase - $KAlSi_3O_8$
   c. zeolites – SiO_4 and AlO_4 tetrahedra are joined together in an open structure that has a three-dimensional network of cavities linked by channels
      i. act as molecular sieves for separating small molecules from larger ones
      ii. used as catalyst in the manufacture of gasoline

5⊠ **Germanium, Tin, and Lead**
   A. Relatively low abundances in the earth's crust.
      1. Sn and Pb.
         a. concentrated in workable deposits
         b. soft, malleable, low-melting metals
   B. Sn - obtained from purified cassiterite (SnO_2) by reduction with carbon.
      1. Protective coating over steel in making tin cans.
      2. Important component of alloys.
      3. Two allotropic forms.
         a. silver-white metallic form (white tin)
         b. brittle, semiconducting form with diamond structure (gray tin)
      4. Tin disease - the conversion of white tin to gray tin below 13°C.
   C. Pb - obtained from galena (PbS) by roasting in air ($\rightarrow$ PbO) and reducing PbO with CO.
      1. Used in making pipes, cables, pigments, and electrodes for storage batteries.
   D. Germanium - one of the holes in Mendeleev's periodic table.
      1. Used in making transistors and special glasses for infrared devices.
      2. High-melting semiconductor.
      3. Same crystal structure as diamond and silicon.

9⊠ **The Group 5A Elements**
   A. Down the periodic group - increasing atomic size, decreasing ionization energy, and decreasing electronegativity.
   B. Increasing metallic character of heavier elements - evident in the acid-base properties of their oxides.
      1. N and P oxides - acidic.
      2. As and Sb oxides - amphoteric.
      3. Bi_2O_3 oxides - basic.
   C. Electron configuration: ns^2np^3.
      1. Maximum oxidation state: +5.
      2. Minimum oxidation state: -3.
      3. N and P exhibit all oxidation states between -3 and +5.
      4. As and Sb - most important oxidation states: +3 and +5.
   D. Metallic character increases down the group.
      1. Sb^{3+} and Bi^{3+} - found in salts.
      2. No simple cations in compounds of N or P.
   E. As, Sb, and Bi are found in sulfide ores and are used in making various metal alloys.

Nitrogen
   A. Colorless, odorless, tasteless gas.
      1. Makes up 78% of earth's atmosphere.
   B. Separated out of liquid air by fractional distillation.
   C. Uses of N_2.
      1. Gas - protective inert atmosphere in manufacturing processes.
      2. Liquid - refrigerant.

        3. Most important use - Haber process (manufacture of NH_3).

    D. N_2 unreactive due to N-N triple bond.

        1. Reactions involving N_2 have a high activation energy and/or an unfavorable equilibrium constant.

        2. Haber process requires high temperatures, high pressures, and a catalyst.

          a. $N_2 (g) + 3 H_2 (g) \rightarrow 2 NH_3 (g)$

10⊠   E. Classify nitrogen compounds by oxidation state (see Table 19.6, page 797 in text).

11⊠   F. Ammonia - gateway to nitrogen chemistry.

        1. Starting material for industrial synthesis of other important nitrogen compounds.

        2. Colorless, pungent-smelling gas.

        3. Polar, trigonal pyramidal NH_3 molecules.

        4. Very soluble in water (due to hydrogen bonding).

        5. Easily condensed.

        6. Excellent solvent for ionic compounds.

        7. Reacts with acids to yield ammonium salts.

          a. resemble alkali metal salts in their solubility

        8. Brønsted-Lowry base

          a. produces weakly alkaline aqueous solutions.

11⊠   G. Hydrazine (H_2NNH_2) - derivative of NH_3.

        1. Replace one H atom with an NH_2 group.

        2. Prepared by reaction of ammonia with OCl^-.

        3. Poisonous, colorless liquid.

        4. Explosive in the presence of air or other oxidizing agents.

        5. Used as a rocket fuel.

        6. Weak base and versatile reducing agent in aqueous solutions.

11⊠   H. Oxides of nitrogen.

        1. N_2O - colorless, sweet-smelling gas.

          a. laughing gas - small doses are mildly intoxicating

          b. used as a dental anesthetic

          c. propellant for dispensing whipped cream

        2. NO - colorless gas.

          a. produced in laboratory by reacting Cu with dilute HNO_3

          b. large quantities prepared by catalytic oxidation of NH_3

          c. important in biological processes

        3. NO_2 - highly toxic, reddish brown gas that forms rapidly when NO is exposed to air

          a. paramagnetic

          b. dimerizes to form N_2O_4

            i. N_2O_4 predominates at lower temperatures and NO_2 predominates at higher temperatures

        4. HNO_2: produced when NO_2 reacts with water.

          a. disproportionation reaction

          b. $2 NO_2 (g) + H_2O (l) \rightarrow HNO_2 (aq) + H^+ (aq) + NO_3^- (aq)$

   I. Nitric acid - one of the most important inorganic acids.

        1. Used in making ammonium nitrate for fertilizers.

        2. Used in manufacturing explosives, plastics, and dyes.

        3. Produced by the Ostwald process.

          a. air oxidation of NH_3 to nitric oxide

          b. oxidation of nitric oxide to nitrogen dioxide

          c. disproportionation of NO_2 in water

        4. Yellow color of concentrated HNO_3.

          a. due to presence of NO_2 produced by a slight amount of decomposition

5. Strong acid - 100% dissociated in water.
6. Strong oxidizing agent.
    a. stronger oxidizing agent than H^+ (aq)
    b. can oxidize inactive metals
    c. product depends upon the nature of the reducing agent and the reaction conditions
7. Aqua regia - mixture of concentrated HCl and concentrated HNO_3 (3:1).
    a. more potent oxidizing agent than HNO_3

Phosphorus

A. Found in phosphate rock $Ca_3(PO_4)_2$ and in fluorapatite $Ca_5(PO_4)_3F$.
    1. Apatites - phosphate minerals with the formula $3\ Ca_3(PO_4)_2 \cdot CaX_2$ ($X^- = F^-$ or OH^-).
    2. Sixth most abundant element in human body.
        a. bones - $Ca_3(PO_4)_2$
        b. tooth enamel - $Ca_5(PO_4)_3OH$
        c. found in DNA and RNA
B. Exists in two common allotropic forms.
    1. White phosphorus - toxic, waxy, white solid.
        a. discrete tetrahedral P_4 molecules
        b. low melting point
        c. soluble in nonpolar solvents
        d. highly reactive - due to unusual bonding
            i. P-O bonds are "bent," relatively weak, and highly reactive
    2. Red phosphorus - nontoxic; has a polymeric structure.

12⊠ C. Phosphorus compounds - most important oxidation states: +3 and +5.
    1. More likely to be found in a positive oxidation state because of its electronegativity.
D. Phosphine (PH_3) - colorless, extremely poisonous gas.
    1. Most important hydride of phosphorus.
    2. Neutral aqueous solutions.
        a. poor proton acceptor
    3. Easily oxidized: burns in air to form H_3PO_4.
E. Phosphorous halides (PX_3 or PX_5).
    1. Product depends on the relative amounts of the reactants.
    2. Gases, volatile liquids or low-melting solids.

13⊠ F. Oxides and oxoacids of Phosphorus.
    1. Burn P in presence of oxygen → P_4O_6 or P_4O_{10}.
        a. molecular compounds with a tetrahedral array of P atoms
        b. acidic oxides - react with water → H_3PO_3 or H_3PO_4
        c. P_4O_{10} - used as a drying agent
    2. H_3PO_3 - weak diprotic acid.
        a. only two of the three H atoms are bonded to oxygen.
    3. H_3PO_4 - low-melting, colorless, crystalline solid.
        a. phosphoric acid used in the laboratory - syrupy aqueous solution (82% H_3PO_4 by mass)
        b. manufacturing method depends upon use
            i. food additive - begin with molten phosphorus
            ii. fertilizers - begin with $Ca_3(PO_4)_2$
        c. sometimes called orthophosphoric acid
    4. Other phosphoric acids.
        a. diphosphoric acid (pyrophosphoric acid) $H_4P_2O_7$
        b. triphosphoric acid, $H_5P_3O_{10}$
        c. polymetaphosphoric acid $(HPO_3)_n$ - infinitely long chain of phosphate groups

5. $Na_5P_3O_{10}$ - component of some synthetic detergents.
   a. $P_3O_{10}^{5-}$ acts as a water softener

14⊠ **The Group 6A Elements**
   A. Periodic trends.
      1. Oxygen and sulfur - typical nonmetals.
      2. Gray selenium - most stable allotrope of selenium.
         a. lustrous semiconducting solid
      3. Tellurium - semiconductor; classified as a semimetal.
      4. Polonium - radioactive element that occurs in trace amounts in uranium ores, is a silvery white metal.
   B. Electron configuration: ns^2np^6.
      1. Common oxidation state: -2.
         a. stability of state decreases with increasing metallic character
            i. oxygen - powerful oxidizing agent
            ii. H_2Se and H_2Te - reducing agents
      2. S, Se and Te - less electronegative than oxygen.
         a. found in positive oxidation states
         b. common oxidation state: +4 and +6
   C. Commercial uses of Se, Te, and Po - limited.
      1. Se - making red-colored glass and in photocopiers.
      2. Te- used in alloys.
      3. Po - heat source in space equipment and as a source of alpha particles in research.

Sulfur
   A. Many allotropic forms.
      1. Rhombic sulfur - yellow crystalline solid.
         a. most stable form of sulfur
         b. contains crown-shaped S_8 rings
         c. above 95°C convert to monoclinic sulfur (cyclic S_8 molecules pack differently in the crystal)
         d. melts at 113°C when heated at an ordinary rate
      2. Above melting point, sulfur is a fluid, straw-colored liquid.
      3. Between 160 and 195° C - dark, reddish brown, viscous liquid.
         a. S_8 ring opens and forms S_8 chains that form long polymers with more than 200,000 S atoms in the chain.
      4. Above 195°C - more fluid liquid.
      5. Boils at 445°C.
   B. Hydrogen sulfide - colorless gas with the strong, foul odor of rotten eggs.
      1. Extremely toxic.
      2. Prepared by treating FeS with dilute H_2SO_4.
      3. Generated in solution by hydrolysis of thioacetamide for use in qualitative analysis.
      4. Weak diprotic acid and a mild reducing agent.
14⊠   C. Oxides and oxoacids of sulfur.
      1. SO_2 - colorless, toxic gas formed when sulfur burns in air.
         a. slowly oxidized in the atmosphere to SO_3
         b. SO_3 dissolves in rainwater to give sulfuric acid
         c. aqueous solutions contain dissolved SO_2 and little H_2SO_3
15⊠   D. Sulfuric acid - world's most important industrial chemical.
      1. Manufactured by contact process.
         a. S burns in air to give SO_2
         b. SO_2 is oxidized to SO_3

328

　　　c. SO_3 reacts with water to give H_2SO_4
　2. Strong acid in dissociation of first proton.
　3. Forms two series of salts: hydrogen sulfates and sulfates.
　4. Oxidizing properties depend on its concentration and the temperature.
　　　a. dilute solutions at room temperatures - behaves like HCl
　　　b. hot, concentrated H_2SO_4 can oxidize metals not oxidized by H^+
　5. Used in manufacturing soluble phosphate and ammonium sulfate fertilizers.

The Halogens: Oxoacids and Oxoacid Salts
　A. Electron configuration: ns^2np^5.
　　1. Most electronegative group of elements in the periodic table.
　　2. Gain an electron when forming ionic compounds.
　　3. Share an electron with nonmetals to form molecular compounds.

16⊠　B. Oxoacids and oxoacid salts - most important compounds of halogens in positive oxidation states.
　　1. General formula - HXO_n.
　　2. Oxidation state = +1, +3, +5, or +7 (depends on value of n).
　C. Acid strength increases with increasing oxidation state of the halogen.
　　1. Acidic hydrogen bonded to oxygen.
　　2. Strong oxidizing agents.
　D. Hypohalous acids formed from halogen disproportionation reactions.
　　1. X_2 (g, l, or s) + H_2O (l) $\rightleftarrows$ HOX (aq) + H^+ (aq) + X^- (aq)
　　　a. equilibrium lies to the left
　　　b. shift right in basic solution
　　2. NaOCl - strong oxidizing agent.
　　　a. 5% solution - chlorine bleach
　E. React Cl_2 with hot NaOH $\rightarrow$ $NaClO_3$.
　　1. Used as weed killers and strong oxidizing agent.
　F. Anhydrous perchloric acid - colorless, shock-sensitive liquid that decomposes explosively on heating.
　　1. Powerful and dangerous oxidizing agent.
　　2. Perchlorate salts - strong oxidants.
　G. Iodine forms more than one perhalic acid.
　　1. Paraperiodic acid, H_5IO_6 - white crystals obtained from evaporating HIO_4 solutions.
　　　a. weak, polyprotic acid
　　2. Metaperiodic acid, HIO_4 (s), - produced when H_5IO_6 loses water.
　　　a. strong monoprotic acid

Self-Test

This section is intended to test your knowledge of the material covered in this chapter. Think through these problems and make certain you understand what is going on. Ask yourself if your answer makes sense. Many of these questions are linked to the chapter learning goals. Therefore, successful completion of these problems indicates you have mastered the learning goals for this chapter. You will receive the greatest benefit from this section if you use it as a mock exam. You will then discover which topics you have mastered and which topics you need to study in more detail.

Fill-in-the-Blank

1.　The second-row elements are unable to form more than four bonds because they lack

　　_____.

2. Boron forms _____ compounds.

3. Boranes are electron-deficient molecules that contain _____
   bonds.

4. The Ostwald process is used to produce_____.

5. Sulfuric acid is manufactured by the _____.

6. The three primary allotropes of carbon are _____.

7. The basic structural building block for silicates is _____.

8. Binary compounds of carbon with a negative oxidation state are referred to as _____.

9. Ammonia is produced by the _____.

10. The two common allotropic forms of phosphorus are _____.

General Questions

1. For the following pairs of elements, determine which has the more metallic character:
   a. Ca or Ba   b. Al or Ga   c. Rb or Sr

2. For the following pairs of elements, determine which has the more ionic hydride:
   a. Li or Na   b. Sn or Sb   c. Sb or Bi

3. What are the general trends down a group in the periodic table for ionization energy, atomic radius, electronegativity, and basicity of oxides.

4. Carbon has two common allotropes: diamond and graphite. Silicon is found only in the diamond structure. Explain this difference.

5. With the exception of boron, the group 3A elements are metals. Why does boron exhibit nonmetallic character?

6. Why are boron halides able to behave as Lewis acids?

7. Give the valence electron configuration for the group 3A. Give the stable oxidation states for these elements.

8. Describe the bonding in diborane.

9. Give the valence electron configuration and oxidation states for the group 4A elements.

10. Explain why graphite has a slippery feel and can be used as a lubricant.

11. Give some of the industrial uses of CO_2.

12. Describe the steps used for the purification of silicon. Include chemical equations in your description.

13. Determine the charge, number of shared oxygens per silicon and the structure of the silicates in the following minerals:
    a. beryl   b. asbestos   c. talc   d. quartz   e. orthoclase

14. Describe the acid-base properties of the oxides of group 5A.

15. What are the possible oxidation states for the group 5A elements?

16. Give an example of a nitrogen-containing compound for each common oxidation state exhibited by nitrogen.

17. Using chemical equations, show how the phosphoric acids can be interconverted by removing or adding water molecules.

18. What are the common oxidation states for the group 6A elements?

19. Give the name of the halogen oxoacid, HXO_n (X = Cl, Br, or I and n = 1, 2, 3, and 4).

20. Write chemical equations describing the actions of washing soda and baking soda.

21. Write a balanced equation for the reaction of $Ag(CN)_2^-$ by zinc.

22. Give the chemical equation for the production of NO_2 from nitric acid and copper.

23. Write the chemical reaction for each step in the Ostwald process.

24. Write chemical equations for each step in the contact process.

25. Write the chemical equation for the formation of hypoiodous acid from I_2 (s).

Solutions

Fill-in-the-Blank
1. valence *d* electrons
2. molecular
3. three-center, two-electron bonds
4. nitric acid
5. contact process
6. diamond, graphite, fullerene
7. SiO_4 tetrahedra
8. carbides
9. Haber process
10. white phosphorus and red phosphorus

General Questions
1. a) Ba   b) Ga   c) Rb

2. a) Na   b) Sn   c). Bi

3. Ionization energy - decreases down a group; atomic radius - increases down a group; electronegativity - decreases down a group; basicity of oxides - increases down a group

4. The diamond structure for carbon and silicon involves the use of sp^3 orbitals and a tetrahedral array of σ bonds. Graphite, on the other hand, involves the use of sp^2 orbitals to form σ bonds along

331

with a p orbital perpendicular to the plane which forms π bonds. It is the small size of carbon which allows for the overlap of p orbitals to form π bonds. The $3p$ orbitals in silicon are more diffuse, and therefore are unable to form π bonds. As a consequence, silicon cannot form an allotrope with the graphite structure.

5. It is the much higher electronegativity and much smaller atomic radius of boron that distinguishes the chemistry of this element from the other elements in group 3A.

6. The boron atom in boron halides does not have a complete octet surrounding it. Therefore, it is able to accept a pair of electrons from a Lewis base.

7. ns^2np^1; Stable oxidation states - Ga & In: +3; Tl: +1.

8. Each boron in diborane uses an sp^3 hybrid orbital to overlap with a terminal hydrogen $1s$ orbital. Each bridging H atom is joined to both boron atoms through a three-center, two-electron bond in which the two electrons in the B-H-B bridge are spread out over three atoms.

9. ns^2np^2; common oxidation states - +4; Sn and Pb: +2.

10. The carbon sheets in graphite are held together by London forces which can easily slide over each other, giving it a slippery feel and allowing it to be used as a lubricant.

11. Beverages and fire extinguishers.

12. Silicon is first converted to $SiCl_4$: $Si\ (s)\ +\ 2\ Cl_2\ (g)\ \rightarrow\ SiCl_4\ (g)$ The $SiCl_4$ is removed by fractional distillation and then converted back to silicon by reduction with hydrogen: $SiCl_4\ (g)\ +\ 2\ H_2\ (g)\ \rightarrow\ Si\ (s)\ +\ 4\ HCl\ (g)$. Further purification is accomplished with zone refining.

13. a. $Si_6O_{18}^{12-}$; number of shared oxygens per silicon = 2; cyclic anion
    b. $Si_4O_{11}^{6-}$; number of shared oxygens per silicon = 2.5; double-stranded chain anion
    c. $Si_4O_{10}^{4-}$; number of shared oxygens per silicon = 3; infinitely extended 2-dimensional layer anion
    d. SiO_2; number of shared oxygens per silicon = 4; infinitely extended 3-dimensional anion
    e. $AlSi_3O_8^-$; number of shared oxygens per silicon = 4; extended 3-dimensional anion

14. N and P oxides - acidic; As and Sb oxides - amphoteric; Bi_2O_3 - basic

15. -3 → +5

16. ox. state = -3, NH_3; ox. state = -2, N_2H_4; ox. state = -1, NH_2OH; ox. state = +1, N_2O; ox. state = +2, NO; ox. state = +3, HNO_2, ox. state = +4, NO_2; ox. state = +5, HNO_3

17. $2\ H_3PO_4\ \rightarrow\ H_4P_2O_7\ +\ H_2O$ (diphosphoric acid)
    $H_4P_2O_7\ +\ H_3PO_4\ \rightarrow\ H_5P_3O_{10}\ +\ H_2O$ (triphosphoric acid)

18. O: -2; S, Se, Te: -2, +4, +6

19. $HClO_4$ - perchloric acid, $HClO_3$ - chloric acid, $HClO_2$ - chlorous acid, HClO - hypochlorous
    $HBrO_4$ - perbromic acid; $HBrO_3$ - bromic acid, $HBrO_2$ - bromous acid, HBrO - hypobromous
    HIO_4 - periodic acid; HIO_3 - iodic acid, HIO_2 - iodous acid, HIO - hypoiodous acid

20. Washing soda:
$$Ca^{2+} (aq) + CO_3^{2-} (aq) \rightarrow CaCO_3 (s)$$

$$CO_3^{2-} (aq) + H_2O (l) \rightleftarrows HCO_3^- (aq) + OH^- (aq)$$

Baking soda:
$$NaHCO_3 (s) + H^+ (aq) \rightarrow Na^+ (aq) + CO_2 (g) + H_2O (l)$$

21. $2 Ag(CN)_2^- (aq) + Zn (s) \rightarrow 2 Ag (s) + Zn(CN)_4^{2-} (aq)$

22. $Cu (s) + 2 NO_3^- (aq) + 4 H^+ (aq) \rightarrow 2 NO_2 (g) + 2 H_2O (l)$

23. a) air oxidation of NH_3 to nitric oxide: $4 NH_3 (g) + 5 O_2 (g) \rightarrow 4 NO (g) + 6 H_2O (g)$
    b) oxidation of nitric oxide to nitrogen dioxide: $2 NO (g) + O_2 (g) \rightarrow 2 NO_2 (g)$
    c) disproportionation of NO_2 in water: $3 NO_2 (g) + H_2O (l) \rightarrow 2 HNO_3 (l) + NO$

24. a) S burns in air to give SO_2: $S (s) + O_2 (g) \rightarrow SO_2 (g)$
    b) SO_2 is oxidized to SO_3: $2 SO_2 (g) + O_2 (g) \rightarrow 2 SO_3 (g)$
    c) SO_3 reacts with water to H_2SO_4: $SO_3 (g) + H_2O$ (in conc. H_2SO_4) $\rightarrow H_2SO_4 (l)$

25. $I_2 (s) + H_2O (l) \rightleftarrows HOI (aq) + H^+ (aq) + X^- (aq)$

CHAPTER 20

TRANSITION ELEMENTS AND COORDINATION CHEMISTRY

Chapter Learning Goals

1⊠ Write valence electron configurations for transition-metal atoms and ions.
2⊠ Compare the properties of the first-transition-series elements. Include appearance; valence electron configurations for atoms and ions; common oxidation states; and trends in melting points, atomic radii, densities, ionization energies, and standard oxidation potentials.
3⊠ Balance equations for redox reactions of chromium, iron, and copper.
4⊠ Write formulas of coordination complexes. Identify the ligands and their donor atoms. Determine the coordination number and the oxidation state of the metal and the charge on any complex ion.
5⊠ Given the electron dot structure of a molecule or ion, determine whether it can serve as a chelate ligand.
6⊠ Name coordination compounds.
7⊠ Identify linkage isomers and ionization isomers.
8⊠ Determine the number and structures of diastereoisomers possible for a given coordination complex.
9⊠ Determine which coordination complexes are chiral and draw the structure of the enantiomers.
10⊠ Give a valence bond theory description of a coordination complex and show the number of unpaired electrons and the hybrid orbitals used by the metal ion.
11⊠ Give a crystal field theory description of a coordination complex and show the number of unpaired electrons.

Chapter in Brief

This chapter examines the properties and chemical behavior of transition-metal compounds with particular emphasis on coordination compounds. You begin with a review of the electronic configurations of the transition elements and how these configurations affect the properties and oxidation states. These concepts are then illustrated with the chemistry of chromium, iron and copper. You will then explore the area of coordination chemistry including the structural properties, color, and magnetism of these compounds. Finally, you will examine the different bonding theories that are used to explain the color and magnetism of these complexes.

1⊠ **Electron Configurations**
    A. $4s$ subshell filled first; $3d$ subshell filled according to Hund's rule.
        1. Add one electron to each of the five $3d$ orbitals before adding a second electron.
        2. Two exceptions: Cr and Cu.
            a. Cr: $4s^1 3d^5$
            b. Cu: $4s^1 3d^{10}$
    B. Electron configurations depend on both orbital energies and electron-electron repulsions.
        1. If two valence subshells have similar energies, can't always predict the configurations.
        2. Exceptions from the expected orbital filling pattern result in either half-filled or completely filled subshells.
        3. For Cr and Cu: $3d$ and $4s$ have similar energies and as a consequence, one electron shifts from $4s$ to $3d$
            a. shift decreases the electron – electron repulsions
            b. gives either a filled or half-filled subshell
    C. Easy to predict electron configurations of transition-metal cations.
        1. All the valence electrons occupy the d orbitals.

2. Neutral atom loses one or more electrons; Z_{eff} increases.
   a. remaining electrons more strongly attracted to the nucleus
   b. orbital energies decrease
   c. $3d$ orbitals experience a steeper drop in energy with increasing Z_{eff} than does the $4s$
      i. $3d$ orbitals in cations lower in energy than the $4s$ orbitals

2⊠ **Properties of Transition Elements**
   A. Properties can be understood in terms of electron configurations.
   B. Metallic properties.
      1. Malleable, ductile, lustrous, and good conductors of heat and electricity.
      2. Sharing of d, as well as s, electrons gives rise to stronger metallic bonding.
         a. transition metals are harder, have higher melting and boiling points, and are more dense than the group 1A and 2A metals
      3. Melting points increase as the number of unpaired d electrons available for metallic bonding increases and then decrease as the d electrons pair up and become less available for bonding.
   C. Atomic radii and densities.
      1. Decrease in radii with increasing atomic number.
         a. added d electrons only partially shield the added nuclear charge
         b. upturn in radii toward the end of each series
            i. due to more effective shielding and increasing electron-electron repulsion as d orbitals become doubly occupied
      2. Similar radii - accounts for ability to blend together in forming alloys.
      3. Second and third series - group 4B on have nearly identical radii.
         a. smaller than expected size of the third-series atoms due to lanthanide contraction
      4. Lanthanide contraction - the general decrease in atomic radii of the f-block lanthanide elements between the second and third transition series.
         a. increase in effective nuclear charge as $4f$ subshell is filled
         b. size decrease due to a larger Z_{eff} almost exactly compensates for the expected size increase due to an added quantum shell of electrons
         c. atoms of third series have radii very similar to those of the second series
      5. Densities - inversely related to atomic radii.
         a. second and third series have nearly the same atomic volume
            i. third series has unusually high densities
   D. Ionization energies and standard oxidation potentials.
      1. Ionization energies - increase left to right across a transition series.
         a. due to increase in Z_{eff} and a decrease in atomic radius
      2. $E°$ for **oxidation** potential of first series are positive (except Cu).
         a. the solid metal is oxidized to the aqueous cation more readily than H_2 gas is oxidized to H^+
         b. $M(s) + 2H^+(aq) \rightarrow M^{2+}(aq) + H_2(g)$
         c. better reducing agents than H_2
         d. can be oxidized by "nonoxidizing" acids (lack an oxidizing anion)
         e. Cu (s) requires a stronger oxidizing agent
      3. General trend for $E°$ correlates with the general trend in ionization energies.
         a. ease of oxidation of the metal decreases as the ionization energies increase across the first series

2⊠ **Oxidation States of Transition Elements**
   A. Exhibit a variety of oxidation states.
      1. Have oxidation states less than their group number.
   B. First series form +2 cation (except Sc).

   1. Due to loss of the two $4s$ electrons.
  C. First series - can also lose $3d$ electrons.
   1. $3d$ and $4s$ energies are similar
   2. Energy required to remove the third electron is provided by the larger $\Delta G°$ of hydration of the more highly charged +3 cation.
  D. Highest oxidation state for the group 3B-7B metals is the group number.
   1. Corresponds to loss of all the valence s and d electrons.
   2. Later transition metals - loss of all valence electrons is prohibited by the increasing value of Z_{eff}.
  E. Transition metal ions in a high oxidation state - good oxidizing agents.
  F. Early transition metal ions in low oxidation state - good reducing agents.
  G. Divalent ions of the later metals - poor reducing agents because of the larger value of Z_{eff}.
  H. Stability of the higher oxidation states increases down a periodic group.

Chemistry of Selected Transition Metals

  A. Chromium:
   1. Obtained from chromite ($FeO \cdot Cr_2O_3$).
    a. reduce chromite with carbon $\rightarrow$ ferrochrome (Fe (s) + Cr (s))
     i. used in making stainless steel
   2. Pure form obtained from reduction of Cr_2O_3 with Al.
   3. Used to electroplate metallic objects with an attractive, protective coating.
   4. Hard, lustrous, takes a high polish.
   5. Resistant to corrosion.
    a. microscopic film of Cr_2O_3 protects surface
   6. Systemize aqueous chemistry by oxidation states and the species that exists under acidic and basic conditions. (see Table 20.3, page 833 in text)
   7. Cr^{2+}.
    a. in aqueous solution - $Cr(H_2O)_6^{2+}$ is beautiful blue solution
    b. rapidly oxidized by O_2 to Cr^{3+}
   8. Cr^{3+}.
    a. $Cr(H_2O)_6^{3+}$ - violet
    b. Cr^{3+} solutions are green
     i. anions replace some of the bound water molecules to give green complex ions
   9. $Cr(OH)_n$ - acid strength increases with increasing polarity of O-H bond.
    a. polarity increases with increasing oxidation state of Cr atom
     i. $Cr(OH)_2$ - basic
     ii. $Cr(OH)_3$ - amphoteric
     iii. $CrO_2(OH)_2$ – chromic acid (H_2CrO_4) – acidic
   10. +6 oxidation state.
    a. acidic solution - $Cr_2O_7^{2-}$
     i. powerful oxidizing agent
     ii. used as an oxidant in analytical chemistry
    b. basic solution - CrO_4^{2-}
     i. weaker oxidizing agent
  B. Iron - immensely important in human civilization and in living systems.
   1. Relatively soft and easily corroded.
    a. combine with carbon and other metals to make alloys
   2. Most important ores - hematite (Fe_2O_3) and magnetite (Fe_3O_4).
    a. reduce with coke in a blast furnace $\rightarrow$ Fe (s)
   3. Most important oxidation states: +2 and +3.

      4. Oxidize with an acid which lacks a nonoxidizing anion in absence of air:

          $Fe\ (s) \rightarrow Fe^{2+}\ (aq)$

          a. oxidation of Fe^{2+} has negative standard potential

          b. in presence of air: Fe^{2+} oxidized to Fe^{3+}

      5. Oxidize with an acid that has an oxidizing anion: $Fe\ (s) \rightarrow Fe^{3+}\ (aq)$.

3⊠   6. $Fe^{3+}\ (aq)\ +$ base $\rightarrow Fe(OH)_3\ (s)$

          a. very insoluble

          b. forms if $pH > 2$

          c. reaction accounts for red-brown rust stains in sinks

   C. Copper - found in the elemental state.

      1. Most important ores - sulfides.

          a. chalcopyrite, $CuFeS_2$

      2. High electrical conductivity and negative oxidation potential.

          a. used to make electrical wiring and corrosion-resistant water pipes

      3. Less reactive than other first-series transition metals.

      4. Prolonged exposure to moist air and CO_2: $Cu\ (s) \rightarrow Cu_2(OH)_2CO_3$.

          a. continues to react with acid rain forming $Cu_2(OH)_2SO_4$, the green patina seen on

             bronze monuments

      5. Two common oxidation states: +1 and +2.

3⊠   6. $E^0_{Cu^+/Cu^{2+}}$ less negative than E^0_{Cu/Cu^+}.

          a. any oxidizing agent strong enough to oxidize copper to the copper(I) ion is also able to

             oxidize the copper(I) ion to the Cu(II) ion

          b. $Cu^+\ (aq)$ can undergo a disproportionation reaction

             i. $2\ Cu^+\ (aq) \rightarrow Cu\ (s)\ +\ Cu^{2+}\ (aq)\ E^\circ = +0.37\ V$

             ii. large equilibrium constant, $K = 1.8 \times 10^6$

          c. Cu^+ is not an important species in aqueous solution

      7. Cu^+ exists in solid compounds.

          a. disproportionation equilibrium is reversed in presence of Cl^-

             i. CuCl precipitation shifts the preceding reaction left

      8. Cu^{2+} - more common.

          a. most common compound - blue-colored $CuSO_4 \cdot 5\ H_2O$

             i. 4 H_2O's bound to the Cu^{2+}; 5th H_2O hydrogen-bonded to SO_4^{2-}

             ii. loses color on heating - suggests that blue color is due to the bonding of Cu^{2+} to the

                water molecules

Coordination Compounds

   A. Coordination compound - a compound in which a central metal ion is attached to a group of surrounding molecules or ions by coordinate covalent bonds.

   B. Ligands - the molecule or ions that surround the central metal ion in a complex.

   C. Donor atoms - the atoms that are attached directly to the metal ion.

   D. Complex formation is a Lewis acid-base interaction in which the ligands act as Lewis bases and the central metal ion behaves as a Lewis acid.

   E. Some coordination compounds are salts which contain a complex cation or anion along with enough ions of opposite charge to give a compound that is electrically neutral overall.

      1. Enclose the complex ion in brackets in the formula.

          a. indicates that the complex ion is a discrete structural unit

   F. Metal complex - refers both to neutral molecules and to complex ions.

4⊠   G. Coordination number - the number of ligand donor atoms that surround a central metal ion in a complex.

      1. Most common coordination numbers: 4 and 6.

        2. Depends on the metal ion's size, charge and electron configuration and the size and shape of the ligands.

    H. Characteristic shape of metal complex is determined by the metal ion's coordination number.

        1. Two-coordinate complexes - linear.

        2. Four-coordinate complexes - tetrahedral and square planar.

        3. Six-coordinate complexes - octahedral.

4⊠   I. Charge on a metal complex is equal to the charge on the metal ion plus the sum of the charges on the ligands.

Ligands

    A. All ligands are Lewis bases.

        1. Have at least one unshared pair of electrons.

           a. forms a coordinate covalent bond to a metal ion

4⊠   2. Classified as monodentate or polydentate.

           a. depends on number of ligand donor atoms that bond to the metal

           b. use the electron pair of a single donor atom - monodentate

           c. use electron pairs on more than one donor atom - polydentate

5⊠   B. Chelating agents - polydentate ligands.

        1. Multipoint attachment to a metal ion resembles the grasping of an object by the claws of a crab.

           a. forms a chelate ring

        2. Metal chelate - a complex that contains one or more chelate rings.

4,6⊠ **Naming Coordination Compounds**

    A. If the compound is a salt, name the cation first and then the anion.

    B. In naming a complex ion or a neutral complex, name the ligands first, in alphabetical order, and then the metal.

        1. Anionic ligands end in -o. (see Table 20.5, page 844 in your text).

        2. Complex name is one word.

    C. More than one ligand of a particular type, use Greek prefixes to indicate number.

    D. If the name of a ligand itself contains a Greek prefix, put the ligand name in parenthesis and use an alternate prefix.

    E. Use a Roman numeral in parenthesis immediately following the name of the metal, to indicate the metal's oxidation state.

    F. In naming the metal, use the ending -ate if the metal is in an anionic complex.

EXAMPLE:

Name the following compounds:

$[Co(NH_3)_4Cl_2]NO_3$ $K_2[OsCl_5N]$ $[Co(en)_2CO_3]Br$

SOLUTION: Using the rules for naming coordination complexes, we find

$[Co(NH_3)_4Cl_2]NO_3$ The ligands are ammine and chloro. Since we have 2 Cl⁻ and 1 NO_3^-, we know that the charge on cobalt is +3. The name is tetraamminedichlorocobalt(III) nitrate.

$K_2[OsCl_5N]$ The ligands in this compound are chloro and nitrido. Since we have 2 K⁺, 5 Cl⁻, and 1 N^{3-}, we know that the charge on osmium must be +6. Since the complex is an anion, the ending on osmium is changed to "ate." The compound's name is potassium pentachloronitridoosmate(VI).

[Co(en)$_2$CO$_3$]Br   The ligands in this compound are ethylenediamine and carbonato. Since ethylenediamine contains a Greek prefix, we will use the prefix *bis* to indicate the number of en ligands present. Since we have CO$_3$$^{2-}$ and Br$^-$, we know the charge on cobalt is +3. The compound's name is carbonatobis(ethylenediamine)cobalt(III) bromide.

EXAMPLE:

Write formulas for the following compounds: triammineaquadichlorocobalt(III) chloride, potassium trioxalatochromate(III), sodium hexacyanoferrate(III).

SOLUTION: Using Table 20.5 on page 844 of the text, we can determine the formula for the ligands.

The ammine ligand is NH$_3$ and the aqua ligand is H$_2$O. The formula for this compound is [CoCl$_2$(NH$_3$)$_3$(H$_2$O)]Cl.

The oxalato ligand has the formula C$_2$O$_4$$^{2-}$. Since there are three oxalato ligands with a total charge of -6 and the chromium has a total charge of +3, we know we need 3 K$^+$ to balance the charge. The formula for this compound is K$_3$[Cr(C$_2$O$_4$)$_3$].

The cyano ligand has the formula CN$^-$. Since there are six cyano ligands with a total charge of -6 and the iron has a total charge of +3, we know we need 3 Na$^+$ to balance the charge. The formula for this compound is Na$_3$[Fe(CN)$_6$].

Isomers

A. Isomers - compounds that have the same formula but a different arrangement of their constituent atoms.
  1. Different compounds with different physical and chemical properties.

7⊠

B. Constitutional isomers - isomers that have different connections among their constituent atoms.
  1. Linkage isomers - arise when a ligand can bond to a metal through either of two different donor atoms.
  2. Ionization isomers - isomers that differ in the anion that is bonded to the metal ion.

C. Stereoisomers - isomers that have the same connections among atoms but have a different arrangement of the atoms in space.

8⊠

  1. Diastereoisomers (geometric isomers) - have different relative orientations of their metal-ligand bonds.
    a. *cis* isomer - identical ligands occupy adjacent corners of the square in a square planar complex
    b. *trans* isomer - identical ligands are across from one another in a square planar complex
    c. have different properties
      i. *cis*-Pt(NH$_3$)$_2$Cl$_2$ - polar molecule; more soluble in water
      ii. *trans*-Pt(NH$_3$)$_2$Cl$_2$ - nonpolar; two Pt-Cl and Pt-NH$_3$ dipoles point in opposite directions and cancel out
    d. square planar complexes of the type MA$_2$B$_2$ and MA$_2$BC (M= metal ion; A, B, C = ligands) can exist as *cis-trans* isomers
    e. no *cis-trans* isomers for four coordinate tetrahedral complexes
      i. all four corners of a tetrahedron are adjacent to one another
    f. octahedral complexes of the type MA$_4$B$_2$ can also exist as diastereoisomers
      i. two B ligands can be either on adjacent or on opposite corners of the octahedron
    g. easy to distinguish
      i. various bonds in the *cis* and *trans* isomers point in different directions

9⊠   **Enantiomers and Molecular Handedness**

    A. Enantiomers - molecules or ions that are nonidentical mirror images of one another.
        1. Differ from one another because of their handedness.
           a. chiral - objects that have a handedness to them
           b. achiral - objects that lack handedness
    B. An object is not chiral if it has a symmetry plane cutting through its middle so that one half of the object is a mirror image of the other half.
    C. Certain molecules and ions are chiral.
        1. Tris(ethylenediamine)cobalt(III) ion, $[Co(en)_3]^{3+}$ (see Fig. 20.22, page 852 in text) - two nonidentical mirror image forms.
           a. "right-handed" enantiomer - the three ethylenediamine ligands spiral to the right (clockwise)
           b. "left-handed" enantiomer - the ethylenediamine ligands spiral to the left (counterclockwise)
        2. $[Co(NH_3)_6]^{3+}$ - achiral; has several symmetry planes.
    D. Enantiomers have identical properties except for their reactions with other chiral substances and their effect on plane-polarized light.
        1. Plane-polarized light - light in which the electric vibrations of the lightwave are restricted to a single plane.
        2. Pass ordinary light through a polarizing filter then through a solution of one of the enantiomers.
           a. plane of polarization is rotated either to the right or to the left
           b. pass through the solution of the other enantiomer - plane of polarization is rotated through an equal angle, but in the opposite direction
        3. Optical isomers - another term used for enantiomers because of their effect on plane-polarized light.
        4. Label enantiomers (+) or (-) depending on the direction of rotation of the plane of polarization.
        5. Racemic mixture - 50:50 mixture of the (+) and (-) isomers which produces no net optical rotation.
           a. rotations produced by the individual enantiomers exactly cancel

Color and Magnetic Properties

    A. Color of transition-metal complexes depend on the identity of the metal and the ligands.
    B. Metal complex can absorb light by undergoing an electronic transition from its lowest energy state (E_1) to a high energy state (E_2).
        1. Wavelength of absorbed light depends on the energy separation $\Delta E = E_2 - E_1$ between the two states.
        2. Absorbance - the measure of the amount of light absorbed by a substance.
           a. absorption spectrum - plot of absorbance versus wavelength
           b. color that we see is complementary to the color absorbed

10⊠   **Bonding in Complexes: Valence Bond Theory**

    A. Valence bond theory - bonding results when a filled ligand orbital containing a pair of electrons overlaps a vacant hybrid orbital on the metal ion.
        1. Produces a coordinate covalent bond.
    B. If you know the geometry of a complex, you know which hybrid orbitals the metal ion uses. (see Table 20.7, page 856 in text).
    C. Magnetic behavior of transition metal complexes.
        1. Paramagnetic - substance that contains unpaired electrons and is attracted by magnetic fields.

2. Diamagnetic substance - contains only paired electrons and is weakly repelled by magnetic fields.

3. Number of unpaired electrons in a transition-metal complex can be determined by quantitative measurement of the force exerted on the complex by a magnetic field.

D. High-spin complex - one in which the d electrons are arranged according to Hund's rule to give the maximum number of unpaired electrons.

E. Low-spin complex - one in which the d electrons are paired up to give a maximum number of doubly occupied d orbitals and a minimum number of unpaired electrons.

11⊠ Crystal Field Theory

A. Crystal field theory - a model that views the bonding in complexes as arising from electrostatic interactions and considers the effect of the ligand charges on the energies of the metal ion d orbitals.

1. Accounts for color and magnetic properties of transition-metal complexes.
2. No covalent bonds, no shared electrons, and no hybrid orbitals.
3. Electrostatic interactions within an array of ions.
4. Neutral, dipolar ligand - the electrostatic interactions are of the ion-dipole type.

B. Octahedral complexes.

1. Metal is positively charged and the ligands are negatively charged.
   a. ligands repel each other
   b. minimize repulsion by getting as far apart from one another as possible
   c. metal-ligand attractions > ligand-ligand repulsions
      i. complex is more stable than the separated ions

C. Effect of ligand charges on the energies of d orbitals.

1. Negatively charged d electrons are repelled by the negatively charged ligands.
   a. orbital energies are higher in the complex
   b. not all raised in energy by the same amount
2. $d_{z^2}, d_{x^2-y^2}$ are raised higher in energy than the d_{xy}, d_{xz}, d_{yz} (point between ligands).
3. Crystal field splitting (Δ) - the energy splitting between the two sets of d orbitals. (see Fig. 20.30, page 861 in text)

D. Crystal field splitting energy, Δ - corresponds to λ's in the visible region of the spectrum.

1. Color of complexes due to electronic transitions between the lower and higher energy sets of d orbitals.
2. Size depends on the nature of the ligands.
   a. spectrochemical series - changes in the crystal field splitting as the ligand varies
   b. weak-field ligands - produce a relatively small value of Δ
      i. high-spin complexes
   c. strong-field ligands - produce a relatively large value of Δ
      i. low-spin complexes

E. Spin-state depends on the relative values of Δ and P, the spin-pairing energy.

1. Spin-pairing energy - energy needed to overcome repulsion that exists when an electron is placed into an orbital that already contains another electron.
2. $\Delta > P$; low spin arrangement has lower energy.
3. $\Delta < P$; high spin arrangement has lower energy.

F. Choice of spin-states arises only for complexes d^4-d^7 complexes.

G. Tetrahedral and square planar complexes - have different energy splitting for the d orbitals. (see Fig. 20.31, page 863 in text).

H. Energy splitting in tetrahedral complexes is just the opposite of that in octahedral complexes.

1. Δ(tetrahedral complexes) $\approx \frac{1}{2}\Delta$(octahedral complexes).
   a. none of the orbitals point directly at the ligands
   b. there are only four ligands instead of six
   c. $\Delta < P$

        d. all complexes are high-spin
   I. Square planar complexes - two trans ligands along the z axis are missing.
      1. Large energy gap between the x^2-y^2 orbital and the four lower-energy orbitals.
      2. Most common for metal ions with d^8 electronic configurations.
         a. favors low-spin complexes in which all four lower-energy orbitals are filled and the higher energy x^2-y^2 orbital is vacant

Self-Test

This section is intended to test your knowledge of the material covered in this chapter. Think through these problems and make certain you understand what is going on. Ask yourself if your answer makes sense. Many of these questions are linked to the chapter learning goals. Therefore, successful completion of these problems indicates you have mastered the learning goals for this chapter. You will receive the greatest benefit from this section if you use it as a mock exam. You will then discover which topics you have mastered and which topics you need to study in more detail.

Fill-in-the-Blank

1. Objects that have a handedness to them are called _____.

2. The _____ isomer has identical ligands that occupy adjacent corners of the square in a square planar complex.

3. For the first series transition metals, the solid metal is oxidized to the aqueous cation _____ than H_2 gas is oxidized to H^+.

4. Melting points of transition elements _____ as the number of _____ _____ available for metallic bonding increases.

5. Melting points of the transition elements _____ as the d electrons _____ and become less available for bonding.

6. Early transition metal ions in low oxidation states are good _____ agents.

7. For $Cr(OH)_n$ compounds, acid strength increases with _____ _____.

8. $Cr_2O_7^{2-}$ is a _____ agent in _____ solution.

9. The most important oxidizing states of iron are _____.

10. Cu^+ undergoes _____ reaction to produce Cu^{2+} and Cu.

11. Complex formation is a _____ interaction in which the ligands act as a _____ and the metal ion behaves as a _____.

12. The most common coordination numbers are _____.

13. _____ ligands use electron pairs on more than one donor atom.

14. A _____ is formed by multipoint attachment of a ligand to a metal ion.

15. Ligands form a _____ bond to a metal ion.

16. _____ are compounds that have the same formula but different arrangement of constituent atoms.

17. Square planar complexes are most common for metal ions with _____ electron configurations.

18. Objects that lack handedness are _____.

19. Light in which the electric vibrations of the lightwave are restricted to a single plane is called

    _____.

20. When naming a coordination complex, the number of ligands of a particular type is indicated with

    a _____.

21. Another term used for enantiomers is _____ due to their effect on plane polarized light.

22. A plot of absorbance versus wavelength is called an _____.

23. _____ substances contain unpaired electrons and are attracted by magnetic fields.

24. According to _____, bonding results when a filled ligand orbital containing a pair of electrons overlaps a vacant hybrid orbital on the metal ion.

25. A _____ complex is one in which the d electrons are arranged according to Hund's rule to give the maximum number of unpaired electrons.

26. _____ is a model that views the bonding in complexes as arising from electrostatic interactions and considers the effect of the ligand charges on the energies of the metal ion d orbitals.

27. Low-spin complexes result from a _____ value of Δ.

28. Tetrahedral complexes are _____ spin complexes due to _____.

29. Square planar complexes are _____ spin complexes.

30. The color of the transition metal complexes is due to the _____ between the lower and higher energy sets of d orbitals.

General Questions

1. Write the valence electron configurations for the following atoms and ions: Mn, Pt, Co^{3+}, V^{2+}.

2. How do the d electrons affect the metallic properties of the transition metals?

3. Why do the third series transition metals have unusually high densities?

4. Using only the periodic table, predict which is the better reducing agent, Mn or Fe.

5. Is FeO_4^{2-} a good oxidizing or reducing agent? Why?

6. Which is the more stable oxidation state, Fe^{6+} or Os^{6+}?

7. Write a balanced, net ionic equation in which $Cr_2O_7^{2-}$ in acidic solution reacts with Pb (s) and HSO_4^- (aq) to produce $PbSO_4$ and Cr^{3+}.

8. Write a balanced, net ionic equation showing what happens when Fe (s) is oxidized in the presence of air and a base.

9. Define:
   a. coordination compound
   b. ligand
   c. donor atom
   d. coordination number
   e. chelating agent
   f. linkage isomer
   g. ionization isomers
   h. diastereoisomers
   i. enantiomers
   j. racemic mixture

10. For the following compounds identify the ligands and their donor atoms. Determine the coordination number and the oxidation state of the metal and the charge on any complex ion. Determine if the compound can display geometrical, linkage, or optical isomerism.
    a. $[Co(NH_3)_4(H_2O)_2]Br_3$
    b. $[Co(NH_3)_5(H_2O)]Cl_2$
    c. $[Co(en)_2(Cl)(SCN)]NO_2$
    d. $Na_4[Fe(CN)_6]$
    e. $[Co(en)_3]Cl_3$

11. Name the compounds in Question 10.

12. Write formulas for the following compounds:
    a. tetraquadichlorochromium(III) chloride
    b. pentaamminedinitrogenruthenium(II) chloride
    c. sodium hexachloropalladate(IV)
    d. sodium pentacyanoiodoferrate(II) dihydrate
    e. tetraamminecopper(II) sulfate

13. Use valence bond theory to explain the bonding in $[FeCl_4]^-$ (tetrahedral complex) and $[Ni(CN)_4]^{2-}$ (square planar). What hybrid orbitals are used by the metal? Are the complexes low-spin or high-spin complexes?

14. Using the spectrochemical series in your text (page 862), determine the number of unpaired electrons in $[CoF_6]^{3-}$ and $[Fe(CN)_6]^{3-}$.

15. Why is the crystal field splitting in a tetrahedral complex approximately half of the crystal field splitting in an octahedral complex?

Solutions

Fill-in-the-Blank
1. chiral
2. *cis*

344

3. more readily
4. increase; d electrons
5. decrease; pair-up
6. reducing
7. increasing polarity of the O-H bond
8. powerful oxidizing agent; acidic
9. +2 (ferrous) and +3 (ferric)
10. disproportionation
11. Lewis acid-base; Lewis base; Lewis acid
12. 4 and 6
13. Polydentate
14. metal chelate
15. coordinate covalent
16. Isomers
17. d^8
18. achiral
19. plane polarized light
20. Greek prefix
21. optical isomers
22. absorbance spectrum
23. paramagnetic
24. Valence bond theory
25. high-spin complex
26. Crystal field theory
27. large
28. high; the small value of Δ
29. low
30. electronic transitions

General Questions

1. Mn: $[Ar] 4s^2 3d^5$;   Pt: $[Xe] 6s^2 4f^{14} 5d^8$;   Co^{3+}: $[Ar] 3d^6$;   V^{2+}: $[Ar] 3d^3$

2. The sharing of d, as well as s, electrons gives rise to stronger metallic bonding. As a consequence, transition metals are harder, have higher melting and boiling points and are more dense than the group 1A and 2A metals. The melting points increase as the number of unpaired d electrons available for metallic bonding increases and then decrease as the d electrons pair up and become less available for bonding.

3. As the $4f$ subshell is filled, there is an increase in the effective nuclear charge. However, the size decrease, due to a larger Z_{eff}, is almost exactly compensated for by the expected size increase due to an added quantum shell of electrons in the third transition series. As a consequence, the atoms of the third transition series have atomic radii (or volumes) very similar to the second transition series. The increase in atomic mass of the third transition series coupled with the similarity in atomic volume for the second and third transition series gives rise to unusually high densities for the atoms in the third transition series.

4. The ease of oxidation of the metal decreases as the ionization energies increase across the first transition series. Therefore, the strength of the reducing agent decreases across the first transition series and Mn is a better reducing agent than Fe.

5. Ions that have transition metals in a high oxidation state are good oxidizing agents.

6. Os^{6+} is more stable because the stability of the higher oxidation states increases down a periodic group.

7. In acidic solution, $Cr_2O_7^{2-}$ undergoes the following half-reaction:

$$Cr_2O_7^{2-} (aq) + H^+ (aq) \rightarrow Cr^{3+} (aq)$$

The other half-reaction is:

$$Pb (s) + HSO_4^- (aq) \rightarrow PbSO_4 (s)$$

Using the rules to balance a redox equation, we end up with the following balanced half-reactions:

$$Cr_2O_7^{2-} (aq) + 14 H^+ (aq) + 6 e^- \rightarrow 2 Cr^{3+} (aq) + 7 H_2O (l)$$

$$Pb (s) + HSO_4^- (aq) \rightarrow PbSO_4 (s) + H^+ (aq) + 2 e^-$$

To get the overall, balanced net ionic equation, we need to multiply the bottom equation by 3 and cancel out what is common. This gives:

$$Cr_2O_7^{2-} (aq) + 11 H^+ (aq) + 3 Pb (s) + 3 HSO_4^- (aq) \rightarrow 2 Cr^{3+} (aq) + 7 H_2O (l) + 3 PbSO_4 (s)$$

8. $Fe (s) + 2 H^+ (aq) \rightarrow Fe^{2+}(aq) + H_2 (g)$

$$4 Fe^{2+} (aq) + O_2 (g) + 4 H^+ (aq) \rightarrow 4 Fe^{3+} (aq) + 2 H_2O (l)$$

$$Fe^{3+} (aq) + 3 OH^- (aq) \rightarrow Fe(OH)_3 (s)$$

9. a. coordination compound - a compound in which a metal ion is attached to a group of surrounding molecules or ions by coordinate covalent bonds.
   b. ligand - the molecule or ions that surround the central metal ion in a complex.
   c. donor atom - the atoms that are attached directly to the metal ion.
   d. coordination number - the number of ligand donor atoms that surround a central metal ion in a complex.
   e. chelating agent - a polydentate ligand in which electron pairs on more than one donor atom can bond to the metal ion.
   f. linkage isomers - isomers that result from a ligand's ability to bond to a metal through either of two different donor atoms.
   g. ionization isomers - isomers that differ in the ion that is bonded to the metal ion.
   h. diastereoisomers - isomers which have different relative orientations of their metal-ligand bonds.
   i. enantiomers - molecules or ions that are nonidentical mirror images of one another.
   j. racemic mixture - a 50:50 mixture of the (+) and (-) isomers which produces no net optical rotation.

10. a. Ligands: NH_3 (donor atom - N); H_2O (donor atom - O)
       Coordination number = 6; oxidation state of metal = +3; Has the general formula MA_2B_4; therefore can be geometrical isomers.
    b. Ligands: NH_3 (donor atom - N); H_2O (donor atom - O)
       Coordination number = 6; oxidation state of metal = +3.
    c. Ligands: ethylenediamine (donor atoms - N), Cl^-, SCN^- (donor atom either N or S)
       Coordination number = 6 (en is a bidentate ligand); oxidation state of metal = +3; Because

SCN⁻ can bond through either the N or S, this compound can display linkage isomerism. The en ligand can spiral either to the right or to the left, indicating the compound can exist as enantiomers. It is also possible for this complex to have ionization isomers with the NO_2^- changing places with either the Cl⁻ or SCN⁻ ligands. It can also have *cis* and *trans* diastereoisomers.

   d. Ligands: CN⁻ (donor atom - C)
      Coordination number = 6; oxidation state of metal = +2; Charge on complex ion = - 4

   e. Ligands: en (donor atoms - N)
      Coordination number = 6 (en is a bidentate ligand); oxidation state of metal = +3; The en ligand can spiral either to the right or to the left, indicating the compound can exist as enantiomers.

11. a. tetraaminediaquacobalt(III) bromide
   b. pentaammineaquocobalt(III) chloride (linkage through O) depending on the linkage isomer.
   c. chlorobis(ethylenediamine)thiocyanatocobalt(III) nitrate or chlorobis(ethylenediamine)isothiocyanatocobalt(III) nitrate depending on the linkage isomer.
   d. sodium hexacyanoferrate(II)
   e. tris(ethylenediamine)cobalt(III) chloride

12. a. $[CrCl_2(H_2O)_4]Cl$
   b. $[Ru(NH_3)_5(N_2)]Cl_2$
   c. $Na_2[PdCl_6]$
   d. $Na_4[Fe(CN)_5I] \cdot 2\,H_2O$
   e. $[Cu(NH_3)_4]SO_4$

13. a. Since $[FeCl_4]^-$ is tetrahedral, we know that the hybridization on the metal ion is sp^3. The Fe^{3+} ion has the electron configuration [Ar] $3d^5$. The orbital diagram for this ion is:

[Ar] ↑ ↑ ↑ ↑ ↑   __ __ __ __
     3d   4s   4p

To accept a share in the 4 electron pairs provided by the chloride ion, Fe^{3+} must use the empty 4s and 4p orbitals to form hybrid orbitals. The orbital diagram for the complex is:

[Ar] ↑ ↑ ↑ ↑ ↑   ↑↓ ↑↓ ↑↓ ↑↓
     3d sp^3

This complex would be high-spin since it gives the maximum number of unpaired electrons.

   b. $[Ni(CN)_4]^{2-}$ is square planar; therefore, the hybrid orbitals being used by the Ni^{2+} ion are dsp^2. The Ni^{2+} ion has the electron configuration [Ar] $3d^8$. The orbital diagram for this ion is:

[Ar] ↑↓ ↑↓ ↑↓ ↑ ↑   __   __ __ __
     3d   4s   4p

For Ni^{2+} to use dsp^2 hybrid orbitals, the two unpaired electrons must first pair up. Ni^{2+} can then accept a share in the 4 electron pairs provided by the CN⁻ ion. The orbital diagram for the complex is:

[Ar] $\underline{\uparrow\downarrow}$ $\underline{\uparrow\downarrow}$ $\underline{\uparrow\downarrow}$ $\underline{\uparrow\downarrow}$ $\quad$ $\underline{\uparrow\downarrow}$ $\underline{\uparrow\downarrow}$ $\underline{\uparrow\downarrow}$ $\underline{\uparrow\downarrow}$ $\quad$ $\underline{}$

$\quad\quad\quad\quad$ $3d$ $\quad\quad\quad\quad\quad\quad$ dsp^2 $\quad\quad\quad$ $4p$

The complex is low-spin since there are no unpaired electrons.

14. From the spectrochemical series, we know that F^- is a weak-field ligand and that CN^- is a strong field ligand. Therefore, the Δ for $[CoF_6]^{3-}$ is small whereas the Δ for $[Fe(CN)_6]^{3-}$ is large. The Co^{3+} ion has the electron configuration [Ar] $3d^6$. These six electrons will first fill all of the d orbitals before pairing up since $\Delta < P$. Therefore, the number of unpaired electrons equals 4. The Fe^{3+} ion has the electron configuration [Ar] $3d^5$. Since $\Delta > P$, the electrons will pair up in the lower energy d orbitals in the complex, leaving only one unpaired electron.

For $[CoF_6]^{3-}$:

$\quad\quad\quad$ $\underline{\uparrow}$ $\quad$ $\underline{\uparrow}$

$\quad\quad\quad$ d_{z^2} $\quad$ $d_{x^2-y^2}$

$\quad\quad\quad\quad$ $\uparrow$

$\quad\quad\quad\quad$ Δ

$\quad\quad\quad\quad$ $\downarrow$

$\underline{\uparrow\downarrow}$ $\quad$ $\underline{\uparrow}$ $\quad$ $\underline{\uparrow}$

d_{xy} $\quad\quad$ d_{xz} $\quad$ d_{yz}

For $[Fe(CN)_6]^{3-}$:

$\quad\quad\quad$ $\underline{}$ $\quad$ $\underline{}$

$\quad\quad\quad$ d_{z^2} $\quad$ $d_{x^2-y^2}$

$\quad\quad\quad\quad$ $\uparrow$

$\quad\quad\quad\quad$ Δ

$\quad\quad\quad\quad$ $\downarrow$

$\underline{\uparrow\downarrow}$ $\quad$ $\underline{\uparrow\downarrow}$ $\quad$ $\underline{\uparrow}$

d_{xy} $\quad\quad$ d_{xz} $\quad$ d_{yz}

15. The crystal field splitting in tetrahedral complexes is half of the crystal field splitting in octahedral complexes because none of the d orbitals point directly at the ligands, and there are only four ligands as opposed to six in an octahedral complex.

CHAPTER 21

METALS AND SOLID-STATE MATERIALS

Chapter Learning Goals

1⊠ Predict whether a given metal is likely to be found in nature as an oxide, a sulfide, a carbonate, a silicate, a chloride, or a free metal.

2⊠ Describe the three steps in the metallurgical processes for isolating and purifying a metal from its ore.

3⊠ Describe the Mond process.

4⊠ Describe the processes by which iron ore is converted to steel.

5⊠ Explain the properties of a metal according to the electron-sea and the band theory models.

6⊠ Use band theory to explain transition-metal melting point trends.

7⊠ Classify doped semiconductors as *n*-type or *p*-type.

8⊠ Define superconducting transition temperature (T_c) and the Meissner effect.

9⊠ Define the terms ceramics, sintering, sol, and gel.

10⊠ Describe the sol-gel method for the production of ceramic powders.

11⊠ Classify composite materials as ceramic-ceramic, ceramic-metal, or ceramic-polymer.

Chapter in Brief

This chapter explores the chemistry behind metals and solid-state materials. You will examine the natural sources of the metallic elements, the methods used to obtain metals from their ores, and the models used to describe the bonding in metals. You will also study the structure, bonding, properties, and applications of semiconductors, superconductors, ceramics, and composites.

Sources of the Metallic Elements

    A. Minerals - the crystalline, inorganic constituents of the rocks that make up the earth's crust.
        1. Source of most metals.
    B. Most abundant minerals - silicates and aluminosilicates.
        1. Difficult to concentrate and reduce.
        2. Not important sources of metals.
    C. More important minerals - sulfides and oxides.
        1. Hematite (Fe_2O_3), rutile (TiO_2), and cinnabar (HgS).
        2. Ores - mineral deposits from which metals can be produced economically.

1⊠   D. Chemical composition of the most common ores correlates with the location of the metal in the periodic table.
        1. Early transition metals - occur as oxides.
            a. less electronegative metals form compounds by losing electrons to highly electronegative nonmetals
        2. Late transition metals - occur as sulfides.
            a. the more electronegative metals tend to form compounds with more covalent character by bonding to the less electronegative nonmetals
        3. Electronegative *p*-block metals - sulfide ores.
        4. *s*-block metals - occur as carbonates, silicates, and chlorides (for Na and K).
            a. *s*-block oxides are strongly basic and too reactive to exist with acidic oxides

Metallurgy

    A. Ore - a complex mixture of a metal-containing mineral and economically worthless material called gangue.

1. Gangue - consists of sand, clay, and other impurities.
B. Metallurgy - the science and technology of extracting metals from their ores.

2⊠
    1. Three-step process.
      a. concentration of the ore and chemical treatment prior to reduction
      b. reduction of the mineral to the free metal
      c. refining or purification of the metal
    2. Used for making alloys - metallic materials composed of two or more elements.
C. Concentration and chemical treatment of ores.
    1. Concentrate by separating the mineral from the gangue.
      a. have different properties which are exploited in various separation methods
    2. Metal sulfides ores - concentrated by flotation.
      a. process exploits the differences in the ability of water and oil to wet the surfaces of the mineral and the gangue
    3. Bayer process - uses chemical treatment to concentrate the mineral.
      a. separate Al_2O_3 in bauxite from Fe_2O_3 impurities by treatment with hot aqueous NaOH
    4. Roasting - a process that involves heating the mineral in air.
      a. used to convert minerals to compounds that are more easily reduced
D. Reduction - the concentrated ore is reduced to the free metal, either by chemical reduction or by electrolysis.
    1. Method used depends on the activity of the metal as measured by E^o for reduction (see Table 21.2, page 877 in text).
      a. most active metals - most negative standard reduction potentials
        i. the most difficult to reduce
      b. the least active metals have the most positive standard reduction potentials
        i. easiest to reduce
    2. Au and Pt - inactive; found in nature in uncombined form.
    3. Cu, Ag, and Hg - occur as sulfide ores.
      a. easily reduced by roasting
    4. Cr, Zn, and W - more active metals.
      a. reduce metal oxides with a chemical reducing agent (C, H_2 or a more electropositive metal)
    5. Most active metals - produced by electrolytic reduction.
      a. no chemical reducing agent strong enough to reduce compounds
E. Refining - purification methods used include distillation, chemical purification, and electrorefining.
    1. Zn - volatile enough to be refined by distillation.

3⊠
    2. Ni - purified by the Mond process.
      a. a chemical method involving formation and subsequent decomposition of $Ni(CO)_4$
    3. Cu - purified by electrorefining.
      a. an electrolytic process in which Cu is oxidized to Cu^{2+} at an impure Cu anode, and Cu^{2+} from aqueous $CuSO_4$ is reduced to Cu at a pure Cu cathode

Iron and Steel

A. Metallurgy of iron.
    1. Special technological importance.
      a. major constituent of steel
B. Production of iron - carbon monoxide reduction of Fe ore in a blast furnace (see Fig. 21.5, page 880 in text)
    1. Overall reaction: Fe_2O_3 (s) + 3 CO (g) → 2 Fe(l) + 3 CO_2 (g).
C. Cast iron (pig iron) obtained from a blast furnace; brittle material containing 4% elemental carbon and smaller amount of other impurities.

4⊠
D. Basic oxygen process - used for the purification and conversion of iron to steel.

1. Expose molten iron from the blast furnace to a jet of pure oxygen gas in a furnace lined with basic oxides.
   a. oxidizes impurities
   b. acidic oxides produced react with CaO and form a slag that is poured off
2. Produces steels with about 1% carbon and very small amounts of P and S.

F. Composition of liquid steel - monitored by chemical analysis.

Bonding in Metals

A. Metals - malleable, ductile, lustrous, and good conductors of heat and electricity.
   1. Malleable - can be hammered into sheets.
   2. Ductile - can be drawn into wires.
   3. Luster - shiny appearance.

B. Can understand properties by examining the bonding in metals.

C. Valence electrons can't be localized in a bond between any particular pair of atoms.
   1. Delocalized and belong to the crystal as a whole.

5⊠   D. Electron-sea model - the crystal is a 3-dimensional array of metal cations immersed in a sea of delocalized electrons that are free to move throughout the crystal.
   1. Delocalized, mobile valence electrons act as electrostatic glue that holds the metal cations together.
   2. Simple qualitative explanation for the electrical and thermal conductivity of metals.
      a. mobile electrons are free to move from a negative electrode to a positive electrode when subjected to an electrical potential
      b. mobile electrons conduct heat by carrying kinetic energy from one part of the crystal to another
      c. electrons extend in all directions, therefore, when a metallic crystal is deformed, no localized bonds are broken
         i. explains malleability and ductility

5⊠   E. Molecular orbital theory for metals.
   1. The number of molecular orbitals formed is the same as the number of atomic orbitals combined.
   2. The difference in energy between successive MOs decreases as the number of atoms increases.
      a. MOs merge into an almost continuous band of energy levels
      b. MO theory for metals is called band theory
      c. bottom half of band consists of bonding MOs
      d. top half of band consists of antibonding MOs
   3. Each electron in a metal has a discrete kinetic energy and a discrete velocity.
      a. depend on the particular MO
      b. increase from the bottom to the top of a band
   4. One-dimensional metal wire.
      a. energy levels within a band occur in degenerate pairs
         i. one - electrons moving to the right
         ii. other set - electrons moving to the left
      b. no electrical potential - no net electric current in either direction
         i. two sets are equally populated
      c. electrical potential - electrons moving right (toward the + battery terminal) are accelerated while electrons moving left (toward the - battery terminal) are decelerated
         i. electrons moving very slowly left change direction
         ii. number of electrons moving right > number of electrons moving left
         iii. net electric current
   5. Electrical potential can shift electrons from one set of energy levels to the other only if the band is partially filled.

      a. electrical insulators - materials that have only completely filled bands

      b. metals - materials that have partially filled bands

      c. can have *s* and *p* subshells sufficiently close in energy so that the *s* and *p* bands overlap, resulting in a partially filled composite band

    6. Transition metals - *d* band overlaps the *s* band.

      a. composite band has six MO's per metal atom

      b. maximum bonding for metals with six valence electrons per metal atom

6⊠   c. melting points of transition metals go through a maximum at group 6B

Semiconductors

    A. Semiconductor - a material that has an electrical conductivity intermediate between that of a metal and that of an insulator.

    B. Valence band - the bonding MOs.

    C. Conduction band - the higher-energy, antibonding MOs.

    D. Band gap - the separation of the valence band and the conduction band in terms of energy.

    E. Insulator - no vacant MOs in the valence band; large band gap.

      1. Can't excite electrons within the valence band.

      2. Can't excite electrons to vacant MOs in the conduction band due to the large band gap.

    F. Metal conductors - no energy gap between the highest occupied and lowest unoccupied MOs.

    G. Semiconductor - smaller band gap than insulators.

      1. A few electrons can jump the gap and occupy the higher-energy conduction band.

      2. Partially filled conduction band and valence band .

        a. unoccupied MOs in valence band result from electrons that jumped the gap

      3. Electrical potential can accelerate the electrons in the partially filled bands.

        a. semiconductor conducts a small current

      4. Conductivity increase with increasing temperature.

        a. electrons with enough energy to jump the band gap increases

7⊠   H. Doping - a process in which the conductivity of a semiconductor is increased by adding certain impurities in small amounts.

      1. Add an impurity with more electrons than the semiconductor - *n*-type semiconductor.

        a. charge carriers are negative electrons

      2. Example of *n*-type semiconductor: Si doped with P.

        a. each P atom occupies a Si position and introduces an extra electron

        b. extra electron occupies the conduction band

        c. more electrons in conduction band causes the conductivity to be higher

      3. Add an impurity with less electrons than the semiconductor - *p*-type semiconductor.

        a. creates positive holes in the valence band

Superconductors

8⊠   A. Superconductor - a material that loses all electrical resistance below a characteristic temperature, the superconducting transition temperature (T_c).

      1. Below T_c, once an electric current is started i will flow indefinitely without loss of energy.

    B. 1986 - $Ba_xLa_{2-x}CuO_4$ was discovered to have a T_c = 35 K.

      1. Soon after, other copper-containing oxides were discovered with even higher T_c.

      2. Ceramics - nonmetallic inorganic solids.

        a. most are electrical insulators

    C. No generally accepted theory of superconductivity in ceramic superconductors.

      1. Infinitely extended layers of Cu and O and the fractional oxidation number of Cu appear to play a role in the current flow.

    D. Superconductors can levitate a magnet.

      1. Cool a superconductor below T_c and lower a magnet toward it, the magnet and superconductor repel each other.

E. Present applications of conventional superconductors require the use of liquid helium (expensive and requires cryogenic equipment).
   1. Can use liquid N_2 with new, higher temperature superconductors.
      a. abundant refrigerant that is cheaper than milk
F. Problems with high-temperature superconductors.
   1. Brittle powders with high melting points.
   2. Not easily made into wires and coils needed for electrical equipment.
   3. Currents carried at 77 K (boiling point of liquid N_2) are too low for practical applications.
G. High-temperature superconductors based on fullerene - the allotrope of carbon that contains C_{60} molecules.
   1. K_3C_{60} - metallic conductor at room temperature; superconductor at 18 K.
   2. Fullerides may prove to be better materials than the Cu-O ceramics for making superconducting wires.
      a. fullerides are 3-dimensional superconductors

Ceramics

9⊠
A. Ceramics - inorganic, nonmetallic, nonmolecular solids, including both crystalline and amorphous materials, such as glasses.
   1. Traditional silicate ceramics - made by heating aluminosilicate clays to high temperatures.
   2. Advanced ceramics - materials that have high-tech engineering, electronic, and biomedical applications.
      a. oxide ceramics - alumina (Al_2O_3)
      b. nonoxide ceramics - silicon carbide and silicon nitride
B. Properties of ceramics - superior to those of metals.
   1. Higher melting points.
   2. Stiffer, harder, and more resistant to wear and corrosion.
   3. Maintain their strength at high temperatures.
   4. Less dense than steel.
      a. attractive lightweight, high-temperature materials for replacing metal components in aircraft, space vehicles, and automobiles
C. Strong chemical bonding.
   1. Covalent network solid.
      a. highly directional covalent bonds prevent planes of atoms from sliding over one another when the solid is subjected to the stress of a load or an impact
      b. solid can't deform to relieve the stress
      c. maintains it shape up to a point, but then the bonds give way suddenly
   2. Oxide ceramics - bonding is largely ionic.
      a. behave similarly to covalent network solids
   3. Metals - deform under stress because their planes of metal cations can slide easily in the electron sea.
   4. Metals dent, ceramics shatter.
D. Ceramic processing - the series of steps that leads from raw material to the finished ceramic object.
   1. Determines the strength and the resistance to fracture of the product.
9⊠
   2. Sintering - a process in which the particles of the powder are welded together without completely melting.
      a. occurs below the melting point
      b. crystal grains grow larger and the density of the material increases as the void spaces between particles disappear.
      c. need high-purity, fine powders that are tightly compacted prior to sintering
         i. Impurities and remaining voids can lead to microscopic cracks

10⊠   E. Sol-gel method - method used for preparing high-purity, fine powders that can be tightly compacted.
         1. Synthesis of a metal oxide powder from a metal alkoxide (a compound derived from a metal and an alcohol).

9⊠   2. Sol - a colloidal dispersion consisting of extremely fine particles having a diameter of only 0.001 to 0.1 μm.

9⊠   3. Gel - a rigid, gelatin-like material.

Composites

     A. Ceramic composite - a hybrid material in which a ceramic powder is mixed, prior to sintering, with fibers of a second ceramic material.
         1. Combines the advantageous properties of both components.
         2. Whiskers - tiny fiber-shaped particles that are very strong because they are single crystals.
     B. Fibers and whiskers increase the strength and fracture toughness of composite materials.
         1. Most of the chemical bonds are aligned along the fiber axis giving the fibers great strength.
         2. Can deflect cracks.
             a. prevents cracks from moving cleanly in one direction
         3. Can bridge cracks.
             a. holds two sides of a crack together

11⊠   C. Composites in which the two phases are different types of materials.
         1. Ceramic-metal composites (cermets).
         2. Ceramic-polymer composites.
         3. High strength-to-weight ratios.
     D. Ceramic fibers used in composites are usually made by high-temperature methods.

Self-Test

This section is intended to test your knowledge of the material covered in this chapter. Think through these problems and make certain you understand what is going on. Ask yourself if your answer makes sense. Many of these questions are linked to the chapter learning goals. Therefore, successful completion of these problems indicates you have mastered the learning goals for this chapter. You will receive the greatest benefit from this section if you use it as a mock exam. You will then discover which topics you have mastered and which topics you need to study in more detail.

True/False

1. The early transition metals occur as sulfides in nature.

2. The *s*-block elements occur as oxides in nature.

3. The most active metals have the most negative standard reduction potentials and are the most difficult to reduce.

4. Insulators have small band gaps.

5. The electrons in a metal extend in all directions and therefore, the metal breaks when it is hammered.

6. The difference in energy between successive MOs in a metal decreases as the number of atoms increases.

7. The valence band in a semiconductor contains the higher-energy, antibonding MOs.

8. When an impurity with more electrons than the semiconductor is added to the semiconductor, an n-type semiconductor is created.

9. Most ceramics are electrical conductors.

10. One of the advantages of the new superconductors that have been synthesized is the ability of these materials to conduct at liquid nitrogen temperatures.

Matching

Minerals

a. the science and technology of extracting metals form their ores

Ore

b. a material that loses all electrical resistance below a characteristic temperature

Metallurgy

c. inorganic, nonmetallic, nonmolecular solids, including both crystalline and amorphous materials

Bayer process

d. tiny fiber-shaped particles that are very strong because they are single crystals

Semiconductor

e. a hybrid material in which a ceramic powder is mixed, prior to sintering

Doping

f. a process in which the particles of the powder are welded together without completely melting

Superconductor

g. method used for preparing high-purity, fine powders that can be tightly compacted

Ceramics

h. a material that has an electrical conductivity intermediate between that of a metal and that of an insulator

Sintering

i. the crystalline, inorganic constituents of the rocks that make up the earth's crust

Sol-gel method

j. a process in which the conductivity of a semiconductor is increased by adding certain impurities in small amounts

Composite

k. a complex mixture of a metal-containing mineral and economically worthless material called gangue

Whisker

l. a chemical treatment used to separate Al_2O_3 in bauxite from impurities by treatment with hot aqueous NaOH

Fill-in-the-Blank

1. The chemical composition of the most common ores correlates with the _____ _____.

2. Roasting is a process that involves _____ and is used to convert minerals to compounds that are _____.

3. The three steps of the metallurgical process for producing metals are _____ _____.

4. The most active metals are produced by electrolytic reduction because _____ _____.

5. The overall reaction for the production of iron using a blast furnace is _____ _____.

6. The electron-sea model for bonding in metals assumes that the crystal is a 3-dimensional array of _____ immersed in a sea of _____ that are free to _____.

7. Based on the electron-sea model, thermal conductivity of metals results from the mobile electrons _____.

8. According to the band theory for bonding in metals, an electrical potential can shift electrons from one set of energy levels to the other only if the _____.

9. Electrical insulators are materials that have only _____.

10. Metals are materials that have _____.

11. The separation of the valence band and the conduction band in terms of energy is called the _____.

12. An insulator is unable to conduct electricity because it can't excite electrons within the filled _____ and the band gap is too _____ to excite electrons to the vacant MOs in the _____.

13. The conductivity of a semiconductor increases with increasing temperature because the number of electrons with enough energy to jump the _____ increases.

14. *p*-type semiconductors are _____ with impurities that create _____ in the valence band.

15. The fullerides may prove to be better materials than the Cu-O ceramics for making superconducting wires because _____.

16. A ceramic breaks when subjected to a stress because _____

_____.

17. _____ increase the strength and fracture toughness of composite

materials.

Solutions

True/False
1. F. The early transition metals occur in nature as oxides because the less electronegative metals form compounds by losing electrons to highly electronegative nonmetals.
2. F. The *s*-block metals are strongly basic and too reactive to exist with acidic oxides. The *s*-block metals occur as carbonates, silicates, and chlorides (for Na and K).
3. T
4. F. Insulators have very large band gaps, which is why they are unable to conduct an electrical current.
5. F. Because the electrons in a metal extend in all directions, no localized bonds are broken when the metal is hammered and therefore, it doesn't break.
6. T
7. F. The valence band contains the lower-energy, bonding MOs.
8. T
9. F. Most ceramics are electrical insulators.
10. T

Matching
    Minerals - i
    Ore - k
    Metallurgy - a
    Bayer process - l
    Semiconductor - h
    Doping - j
    Ceramics - c
    Superconductors - b
    Sintering - f
    Sol-gel method - g
    Composite - e
    Whiskers - d

Fill-in-the-Blank
1. location of the metal in the periodic table
2. heating the mineral in air; are more easily reduced
3. concentration of the ore and chemical treatment prior to reduction; reduction of the mineral to the free metal; refining or purification of the metal
4. no chemical reducing agent is strong enough to reduce the compounds.
5. $Fe_2O_3 \, (s) + 3 \, CO \, (g) \rightarrow 2 \, Fe \, (l) + 3 \, CO_2 \, (g)$
6. metal cations; delocalized electrons; move about the crystal
7. ability to carry kinetic energy from one part of the crystal to another
8. the band is partially filled
9. completely filled bands
10. partially filled bands
11. band gap

12. valence band; large; conduction band
13. band gap
14. doped; positive holes
15. they are 3-dimensional superconductors
16. the highly directional covalent bonds prevent the planes of atoms from sliding over one another
17. Fibers and whiskers

CHAPTER 22

NUCLEAR CHEMISTRY

Chapter Learning Goals

1⊠ Summarize the differences between nuclear reactions and chemical reactions.

2⊠ Write balanced equations for nuclear reactions, identifying the types of radiation and nuclides involved.

3⊠ Use the integrated first-order rate law, solving for half-life, decay constant, or ratio of nuclei initially present to nuclei present at time t.

4⊠ Use the neutron/proton plot for stable isotopes to determine whether a given nuclide is expected to be stable or unstable.

5⊠ Calculate mass defects and binding energies for nuclides. Use values of binding energy per nucleon to compare the relative stabilities of two nuclides.

6⊠ Classify nuclear reactions as fission or fusion. Calculate the energy released by a nuclear fission or fusion reaction.

7⊠ Write balanced equations for nuclear transmutations.

8⊠ Show how radiocarbon dating is used to determine the age of an object.

Chapter in Brief

Nuclear chemistry is the study of the properties and reactions of atomic nuclei. In this chapter you will learn about the characteristics of nuclear reactions, the different types of radioactive particles involved in these reactions, the first-order decay rates for these reactions, as well as the energy changes that occur during reaction. You will examine the factors that tend to induce nuclear stability and the phenomena of nuclear fission and fusion, and nuclear transmutation. You will explore how radioactivity is detected and measured along with the biological effects of radiation. Finally, you will study some of the applications of nuclear chemistry.

Nuclear Reactions and Their Characteristics

    A. An atom is characterized by its atomic number, Z and its mass number, A.

        1. Z - written as a subscript to the left of the element symbol; gives the number of protons in the nucleus.

        2. A - written as a superscript to the left of the element symbol, gives the total number of nucleons.

           a. nucleons - a general term for both protons (p) and neutrons (n)

    B. Nuclides (isotopes) - atoms with identical atomic numbers, but different mass numbers.

    C. Nuclear reactions - reactions that change the nucleus.

        1. Indicate the electrons as $_{-1}^{0}e$.

           a. superscript 0 - the mass of an electron is essentially zero when compared with that of a proton or neutron

           b. subscript (-1) - the charge of an electron is -1

1⊠   D. Differences between nuclear reactions and chemical reactions.

        1. Nuclear reactions - change in an atom's nucleus; chemical reaction - a change in distribution of the outer shell electrons around an atom.

        2. Different nuclides of an element have essentially the same behavior in chemical reactions, but have different behavior in nuclear reactions.

        3. Changes in temperature or pressures or the addition of a catalyst, do not affect the rate of a nuclear reaction.

      4. The nuclear reaction of an atom is essentially the same regardless of whether the atom is in an element or a compound.

      5. The energy changes accompanying nuclear reactions are far greater than those accompanying chemical reactions.

2⊠   Nuclear Reactions and Radioactivity

A. Radioactive - nuclei that spontaneously emit radiation.
    1. Three common types of radiation with different properties.
      a. alpha (α)
      b. beta (β)
      c. gamma (γ)

B. Alpha (α) radiation - a stream of particles that consist of two protons and two neutrons.
    1. Repelled by a positively charged electrode.
    2. Attracted by a negatively charge electrode.
    3. Mass-to-charge ratio as $^{4}_{2}He^{2+}$.
    4. Emission of an α particle reduces the mass number of the nucleus by four and reduces the atomic number by two.
      a. $^{238}_{92}U \rightarrow {}^{4}_{2}He + {}^{234}_{90}Th$
    5. Common for heavy radioactive isotopes (radionuclides).

C. Balanced nuclear reaction.
    1. Sums of the nucleons on both sides are equal.
    2. Sums of the charges on both sides are equal.
    3. Not concerned with ionic charges on atoms.
      a. irrelevant to nuclear disintegration
      b. disappear

D. Beta (β) radiation - occurs when a neutron in the nucleus spontaneously decays into a proton plus an electron ($^{0}_{-1}e$, or β^-), which is then ejected.
    1. Product nucleus - same mass number but a higher atomic number (creates a proton).

E. Gamma (γ) radiation - electromagnetic radiation of very high energy and short wavelength.
    1. Stream of high energy photons.
    2. Almost always accompanies α and β emission.
      a. mechanism for release of energy
    3. γ emission is often not shown in nuclear equations.
      a. doesn't change either the mass number or the atomic number of the product nucleus

F. Positron emission and electron capture.
    1. Positron emission - the conversion of a proton in the nucleus into a neutron plus an ejected positron.
      a. positron ($^{0}_{1}e$, or β^+) - a particle with the same mass as an electron but opposite charge
      b. decrease in atomic number of the product nucleus
      c. no change in the mass number
    2. Electron capture - a proton in the nucleus captures an inner-shell electron which is converted into a neutron.
      a. no change in the mass number of the product nucleus
      b. atomic number decreases by one

3⊠   Radioactive Decay Rates

A. Radioactive decay is kinetically a first-order process whose rate is proportional to the number of radioactive nuclei N in a sample:
    1. Decay rate = $k \times N$.
    2. k = first-order rate constant called the decay constant.

3. Integrated rate law

$$\ln\left(\frac{N}{N_\circ}\right) = -kt \; .$$

   a. N_0 = number of radioactive nuclei originally present; N = number remaining at time t

B. Radioactive decay is characterized by a half-life, $t_{1/2}$.
   1. Time required for the number of radioactive nuclei in a sample to drop to one-half of the initial value.
   2. Each passage of a half-life causes the decay of one-half of whatever sample remains.
   3. The half-life is the same no matter what the size of the sample, the temperature, or any other external condition.
   4. $t_{1/2} = \dfrac{0.693}{k}$ and $k = \dfrac{0.693}{t_{1/2}}$.
   5. To calculate the ratio of remaining and initial amounts of radioactive sample N/N_0 at any time t.

      a. $\ln\left(\dfrac{N}{N_\circ}\right) = -0.693\left(\dfrac{t}{t_{1/2}}\right)$

EXAMPLE:

The radioactive decay of Tl-206 to Pb-206 has a half-life of 4.20 min. Starting with 0.250 g of Tl-206, calculate the grams of Tl-206 left after 60.00 min.

SOLUTION:

We can use the equation $\ln\left(\dfrac{N}{N_\circ}\right) = -0.693\left(\dfrac{t}{t_{1/2}}\right)$, and solve for the mass of Tl-206 by realizing that the mass of a sample is proportional to the number of nuclei.

$$\frac{m}{m_0} = \frac{N}{N_0}$$

$$\ln\left(\frac{m}{0.250 \text{ g}}\right) = -0.693\left(\frac{60.00}{4.20}\right) = -9.900 \; ; \qquad \frac{m}{0.250 \text{ g}} = 5.017 \times 10^{-5}; \quad N = 1.254 \times 10^{-5} \text{ g}$$

Nuclear Stability

   A. Stable radioactive isotope - one that can be prepared and whose half-life can be measured.
      1. Unstable - those isotopes that can't be prepared or that decay too rapidly for their half-lives to be measured.
      2. Nonradioactive (stable indefinitely) - isotopes that do not undergo radioactive decay.
   B. Neutron/proton ratio in the nucleus determines if an isotope is radioactive.
   C. Plot of number of neutrons (y axis) *versus* number of protons (x axis) (see Fig. 22.3, page 915 in text).
      1. Stable nuclides fall in a curved band called the peninsula of nuclear stability.
      2. Sea of instability - area on either side of the peninsula of nuclear stability.
         a. represents the large number of unstable neutron/proton combinations
      3. Island of stability - area predicted to exist for a few superheavy nuclides near 114 protons and 184 neutrons.
   D. Generalizations from plot.
      1. Every element has at least one radioactive isotope.

2. Hydrogen is the only element whose most abundant stable isotope contains more protons than neutrons.
3. The ratio of neutrons to protons gradually increases for elements heavier than calcium.
4. All isotopes beyond bismuth-209 are radioactive.
5. Nonradioactive isotopes generally have an even number of neutrons.
E. Neutrons function as a kind of nuclear glue that holds nuclei together by overcoming proton-proton repulsions.
F. Magic numbers of protons or neutrons - 2, 8, 20, 28, 50, 82, 126.
1. Give rise to particularly stable nuclei.
2. Nucleus with a magic number of either protons or neutrons is unusually stable.
3. Analogous to chemical stability brought about by an octet of electrons.
G. Trends that are apparent from the peninsula of nuclear stability.
1. Elements with an even atomic number have a larger number of nonradioactive nuclides than do elements with an odd atomic number.
2. Radioactive nuclei on the right side of the peninsula undergo nuclear disintegration by positron emission, by electron capture, or by alpha emission.
a. lower neutron/proton ratio
b. processes increase the neutron/proton ratio
3. Nuclei on the left side of the peninsula emit beta particles.
a. higher neutron/proton ratios
b. process decreases the neutron/proton ratio
H. Some nuclides can't reach a nonradioactive nucleus in a single emission.
1. Undergo a decay series of disintegrations.

5⊠   **Energy Changes During Nuclear Reactions**
A. Can calculate the energy change on formation of the nucleus from isolated protons and neutrons.
1. $\Delta E = \Delta mc^2$.
a. relates the energy change of a nuclear process to a corresponding mass change
B. Mass defect - the loss in mass that occurs when protons and neutrons combine to form a nucleus.
1. Lost mass is converted into energy.
a. binding energy - energy that holds the nucleons together
b. calculate from the Einstein equation
c. usually expressed on a per-nucleon basis using the electron volts as the energy unit
i. $1 \text{ eV} = 1.60 \times 10^{-19}$ J
C. Plot binding energy per nucleon.
1. Higher binding energy corresponds to higher stability.
D. Mass and energy are interconvertible.
1. Laws of conservation of mass and conservation of energy must be combined.
a. mass and energy are interconverted in a chemical reaction, but the combination of the two is conserved.

EXAMPLE:

Calculate the nuclear binding energy in MeV/nucleon for $^{174}_{77}\text{Ir}$, atomic mass = 173.966 66 amu.

SOLUTION: First calculate the total mass of the nucleons (97 n + 77 p)

Mass of 97 neutrons	= (97) (1.008 66 amu)	= 97.840 02
Mass of 77 protons	= (77)(1.007 28 amu)	= 77.560 56
Mass of 97 n + 77 p		= 175.400 58 amu

362

Determine the mass of the $^{174}_{77}\text{Ir}$ nucleus by subtracting the mass of 77 e⁻ from the atomic mass.

mass of nucleus = 173.966 66 - 0.042 24 = 173.924 42

Next, solve for the mass defect by subtracting the mass of the ^{174}Ir nucleus from the ^{174}Ir atom.

Mass defect = mass of nucleons - mass of nucleus
= (175.40058 amu) - (173.924 42 amu)
= 1.476 16 amu

Now convert the mass defect from amu to gram to get the mass defect in grams per mole.

$$(1.476\ 16\ \text{amu})\left(1.660\ 54\times10^{-24}\ \frac{\text{g}}{\text{amu}}\right)\left(6.022\times10^{23}\ \text{mol}^{-1}\right)=1.476\ \frac{\text{g}}{\text{mol}}$$

Now, use the Einstein equation to convert the mass defect into the binding energy.

$$\Delta E = \Delta mc^2 = \left(1.476\ \frac{\text{g}}{\text{mol}}\right)\left(10^{-3}\ \frac{\text{kg}}{\text{g}}\right)\left(3.00\times10^8\ \frac{\text{m}}{\text{s}}\right)^2 = 1.33\times10^{14}\ \text{J / mol} = 1.339\times10^{11}\ \text{kJ / mol}$$

We now convert to MeV/nucleon

$$\frac{1.33\times10^{14}\ \dfrac{\text{J}}{\text{mol}}}{6.022\times10^{23}\ \dfrac{\text{nuclei}}{\text{mol}}}\times\frac{1\ \text{MeV}}{1.60\times10^{-13}\ \text{J}}\times\frac{1\ \text{nucleus}}{174\ \text{nucleons}} = 7.93\ \text{MeV / nucleon}$$

6⊠ **Nuclear Fission and Fusion**
    A. Lighter and heavier elements are less stable than midweight elements near iron-56.
        1. Heavy nuclei - gain stability and release energy if they fragment to yield midweight elements.
        2. Light nuclei can gain stability and release energy if they fuse together.
    B. Fission - the fragmentation of heavy nuclei.
        1. Nuclei break into fragments when struck by neutrons.
        2. Doesn't occur in exactly the same way each time.
        3. Uranium-235 - more than 100 different fission pathways.
           a. more frequently occurring pathways
$$^{1}_{0}\text{n} + {}^{235}_{92}\text{U} \rightarrow {}^{142}_{56}\text{Ba} + {}^{91}_{36}\text{Kr} + 3\ {}^{1}_{0}\text{n}$$
           b. three neutrons released induce nine more fissions → 27 neutrons
           c. chain reaction - a reaction that continues to occur even if the supply of neutrons from outside is cut off
           d. small sample size - many of the neutrons escape before initiating additional fission events
           e. critical mass - a sufficient amount of radioactive nuclide that allows the chain reaction to become self-sustaining
        4. Can calculate the amount of energy released during nuclear fission using the mass defect and Einstein equation.
           a. use the masses of the atoms corresponding to the relevant nuclei

C. Nuclear reactors.
   1. Principle of a nuclear reactor - a subcritical amount of uranium fuel is placed in a containment vessel surrounded by circulating coolant, and control rods are added.
      a. control rods - mad of boron and cadmium
         i. absorb and regulate the flow of neutrons
         ii. raised and lowered to maintain fission at a controlled rate.
      b. energy from the controlled fission heats the circulating coolant
      c. heated circulating coolant produces steam to drive a turbine and produce electricity
D. Nuclear fusion - the joining together of light nuclei.
   1. Release enormous amounts of energy.
   2. Fusion of hydrogen nuclei - a potential power source.
      a. hydrogen isotopes are cheap and plentiful
      b. fusion products are nonradioactive and nonpolluting
   3. Technical problems to achieving a practical and controllable fusion method.
      a. to initiate the process, $T = 40 \times 10^6$ K

7⊠ **Nuclear Transmutation**

A. Nuclear transmutation - the change of one element into another brought about by bombardment of an atom with a high-energy particle.
   1. Creates an unstable nucleus which causes a nuclear change to occur.
B. All transuranium elements have been produced by bombardment reactions.

Detecting and Measuring Radioactivity

A. Ionizing radiation - high-energy radiation of all kinds.
   1. Interaction of the radiation with a molecule knocks an electron from the molecule.
      a. Molecule $\xrightarrow{\text{radiation}}$ ion + e⁻
B. Devices used for measuring radiation.
   1. Photographic film badge.
   2. Geiger counter.
   3. Scintillation counter.
C. Expression of radiation intensity depends upon what is being measured.
   1. Becquerel - SI unit for measuring the number of radioactive disintegrations occurring each second in a sample.
      a. 1 Bq = 1 disintegration/s
   2. Curie (Ci): 1 Ci = 3.7×10^{10} disintegrations/s.
   3. Gray - SI unit for measuring the amount of energy absorbed per kilogram of tissue exposed to a radiation source.
      a. 1 Gy = 1 J/kg
      b. 1 rad = 0.01 Gy
   4. Sievert (Sv) - SI unit that measures the amount of tissue damage caused by radiation.
      a. 1 rem (roentgen equivalent for man) = 0.01 Sv

Biological Effects of Radiation

A. The effects of ionizing radiation on the human body depend on:
   1. The kind of radiation and its energy.
   2. The length of exposure.
   3. Whether the source is outside or inside the body.

Applications of Nuclear Chemistry

8⊠ A. Radiocarbon dating - depends on the slow and constant production of radioactive carbon-14 in the upper atmosphere by neutron bombardment of nitrogen atoms.
   1. $^{14}_{7}\text{N} + ^{1}_{0}\text{n} \rightarrow ^{14}_{6}\text{C} + ^{1}_{1}\text{H}$

2. $^{14}C \rightarrow {}^{14}CO_2$ which mixes with $^{12}CO_2$ and is taken up by plants during photosynthesis.
3. Ratio of ^{14}C to ^{12}C in a living organism is the same as that in the atmosphere.
4. $^{14}C/^{12}C$ decreases when the organism dies.
   a. ^{14}C undergoes radioactive decay.
   $$^{14}_{6}C \rightarrow {}^{14}_{7}N + {}^{0}_{-1}e$$
5. For ^{14}C, $t_{1/2} = 5715$ y.
6. Can determine how long ago the organism died by measuring the amount of ^{14}C remaining.
7. Use other radionuclides for dating rocks.
   a. $t_{1/2} = 4.46 \times 10^9$ y for ^{238}U
   b. ^{238}U decays through a series to give ^{206}Pb
   c. measure $^{238}U/^{206}Pb$ ratio can determine the age of uranium-containing rock

B. Medical uses of radioactivity - grouped into four classes.
   1. In vivo procedures.
   2. In vitro procedures.
   3. Radiation therapy.
   4. Imaging procedures.
C. In vivo procedures - studies that take place inside the body.
   1. Assess the functioning of a particular organ or body system.
   2. A radiopharmaceutical agent is administered, and its path in the body is determined by analysis of blood or urine samples.
D. In vitro procedures - studies that take place outside the body.
   1. Radioimmunoassay - techniques which measure small concentrations of substances in body fluids.
   2. Takes advantage of antibodies to bind to antigens.
E. Therapeutic procedures - radiation is used as a weapon to kill diseased tissue.
   1. Involve either external or internal sources of radiation.
   2. Internal radiation therapy is a much more selective technique than external therapy.
F. Imaging procedures - give diagnostic information about the health of body organs by analyzing the distribution pattern of radionuclides introduced into the body.
   1. A radiopharmaceutical agent known to concentrate in a specific tissue or organ is injected into the body and its distribution pattern is monitored by external radiation detectors.
G. Magnetic resonance imaging (MRI) uses radio waves to stimulate certain nuclei in the presence of an extremely powerful magnetic field.
   1. Stimulated nuclei give off a signal that can be measured, interpreted, and correlated with their environment in the body.

Self-Test

This section is intended to test your knowledge of the material covered in this chapter. Think through these problems and make certain you understand what is going on. Ask yourself if your answer makes sense. Many of these questions are linked to the chapter learning goals. Therefore, successful completion of these problems indicates you have mastered the learning goals for this chapter. You will receive the greatest benefit from this section if you use it as a mock exam. You will then discover which topics you have mastered and which topics you need to study in more detail.

General Questions

1. Summarize the differences between nuclear reactions and chemical reactions.

2. Balance the following nuclear reactions:

   a. $^{81}_{36}Kr + ^{0}_{-1}e \rightarrow$

   b. $^{104}_{47}Ag \rightarrow ^{0}_{1}e +$

   c. $^{73}_{31}Ga \rightarrow ^{0}_{-1}e +$

   d. $^{104}_{48}Cd \rightarrow ^{104}_{47}Ag +$

3. Write nuclear equations for:

   a. alpha emission by $^{11}_{5}B$

   b. beta emission by $^{121}_{51}Sb$

   c. neutron emission by $^{70}_{35}Br$

   d. proton emission by $^{41}_{19}K$

4. What is the age of a bone fragment that shows an average of 3.5 disintegrations per minute per gram of carbon? The carbon in living organisms undergoes an average of 15.3 disintegrations per minute per gram, and the half-life of ^{14}C is 5715 y.

5. $t_{1/2}$ for selenium-75 is 120.0 days. What is the decay constant for this radioactive element?

6. The half-life of strontium-90 is 28.1 years. How much of a 0.500 mg sample of strontium-90 remains after 5.00 years?

7. From the neutron/proton plot for stable isotopes, determine if an element with 48 protons and 42 neutrons will be stable. If this element is unstable, what type of nuclear processes can incease the stability?

8. Calculate the mass defect and binding energy for ^{92}Mo (atomic mass = 91.906 91 amu).

9. How much energy in kJ/mol is released in the following reaction:
$^{1}_{0}n + ^{235}_{92}U \rightarrow ^{137}_{52}Te + ^{97}_{40}Zr + 2\,^{1}_{0}n$

   The atomic masses are: ^{235}U (235.0439 amu); ^{137}Te (136.9254 amu); ^{97}Zr (96.991 10 amu); and n (1.008 66 amu); 1 amu = 1.6605×10^{-27} kg

10. Write chemical reactions for the following nuclear transmutations.

    a. Curium-242 reacts with an alpha particle to produce californium-245

    b. ^{14}N reacts with an alpha particle to produce ^{17}O

11. What is the age of a piece of wood that shows an average of 2.2 disintegrations per minute per gram of carbon? The carbon in living organisms undergoes an average of 15.3 disintegrations per minute per gram, and the half-life of ^{14}C is 5715 y.

Solutions

1. Nuclear reactions cause a change in an atom's nucleus; chemical reactions cause a change in the distribution of the outer shell electrons around an atom. Different nuclides of an element have essentially the same behavior in chemical reactions but have different behavior in nuclear reactions. Changes in temperature or pressures or the addition of a catalyst do not affect the rate of a nuclear reaction. The nuclear reaction of an atom is essentially the same whether it is combined with other elements or not. The energy changes accompanying nuclear reactions are far greater than those accompanying chemical reactions.

2. a. $^{81}_{36}Kr + ^{0}_{-1}e \rightarrow ^{81}_{35}Br$

   b. $^{104}_{47}Ag \rightarrow ^{0}_{1}e + ^{104}_{46}Pd$

   c. $^{73}_{31}Ga \rightarrow ^{0}_{-1}e + ^{73}_{32}Ge$

   d. $^{104}_{48}Cd \rightarrow ^{104}_{47}Ag + ^{0}_{1}e$

3. a. $^{11}_{5}B \rightarrow ^{4}_{2}He + ^{7}_{3}Li$

   b. $^{121}_{51}Sb \rightarrow ^{0}_{-1}e + ^{121}_{52}Te$

   c. $^{70}_{35}Br \rightarrow ^{1}_{0}n + ^{69}_{35}Br$

   d. $^{41}_{19}K \rightarrow ^{1}_{1}p + ^{40}_{18}Ar$

4. The ratio of the decay rate at any time t to the decay rate at time $t = 0$ is the same as the ratio of N and N_0.

$$\frac{\text{Decay rate at time } t}{\text{Decay rate at time } t = 0} = \frac{kN}{kN_0} = \frac{N}{N_0}$$

$$\ln\left(\frac{N}{N_o}\right) = -(0.693)\left(\frac{t}{t_{1/2}}\right);$$

Let $N = 3.5$ d/min/g and $N_0 = 15.3$ d/min/g. Solving for t

$$\ln\left(\frac{3.5}{15.3}\right) = -0.693\left(\frac{t}{5715 \text{ y}}\right); \quad t = 12,000 \text{ y}$$

5. $0.693 = kt_{1/2};$ $\quad 0.693 = k(120.0 \text{ days})$ $\quad k = 5.78 \times 10^{-3}$ days^{-1}

6. $\ln\left(\frac{N}{N_o}\right) = -(0.693)\left(\frac{t}{t_{1/2}}\right);$ $\quad \ln\left(\frac{N}{0.500}\right) = -0.693\left(\frac{5.00}{28.1}\right);$ $\quad \ln\left(\frac{N}{0.500}\right) = -0.123$

   Taking the antilog of both sides gives:

$$\frac{N}{0.500} = 0.884 \; ; \qquad N = 0.442 \text{g}$$

7. This particular element will not be stable because the neutron/proton ratio is less than 1. This element can undergo positron emission, electron capture, or alpha emission to increase the neutron/proton ratio and gain stability.

8. Total mass of the nucleons ($50 \, n + 42 \, p$)

Mass of 50 neutrons   = (50) (1.008 66 amu)   = 50.443 00
Mass of 42 protons   = (42) (1.007 28 amu)   = 42.305 76

Mass of 50 neutrons and 42 protons   = 92.738 76

Mass of nucleus = atomic mass - mass of 42 e^- = 91.906 91 - 0.02304 = 91.883 87

Mass defect = mass of nucleons - mass of nucleus
$$= 92.738\ 76 - 91.833\ 87$$
$$= 0.854\ 89$$

$$0.85489 \text{ amu} \times \left(1.660\ 54 \times 10^{-24} \; \frac{\text{g}}{\text{amu}}\right)\left(6.022 \times 10^{23} \text{ mol}^{-1}\right) = 0.8549 \; \frac{\text{g}}{\text{mol}}$$

$$\Delta E = \Delta mc^2 = \left(0.8549 \; \frac{\text{g}}{\text{mol}}\right)\left(10^{-3} \; \frac{\text{kg}}{\text{g}}\right)\left(3.00 \times 10^8 \; \frac{\text{m}}{\text{s}}\right)^2 = 7.69 \times 10^{13} \text{ J / mol} = 7.69 \times 10^{10} \text{ kJ / mol}$$

9. Mass change = (mass of reactants) - (mass of products)

$$= [1.008\ 66 + 235.0439] - [136.9254 + 96.9111 + (2 \times 1.008\ 66)]$$
$$= 0.118\ 74 \text{ amu}$$

$$\Delta E = \Delta mc^2 = \left(0.118\ 74 \text{ amu}\right)\left(1.6605 \times 10^{-27} \; \frac{\text{kg}}{\text{amu}}\right)\left(6.022 \times 10^{23} \text{ mol}^{-1}\right)\left(3.00 \times 10^8 \; \frac{\text{m}}{\text{s}}\right)^2$$

$$\Delta E = 1.07 \times 10^{13} \text{ kg} \cdot \text{m}^2 / \left(\text{s}^2 \cdot \text{mol}\right) = 1.07 \times 10^{13} \text{ J / mol}$$

10. a. $^{242}_{96}\text{Cm} + ^{4}_{2}\text{He} \rightarrow ^{245}_{98}\text{Cf} + ^{1}_{0}\text{n}$

    b. $^{14}_{7}\text{N} + ^{4}_{2}\text{He} \rightarrow ^{17}_{8}\text{O} + ^{1}_{1}\text{p}$

11. $\ln\left(\dfrac{2.2 \text{ d / min / g}}{15.3 \text{ d / min / g}}\right) = -0.693\left(\dfrac{t}{5715 \text{ y}}\right) \; ; \; t = 16{,}000 \text{ y}$

CHAPTER 23

ORGANIC CHEMISTRY

Chapter Learning Goals

1☒ Draw electron dot structures for isomers of simple alkanes.
2☒ Write condensed structures of organic molecules.

3☒ Determine which structures represent different molecules and which are merely different conformations of the same molecule.
4☒ Given the structure of an alkane, determine its IUPAC name and vice versa.
5☒ Given the structure of a cycloalkane, determine its IUPAC name and vice versa.
6☒ Predict the products of and write balanced chemical equations for reactions of alkanes with chlorine.
7☒ Identify functional groups in molecules. Draw structures of molecules containing functional groups listed in Table 23.3 on p. 958 of your text.

8☒ Given the structure of an alkene or alkyne, determine its IUPAC name and vice versa.
9☒ Predict the products and write balanced chemical equations for alkene addition reactions.
10☒ Given the structure of an aromatic compound, determine its IUPAC name and vice versa.
11☒ Predict the products and write balanced chemical equations for aromatic substitution reactions.
12☒ Classify molecules as alcohols, ethers, or amines. State the general properties of these classes of molecules.
13☒ Classify molecules as ketones or aldehydes.
14☒ Classify molecules as carboxylic acids, esters, or amides. State the general properties of these classes of molecules.
15☒ Give systematic names of carboxylic acids, esters, and amides.
16☒ Predict the products and write balanced chemical equations for carbonyl substitution reaction and reactions of carboxylic acids, esters, and amides with water.
17☒ Give the formula for a segment of the polymer formed as a result of (a) an alkene polymerization and (b) a polymerization involving molecules with two different functional groups.

Chapter in Brief

Organic chemistry, once defined as the study of compounds from living organisms, is now defined as the study of carbon compounds. In this chapter you will begin to explore the ability of carbon atoms to bond together, forming a vast array of compounds, such as long chains and rings. You will learn to draw the structures of and apply the IUPAC rules of nomenclature for the simple hydrocarbons. You will classify molecules based upon the functional groups present and learn the general properties of the different classes of molecules. Finally, you will learn how to predict the products and write balanced chemical equations for reactions involving the different classes of molecules.

The Nature of Organic Molecules

    A. Carbon is tetravalent.
    B. Organic molecules have covalent bonds.
    C. Organic molecules have polar covalent bonds when carbon bonds to an element on the right or left side of the periodic table.
    D. Carbon can form multiple covalent bonds by sharing more than two electrons with a neighboring atom.
    E. Organic molecules have specific 3-dimensional shapes, as predicted by the VSEPR model.
        1. C bonded to 4 atoms - bonds point towards corner of a tetrahedron.

　　　　2. C bonded to 3 atoms - bond angle = 120°.
　　　　3. C bonded to 2 atoms - bond angle = 180°.
　　F. Carbon uses hybrid atomic orbitals for bonding to other atoms.
　　　　1. Single bonded carbon (C bonded to 4 atoms) - uses sp^3 hybrids.
　　　　2. Doubly bonded carbons (C bonded to 3 atoms) - uses sp^2 hybrids which point towards the corners of an equilateral triangle.
　　　　　a. unhybridized p orbital which is perpendicular to the plane of sp^2 hybrid orbitals.
　　　　　b. overlap of unhybridized p orbitals on 2 C's forms π bond
　　　　3. Triply bonded carbons (C bonded to 2 atoms) – uses sp hybrid orbitals which are 180° to each other.
　　　　　a. 2 unhybridized p orbitals
　　　　　　i. oriented 90° from each other.
　　　　　　ii. overlap other unhybridized orbitals on 2 C's to form 2 π bonds.
　　G. Covalent bonds give organic compounds properties that are different from ionic salts.
　　　　1. Weak intermolecular forces in organic compounds lead to:
　　　　　a. lower melting and boiling points
　　　　　b. insolubility in water and lack of electrical conductivity

Alkanes and Their Isomers

　　A. Hydrocarbons - molecules that contain only carbon and hydrogen.
　　B. Alkanes - hydrocarbons that have only single bonds.
　　　　1. Saturated hydrocarbons.
　　C. Straight-chain alkanes - compounds with all their carbons connected in a row.
　　D. Branched-chain alkanes - those with a branching connection of carbons.
　　E. Isomers - molecules that have the same molecular formula, but different structures.
　　　　1. Different isomers are different chemical compounds.
　　　　　a. have different structures and different physical properties

Writing Organic Structures

　　A. Condensed structures - carbon-hydrogen and carbon-carbon single bonds aren't shown.
　　　　1. Bonds are understood.
　　　　2. The horizontal bonds between carbons aren't shown.

1,2⊠　　***EXAMPLE:***

　　Draw all possible isomers for C_6H_{14}.

SOLUTION:

$$CH_3CH_2CH_2CH_2CH_2CH_3$$

$$CH_3CHCH_2CH_2CH_3$$
$$| \atop CH_3$$

$$\begin{array}{cc} CH_3 & CH_3 \\ | & | \\ CH_3CHCHCH_3 & CH_3CCH_2CH_3 \\ | & | \\ CH_3 & CH_3 \end{array}$$

$$CH_3CH_2CHCH_2CH_3$$
$$| \atop CH_3$$

3⊠　　**The Shapes of Organic Molecules**

　　A. A molecule can be arbitrarily shown in a great many ways.
　　B. Rotation around carbon-carbon single bonds is possible.
　　　　1. Conformations - a number of possible 3-dimensional structures resulting from the ability of parts of a molecule to spin around a carbon-carbon single bond.

EXAMPLE:

The following molecules have the same formula, C_7H_{16}. Which are isomers and which are the same molecule?

a. $CH_3CH_2CH_2CH_2CH_2CH_2CH_3$

b. $CH_3CHCH_2CH_2CH_3$
$\quad\quad\quad |$
$\quad\quad CH_2CH_3$

c. $CH_3CH_2CH_2CHCH_3$
$\quad\quad\quad\quad\quad |$
$\quad\quad\quad\quad CH_2CH_3$

d. $CH_3CH_2CH_2CH_2$
$\quad\quad\quad\quad\quad\quad |$
$\quad\quad\quad\quad\quad\quad CH_2$
$\quad\quad\quad\quad\quad\quad |$
$\quad\quad\quad\quad\quad\quad CH_2$
$\quad\quad\quad\quad\quad\quad |$
$\quad\quad\quad\quad\quad\quad CH_3$

$\quad\quad\quad\quad\quad CH_3$
$\quad\quad\quad\quad\quad |$
e. $CH_3CH_2CHCHCH_3$
$\quad\quad\quad\quad\quad\quad |$
$\quad\quad\quad\quad\quad CH_3$

SOLUTION:

The isomers are a, b, and e.

4⊠ Naming Alkanes

A. Three parts to a chemical name.
1. Parent name - specifies the overall size of the molecule by telling how many carbon atoms are present in the longest continuous chain.
2. Suffix - identifies what family the molecule belongs to.
3. The prefix - specifies the location of various substituent groups attached to the parent chain.
B. Straight-chain alkanes - named by counting the number of carbon atoms in the chain and adding the family suffix *-ane*.
1. First four compounds.
   a. methane - CH_4
   b. ethane - CH_3CH_3
   c. propane - $CH_3CH_2CH_3$
   d. butane - $CH_3CH_2CH_2CH_3$
2. All other alkanes are named from Greek numbers according to the number of carbons present.
C. Four steps to naming branched-chain alkanes.
1. Name the main chain.
   a. the longest continuous chain of carbons present
   b. use the name of that chain as the parent name
2. Number the carbon atoms in the main chain.
   a. begin at the end near the first branch point
3. Identify and number the branching substituent.
   a. alkyl group - the part of an alkane that remains when a hydrogen is removed
      i. name by replacing the *-ane* ending of the parent alkane with an *-yl* ending
4. Write the name as a single word.
   a. separate different prefixes with hyphens
   b. separate numbers with commas
   c. cite substituent names in alphabetical order
      i. don't use the prefixes for alphabetizing
D. Alkyl groups.
1. Hydrogens in methane and ethane are equivalent.
   a. can only form 1 alkyl group

2. Hydrogens in propane are not equivalent.
    a. 6 H's on each end and 2 H's in the middle
    b. forms two different propyl groups
    c. remove an end H - *n*-propyl
    d. remove a middle H – isopropyl
3. Four different kinds of butyl groups.
    a. butyl and *sec*-butyl:
        i. are derived from straight chain alkanes
        ii. *n*-butyl: remove H from end C
        iii. *sec*-butyl: remove H from a C atom bonded to 2 other C atoms
    b. isobutyl and *tert*-butyl:
        i. derived from branched chain isobutane
        ii. isobutyl: remove H from end C on the branched-chain isobutane
        iii. *tert*-butyl: remove H from a C bonded to three other C atoms
4. Alkyl groups are not compounds.

EXAMPLE:
Name the compounds in the previous example.

SOLUTION:
a and d. *n*-heptane   b and c. 2-ethylpentane   e. 2,3-dimethylpentane

Cycloalkanes
A. Acylic alkanes - open-chain alkanes.
B. Cycloalkanes - alkanes that contain rings of carbon atoms.
C. Structures are represented by polygons.
    1. Carbon atom is understood to be at every junction of lines.
    2. Mentally supply the proper number of hydrogen atoms needed to fill out carbon's valency.
D. Name substituted cycloalkanes using the cycloalkane as the parent name and identifying the positions on the ring where substituents are attached.
    1. Start numbering at the group that has alphabetical priority.
    2. Proceed in the direction that gives the second substituent the lower possible number.

EXAMPLE:
Name the following compounds:

a. 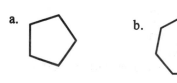   b.   c.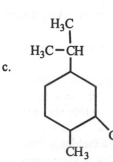

SOLUTION:
a. cyclopentane   b. cycloheptane   c. 1-chloro-3-isopropyl-6-methylcyclohexane

Reactions of Alkanes

A. Alkanes have relatively low chemical reactivity.
B. React with oxygen and with halogens.
    1. Alkanes react with oxygen during combustion.

a. CO_2 and H_2O are products

6⊠   2. Alkanes react with Cl_2 or Br_2 when irradiated with uv light.

7⊠   **Families of Organic Molecules: Functional Groups**

     A. Can classify organic compounds into families according to their structural features.

        1. Chemical behavior of the members of a family is predictable.

     B. Functional group - a part of a larger molecule that is composed of an atom or group of atoms that has characteristic chemical behavior.

        1. A given functional group undergoes the same kinds of reactions in every molecule it's a part of.

     C. The chemistry of an organic molecule, regardless of its size and complexity, is determined by the functional groups it contains.

     D. Most common functional groups - Table 23.3 (page 958 in your text)

EXAMPLE:

     Give the functional group for alcohols, ketones, and amines.

SOLUTION:

Alcohols -

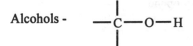

Ketones -

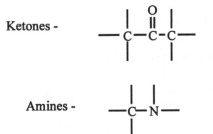

Amines -

Alkenes and Alkynes

     A. Alkenes - hydrocarbons that contain a carbon-carbon double bond.

     B. Alkynes - hydrocarbons that contain a carbon-carbon triple bond.

     C. Unsaturated hydrocarbons - hydrocarbons that have fewer hydrogens per carbon than the related alkanes.

        1. Alkenes and alkynes are unsaturated.

8⊠   D. Name alkenes by counting the longest chain of carbons that contains the double bond and add the family suffix -_ene_.

     E. Alkene isomers that exist because of double-bond position.

        1. For butene and higher - specify the position of the double bond with a numerical prefix.

           a. start numbering from the chain end nearer the double bond

           b. only cite the first of the double-bonded carbons

           c. identify and number the position of any substituent present

           d. if the double bond is equidistant from both ends of the chain, numbering starts at the end nearer the substituent

F. Alkene isomers that exist because of double-bond geometry.
  1. *cis* isomer - functional groups are on the same side of the double bond.
  2. *trans* isomer - function groups are on different sides of the double bond.

G. Alkynes are similar to alkenes.
  1. Names using the family suffix *-yne*.
  2. Simplest alkyne - acetylene.
  3. Isomers are possible depending on the position of the triple bond in the chain.

EXAMPLE:
Name the following compounds:

a. $CH_2=CHCH_2CH_3$ b. $CH_3-C\equiv CCH_3$ c. $CH_3-\underset{\underset{CH_2}{\parallel}}{C}-CH_3$

d. $Br-CH_2-C\equiv C-\underset{\underset{CH_3}{|}}{C}HCH_3$ (with CH_3 above)

SOLUTION:
  a. 1-butene   b. 2-butyne   c. 2-methyl-1-propene

  d. 1-bromo-4-methyl-3-pentyne

EXAMPLE:
Draw structures for:
a. 2,4-dibromo-6-methyl-3-octene
b. 1-bromo-3,4-dimethyl-1-pentyne
c. 2,4-dimethyl-4-pentene

SOLUTION:

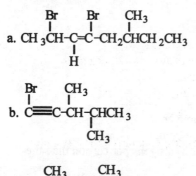

9⊠ **Reactions of Alkenes and Alkynes**
A. Addition reactions - X-Y adds to the multiple bond of the unsaturated reactant to yield a saturated product.
B. Hydrogenation - addition of hydrogen.
  1. Alkene + $H_2 \rightarrow$ alkane
C. Halogenation - addition of Cl_2 and Br_2.
  1. Alkene + $X_2 \rightarrow$ 1,2-dihaloalkane
D. Hydrolysis - alkenes react with water in the presence of H_2SO_4 to produce an alcohol.

EXAMPLE:

Show the products of the reaction of propene with a) H_2, b) Cl_2 and c) H_2O (H_2SO_4 catalyst).

SOLUTION:

   a. $CH_3CH_2CH_3$

   b. $ClCH_2\underset{\underset{Cl}{|}}{CH}CH_3$

   c. $CH_3\underset{\underset{OH}{|}}{CH}CH_3$

Aromatic Compounds and Their Reactions

  A. Aromatic compounds - a class of compounds containing a six-membered ring with three double bonds.

    1. Simplest aromatic - benzene.

  B. Benzene and other aromatic compounds are much less reactive than alkenes and don't normally undergo addition reaction.

  C. Benzene's stability is a consequence of its electronic structure.

    1. six π electrons are spread around the entire ring.

    2. 2 resonance structures

10⊠   D. Name substituted aromatic compounds by using the suffix *-benzene*

  E. Disubstituted aromatic compounds - use one of the prefixes *ortho-*, *meta-*, or *para-*.

    1. *ortho* (*o*) - two substituents in a 1,2 relationship on the ring.

    2. *meta* (*m*) - two substituents in a 1,3 relationship on the ring.

    3. *para* (*p*) - two substituents in a 1,4 relationship on the ring.

  F. Phenyl - the benzene ring itself is a substituent.

11⊠   G. Substitution reactions - a group substitutes for one of the hydrogen atoms on the aromatic ring without changing the ring itself.

    1. All 6 H's are equivalent.

    2. Nitration - substitution of a nitro group ($-NO_2$) in presence of H_2SO_4.

      a. key step in synthesis of explosives

      b. important pharmaceutical agents

      c. nitrobenzene used as starting material for dyes

    3. Halogenation - substitution of either Br or Cl in presence of $FeBr_3$ or $FeCl_3$.

      a. step used in the synthesis of numerous pharmaceutical agents

    4. Sulfonation - substitution of a sulfonic acid group ($-SO_3H$).

      a. result of benzene reacting with H_2SO_4 and SO_3

      b. key step in synthesis of aspirin and the sulfa-drug family of antibiotics

EXAMPLE:
Name the following compounds:

SOLUTION:
iodobenzene, *meta*-dibromobenzene

12⊠   Alcohols, Ethers, and Amines

A. Alcohols can be considered in two ways.
   1. Derivatives of water in which one of the H's is replaced by an organic substituent.
   2. Derivatives of alkanes in which one of the H's is replaced by a -OH group.
   3. Structural resemblance to water.
      a. simple alcohols are water soluble.
   4. Name by:
      a. specifying the point of attachment of the -OH group to the hydrocarbon chain
      b. use the suffix -*ol* to replace the terminal -*e* in the alkane name
      c. number chain at the end nearer to -OH group
   5. Simple alcohols - most important and commonly encountered organic chemicals.
   6. Methanol - wood alcohol.
      a. produced by catalytic reduction of CO with hydrogen gas
      b. important industrial starting material for preparing formaldehyde, acetic acid, and other chemicals
   7. Ethanol - one of the oldest known pure organic chemicals.
      a. alcohol present in wine, beer, and distilled liquors
   8. 2-propanol (rubbing alcohol) - used primarily as a solvent.
   9. Other important alcohols.
      a. ethylene glycol - principal constituent of automobile antifreeze
      b. glycerol - moisturizing agent in many foods and cosmetics
      c. phenol - preparing nylon, epoxy adhesives, and heat-setting resins
B. Ethers - compounds that have 2 organic groups bonded to the same oxygen atom.
   1. Inert chemically - used as reaction solvents.
   2. Diethyl ether - used for many years as a surgical anesthetic agent.
C. Amines - organic derivatives of NH_3.
   1. One or more of the NH_3 hydrogens is replaced by an organic substituent.
   2. Use suffix -*amine* in naming these compounds.
   3. Amines are bases.
      a. can use the lone pair of electrons on nitrogen to accept H^+ from an acid
         i. produce ammonium salts
   4. Ammonium salts are more soluble in water than are neutral amines.
      a. practical consequences in drug delivery

EXAMPLE:
Name the following compounds:

a. $CH_3CHCH_2CH_2-OH$
         |
        CH_3

b. $CH_3CH_2NH_2$

SOLUTION:
  a. 3-methyl-1-butanol   b. ethylamine

Ketones and Aldehydes
  A. Carbonyl group - group with a carbon-oxygen double bond.
  B. Classify into two categories.
    1. Based on their chemical properties.

13⊠   2. Aldehydes and ketones - CO group is bonded to atoms (H and C) that are not strongly electronegative.
      a. bonds to the carbonyl group C are nonpolar.
      b. two groups of compounds have similar properties.

14⊠   3. Carboxylic acids, esters, and amides - CO group is bonded to an atom (O or N) that is strongly electronegative.
      a. bonds to the carbonyl group C are polar
  C. Industrial preparation of simple aldehydes and ketones - involves oxidation of the related alcohol.
  D. Aldehyde and ketone functional groups are present in many biologically important compounds.

14,15⊠ **Carboxylic Acids, Esters, and Amides**
  A. CO group bonded to an atom (O or N) that strongly attracts electrons.
  B. All three families undergo carbonyl-group substitution reactions.
    1. A group substitutes for the -OH, -OC, or -N group of the starting material.
  C. Carboxylic acids - occur widely throughout the plant and animal kingdoms.
    1. Dissociate slightly in aqueous solution to give H_3O^+ and a carboxylate anion.
    2. Undergo acid-catalyzed reaction with an alcohol to yield an ester.
  D. Esters - many uses in medicine, in industry, and in living systems.
    1. Important pharmaceutical agents.
      a. aspirin and benzocaine
    2. Undergo hydrolysis - splits the ester molecule into a carboxylic acid and an alcohol.
    3. Saponification - base-catalyzed ester hydrolysis.
      a. soap - mixture of sodium salts of long-chain carboxylic acids
      b. produced by hydrolysis with aqueous NaOH of the naturally occurring esters in animal fat
    4. Named by identifying the alcohol-related part and the acid-related part using the *-ate* ending.
  E. Amides - amide bond between nitrogen and a carbonyl-group carbon is the fundamental link used by organisms for forming proteins.
    1. Neutral - do not act as proton acceptors and do not form ammonium salts.
    2. Can be prepared by the reaction of a carboxylic acid with ammonia or an amine.
    3. Name by first citing the *N*-alkyl group on the amine part and then identifying the carboxylic acid part using the *-amide* ending.

EXAMPLE:
Name the following compounds:

  a. $CH_3CH_2\!-\!O\!-\!\overset{\displaystyle O}{\underset{\displaystyle \|}{C}}\!-\!CH_3$ b. $CH_3\!-\!\overset{\displaystyle O}{\underset{\displaystyle \|}{C}}\!-\!NH_2$

SOLUTION:
  a. ethyl acetate   b. acetamide

377

EXAMPLE:

Give the products for the following reaction:

$$CH_3-\underset{\underset{O}{\|}}{C}-O-CH_3 \ + H_2O \xrightarrow{\ H^+\ }$$

SOLUTION:

$$CH_3-\underset{\underset{O}{\|}}{C}-OH \ + \ HOCH_3$$

17⊠ **Synthetic Polymers**

    A. Polymers - large molecules formed by the repetitive bonding together of many smaller molecules, called monomers.

        1. Simple alkenes - vinyl monomers undergo polymerization reactions.

            a. fundamental process - addition reaction to the double bond

    B. Second kind of polymerization process - molecules with two functional groups react.

        1. Diacids and diamines react to give polyamides.

        2. Diacids and dialcohols react to give polyesters.

Self-Test

This section is intended to test your knowledge of the material covered in this chapter. Think through these problems and make certain you understand what is going on. Ask yourself if your answer makes sense. Many of these questions are linked to the chapter learning goals. Therefore, successful completion of these problems indicates you have mastered the learning goals for this chapter. You will receive the greatest benefit from this section if you use it as a mock exam. You will then discover which topics you have mastered and which topics you need to study in more detail.

True/False

1. Organic molecules have polar covalent bonds when carbon bonds to hydrogen.

2. Doubly bonded carbons use sp^2 hybrid orbitals to form σ bonds.

3. Hydrocarbons are molecules that contain only C and H.

4. Different isomers have different structures but the same physical properties.

5. Alkyl groups are not compounds.

6. Alkanes are very reactive.

7. Benzene's stability is a consequence of the six π electrons which are spread around the ring.

8. Amines are organic derivatives of H_2O.

9. Aldehyde and ketone functional groups are present in many biological important compounds.

10. Amides are neutral.

Fill-in-the-Blank

1. Carbon is _____ in terms of its bonding.

2. Molecules that have the same molecular formula but different structures are called _____.

3. A number of possible 3-dimensional structures resulting from the ability of parts of a molecule to spin around a carbon-carbon single bond are referred to as _____.

4. Alkanes that contain rings of carbon atoms are known as _____.

5. The chemistry of an organic molecule, regardless of its size or complexity, is determined by the _____ it contains.

6. The simplest aromatic compound is _____.

7. _____ are compounds that have two organic groups bonded to the same oxygen atom and whose reactivity is _____.

8. Organic derivatives of NH_3 are _____.

9. Upon hydrolysis, an ester molecule splits into an _____ and an _____.

10. Polyesters are polymers formed by the reaction of _____ and _____.

General Questions

1. Name the following compounds:

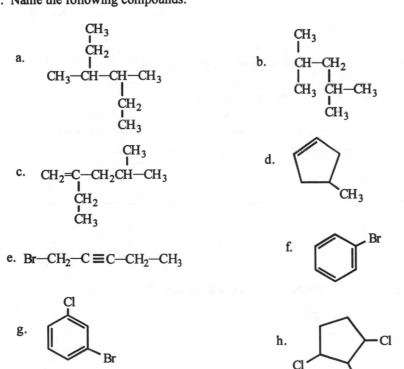

i. 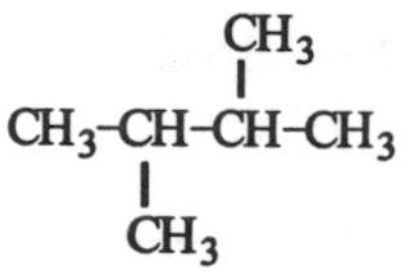

j. 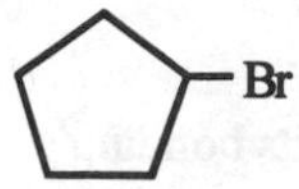

k. 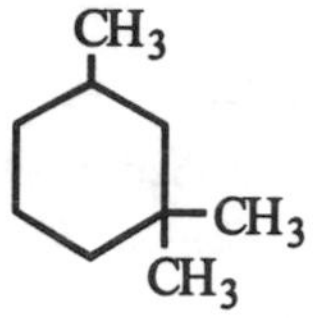

l. 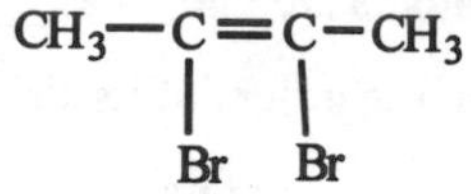

2. Identify the following alkyl groups:

a. CH_3CH_2–

b. 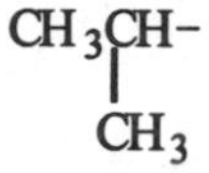

c. 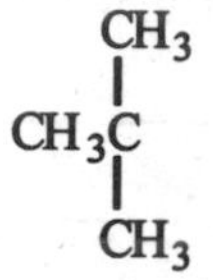

3. Write structures for the following compounds:
   a. 4-methylcyclohexene
   b. 2-pentyne
   c. *para*-chloroiodobenzene
   d. 1-chloro-3-ethyl-3-hexene
   e. 2,3-dimethylhexane
   f. 1,3-dichloro-1-butyne
   g. chlorobenzene

4. Give the first chemical equation for the reaction between ethane and Br_2.

5. Predict the outcome of the following reactions. Name the organic product.
   a. propene and bromine
   b. propene and hydrogen
   c. cyclohexene and water

6. Determine if the following isomers are *cis* or *trans*.

a. 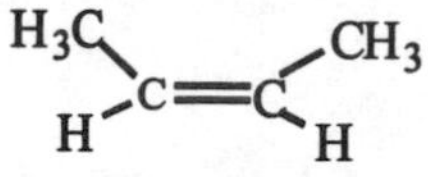

b. 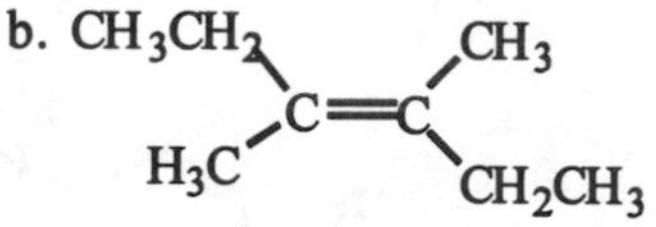

7. How would you prepare the following compounds beginning with benzene?

a. 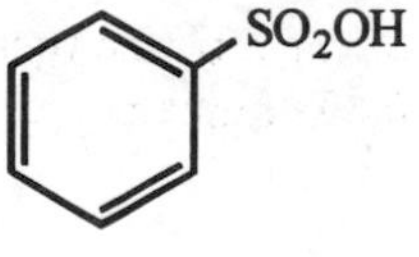

b. 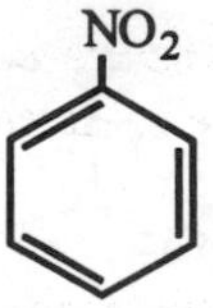

8. Classify the following compounds based on the functional group present. Name these compounds.

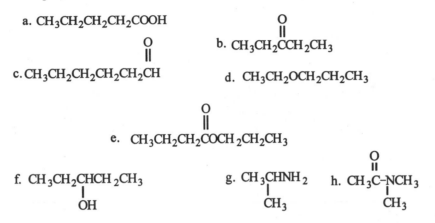

a. $CH_3CH_2CH_2CH_2COOH$

b. $CH_3CH_2\overset{\overset{\displaystyle O}{\|}}{C}CH_2CH_3$

c. $CH_3CH_2CH_2CH_2CH_2\overset{\overset{\displaystyle O}{\|}}{C}H$

d. $CH_3CH_2OCH_2CH_2CH_3$

e. $CH_3CH_2CH_2\overset{\overset{\displaystyle O}{\|}}{C}OCH_2CH_2CH_3$

f. $CH_3CH_2\underset{\underset{\displaystyle OH}{|}}{C}HCH_2CH_3$

g. $CH_3\underset{\underset{\displaystyle CH_3}{|}}{C}HNH_2$

h. $CH_3\overset{\overset{\displaystyle O}{\|}}{C}-\underset{\underset{\displaystyle CH_3}{|}}{N}CH_3$

9. Predict the outcome of the following reactions. Name the organic product.

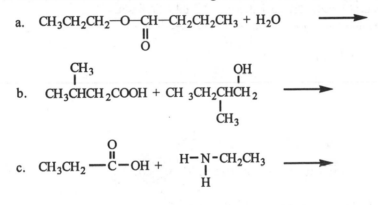

a. $CH_3CH_2CH_2-O-\overset{\overset{\displaystyle }{}}{\underset{\underset{\displaystyle O}{\|}}{C}}H-CH_2CH_2CH_3 + H_2O \longrightarrow$

b. $CH_3\underset{\underset{\displaystyle }{|}}{\overset{\overset{\displaystyle CH_3}{|}}{C}}HCH_2COOH + CH_3CH_2\underset{\underset{\displaystyle CH_3}{|}}{C}H\overset{\overset{\displaystyle OH}{|}}{C}H_2 \longrightarrow$

c. $CH_3CH_2-\overset{\overset{\displaystyle O}{\|}}{C}-OH + \underset{\underset{\displaystyle H}{|}}{H-N-CH_2CH_3} \longrightarrow$

10. Draw structural formulas for the polymers made from:

a. $CH_2=CH-Br$

b. $HO-\overset{\overset{\displaystyle }{}}{\underset{\underset{\displaystyle O}{\|}}{C}}-\text{⬡}-\overset{\overset{\displaystyle }{}}{\underset{\underset{\displaystyle O}{\|}}{C}}-OH + HO-CH_2CH_2-OH$

Solutions

True/False

1. F. Organic molecules have polar covalent bonds when carbon bonds to an element on the right or left side of the periodic table leading to a difference in electronegativity of 0.40 or greater.
2. T
3. T
4. F. Different isomers are different chemical compounds with different physical properties.
5. T
6. F. Alkanes have low reactivity.
7. T

8. F; Amines are organic derivatives of NH_3.
9. T
10. T

Fill-in-the-Blank
1. tetravalent
2. isomers
3. conformers
4. cycloalkanes
5. functional group
6. benzene
7. Ethers; inert
8. amines
9. alcohol; carboxylic acid
10. diacids; dialcohols

General Questions
1. a. 3,4-dimethylhexane; b. 2,4-dimethylpentane; c. 2-ethyl-4-methyl-1-pentene; d. 4-methylcyclopentene; e. 1-bromo-2-pentyne; f. bromobenzene; g. *meta*-bromochlorobenzene; h. 1,3-dichloro-2-methylcyclopentane; i. 2,3-dimethylbutane; j. bromocyclopentane; k. 1,1,3-trimethylcyclohexane; l. 2,2-dibromo-2-butene

2. a. ethyl b. isopropyl c. *tert*-butyl

3.

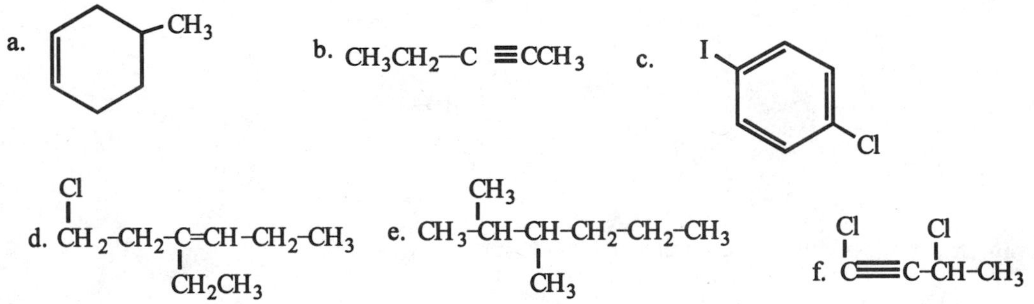
a.
CH3
b. CH3CH2—C ≡CCH3
c.
I
Cl
Cl
d. CH2-CH2-C=CH-CH2-CH3
CH2CH3
CH3
e. CH3-CH-CH-CH2-CH2-CH3
CH3
Cl
Cl
f. C≡C-CH-CH3

4. $CH_3CH_3 + Br_2 \longrightarrow CH_3CH_2Br$

5. a. 1,2-dibromopropane; b. propane; c. cyclohexanol

6. a. *cis* b. *trans*

7. benzene + H_2SO_4 + SO_3 ; benzene + HNO_3 + H_2SO_4

8. a. carboxylic acid: pentanoic acid; b. ketone: 3-pentanone
   c. aldehyde: hexanal; d. ether: ethylpropylether
   e. ester: propylbutanoate; f. alcohol: 3-pentanol
   g. amine: isopropylamine; h. amide: *N,N*-dimethylacetamide

9. a. propanol + butanoic acid; b. 2-methylbutyl-3-methylbutanoate; c. *N*-ethylpropanamide

10.

a

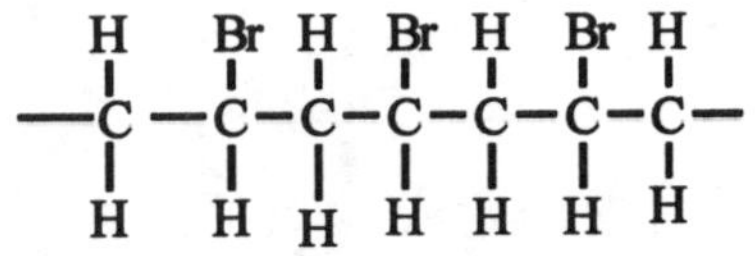

$$-\underset{\underset{H}{|}}{\overset{\overset{H}{|}}{C}}-\underset{\underset{H}{|}}{\overset{\overset{Br}{|}}{C}}-\underset{\underset{H}{|}}{\overset{\overset{H}{|}}{C}}-\underset{\underset{H}{|}}{\overset{\overset{Br}{|}}{C}}-\underset{\underset{H}{|}}{\overset{\overset{H}{|}}{C}}-\underset{\underset{H}{|}}{\overset{\overset{Br}{|}}{C}}-\underset{\underset{H}{|}}{\overset{\overset{H}{|}}{C}}-$$

b. dacron -

$-CH_2-O-\overset{}{\underset{O}{C}}-\bigcirc-C-O-CH_2CH_2-O-\overset{}{\underset{O}{C}}-\bigcirc-C-O-CH_2CH_2-$

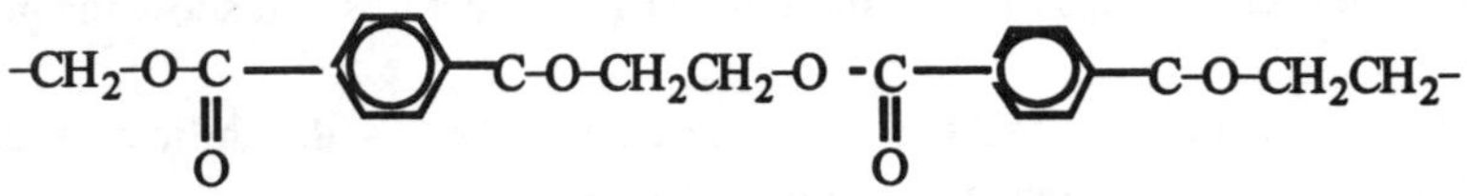

CHAPTER 24

BIOCHEMISTRY

Chapter Learning Goals

1⊠ Explain the metabolic reactions between a molecule containing the –OH functional group and ATP to form a phosphate.

2⊠ Identify functional groups in α-amino acids. Draw structures that show the geometry about each atom.

3⊠ Classify amino acid side chains as acidic, basic, or neutral, hydrophilic or hydrophobic.

4⊠ Determine whether a molecule is chiral or achiral.

5⊠ Draw the structure and give the three-letter shorthand notation for simple proteins.

6⊠ Classify a carbohydrate as simple or complex. Classify a monosaccharide as a ketose or aldose.

7⊠ Draw tetrahedral representations of simple carbohydrates, and determine the number of chiral carbon atoms and the maximum number of isomers of the molecule.

8⊠ Draw the structures of simple fats and oils.

9⊠ Draw the structures of simple nucleic acids.

10⊠ Show the complementary base pairs in strands of DNA.

11⊠ Show the RNA base sequence complementary to a given DNA base sequence.

Chapter in Brief

This chapter begins by looking at biochemical energetics. You then begin an examination of the main classes of biomolecules: proteins, carbohydrates, lipids, and nucleic acids. Your detailed study of these molecules includes the molecular handedness of amino acids and carbohydrates, the levels of protein structure, and the cyclic structures of monosaccharides. You will also examine the lock and key model of enzyme action and the Watson-Crick model for base pairing in DNA.

1⊠ **Biochemical Energy**

    A. Metabolism - the sum of the many organic reactions that go on in cells.

        1. Extract and release energy from food.

        2. End products - carbon dioxide, water and energy.

        3. Occur in either linear or cyclic long sequences.

            a. linear sequence - product of one reaction serves as the starting material for the next

            b. cyclic sequence - a series of reactions regenerates the first reactant

    B. Catabolism - reaction sequences that break molecules apart.

        1. Generally release energy that is used to power living organisms.

        2. First stage - digestion.

            a. bulk food is broken down into small molecules

        3. Second stage - small molecules are broken down into 2-carbon acetyl groups attached to coenzyme A.

            a. produces acetyl coenzyme A - intermediate in the breakdown of all main classes of food molecules

        4. Third stage - citric acid cycle.

            a. acetyl groups are oxidized; produces carbon dioxide and water

            b. releases a great deal of energy which is used in fourth stage

        5. Fourth stage - the respiratory chain.

            a. produces adenosine triphosphate (ATP)

            b. plays a pivotal role in the production of biological energy

C. Anabolism - reaction sequences that put building blocks back together to assemble larger molecules.
  1. Absorb energy.
D. The entire process of energy production revolves around the ATP $\rightleftharpoons$ ADP interconversion
  1. Catabolic reactions "pay off" in ATP by synthesizing it from adenosine diphosphate (ADP) plus hydrogen phosphate ion.
  2. Anabolic reactions "spend" ATP by transferring a phosphate group to other molecules, thereby regenerating ADP.
E. ATP releases a large amount of energy when its P-O-P bonds are broken and a phosphate group is transferred.
  1. Usefulness is due to ATP's ability to drive otherwise unfavorable reactions.
  2. Overall free-energy change for the two reactions together is favorable.

Amino Acids and Peptides
A. Protein - biological polymers made up of many amino acids linked together to form long chains.
  1. A group of biological molecules that are of primary importance to all living organisms.
  2. Different biological functions.
    a. some serve a structural purpose
    b. some act as hormones to regulate specific body processes
    c. some are enzymes
      i. biological catalysts that carry out body chemistry
2⊠ B. Amino acids - molecules that contain two functional groups
  1. a basic amino group ($-NH_2$)
  2. an acidic group (-COOH)
C. Peptide bond - an amide bond that forms when two amino acids link together.
D. Dipeptide - the molecule that results when two amino acids are linked together by formation of a peptide bond between the $-NH_2$ group of one and the -COOH group of the second.
  1. Tripeptide - link three amino acids by two peptide bonds.
  2. Polypeptides - chains of up to 100 amino acids.
  3. Protein - chains with more than 100 amino acids.
E. 20 different amino acids commonly found in proteins.
  1. Refer to each by a three-letter shorthand code (Fig. 24.2 page 993 in text).
  2. α-amino acids - the amino group is connected to the carbon atom alpha to (next to) the carboxylic acid group.
  3. Differ in the nature of the group (side chain, R) attached to the α-carbon.
3⊠ 4. Classified as neutral, basic, or acidic.
    a. depends on side chain
    b. neutral - divide into hydrophobic (nonpolar side chains) and hydrophilic (polar) side chains
    c. acidic - an additional carboxylic acid on side chain
    d. basic - additional amine function on side chain

Amino Acids and Molecular Handedness
A. Chiral molecule - a molecule that is not identical to its reflected mirror image.
  1. Right-handed molecule (D).
  2. Left-handed molecule (L).
4⊠ B. A carbon atom bonded to four different atoms or groups of atoms is chiral.
  1. Achiral - carbon atom is bonded to two or more of the same groups.
C. Enantiomers - two mirror-image forms of a chiral molecule.
  1. 19 of common amino acids are chiral because they have 4 different groups bonded to their α-carbons.

2. Only L-amino acids are found in proteins.

Proteins

A. Residues - individual amino acids which are linked together by peptide bonds to form proteins.

B. Backbone - the repeating chain of amide linkages which the side chains are attached to.

C. The number of possible isomeric peptides increases rapidly as the number of amino acid residues increases.

5⊠   D. All noncyclic proteins have an N-terminal amino acid and a C-terminal amino acid.

    1. N-terminal amino acid - has a free $-NH_2$ group on one end.

    2. C-terminal amino acid - has a free -COOH group on the other end.

    3. Protein is written with the N-terminal residue on the left and the C-terminal residue on the right.

    4. Use three-letter abbreviations to indicate the name. (see Fig. 24.2, page 993 in text.)

E. Fibrous proteins - consist of polypeptide chains arranged side by side in long filaments.

F. Globular proteins - polypeptide chains coiled into compact, nearly spherical shapes.

G. Proteins are also classified by their biological functions. (Table 24.1, page 998 in text.)

Levels of Protein Structure

A. Four levels of structure used to describe proteins.

    1. Primary structure - specifies the sequence in which the various amino acids are linked together.

    2. Secondary structure - - specifies how segments of the protein chain are oriented into a regular pattern.

    3. Tertiary structure - specifies how the entire protein chain is coiled and folded into a specific 3-dimensional shape.

    4. Quaternary structure - how several protein chains aggregate to form a larger unit.

B. Primary protein structure - most important of the four structural levels

    1. Amino acid sequence determines the overall shape and functions.

        a. change 1 amino acid - alter biological properties

C. Secondary protein structure - the regular patterns which result from the orientation of segments of folded proteins.

    1. α-helix - the protein wraps into a helical coil, much like the cord on a telephone.

        a. stabilized by formation of hydrogen bonds between the N-H group of one amino acid and the C=O group of another amino acid four residues away

    2. β-pleated sheet - polypeptide chains line up in a parallel arrangement held together by hydrogen bonds.

        a. not as common as the α-helix

        b. found in proteins where sections of peptide chains double back on themselves

D. Tertiary protein structure - structures which result primarily from interactions of side-chain R groups in the protein.

    1. Stabilization results from the hydrophobic interactions of the hydrocarbon side chains on amino acids.

    2. Neutral, nonpolar side chains congregate on the hydrocarbon-like interior of a protein molecule.

    3. Polar side chains are found on the exterior of the protein.

    4. Other stabilizing factors.

        a. disulfide bridges - covalent S-S bonds formed between nearby cysteine residues

        b. salt bridges - ionic attractions between positively and negatively charged sites on the protein

        c. hydrogen bonds between nearby amino acids

Enzymes

A. Enzymes - large proteins that act as catalysts for biological reactions.
   1. Catalyst - an agent that speeds up the rate of a chemical reaction without itself undergoing change.
   2. Larger, more complicated molecules than simple inorganic catalysts.
   3. More specific in their action.
      a. catalyze only a single reaction of a single compound called the enzyme's substrate
   4. Turnover number - the number of substrate molecules acted on by one molecule of enzyme per unit time.
      a. measures catalytic activity of enzyme

B. Cofactors - small, nonprotein portions of an enzyme, either an inorganic ion or a small organic molecule.
   1. Apoenzyme - the protein part of an enzyme.
   2. Holoenzyme - the entire assembly of apoenzyme plus cofactor.
      a. only holoenzymes are active as catalysts
   3. Coenzyme - small organic molecule which is an enzyme cofactor.

C. Lock-and-key model - model used to explain the specificity and catalytic activity of enzymes
   1. Enzyme - irregularly shaped molecule with a cleft or crevice in its middle.
      a. active site - a small 3-D region inside the crevice with the specific shape necessary to bind the substrate and catalyze the appropriate reaction
         i. lined by various acidic, basic, and neutral amino acid side chains
   2. Forces two reagents into close contact.
   3. Catalytic ability of enzymes is due to their ability to:
      a. bring reagents together
      b. hold reagents at the exact distance and with the exact orientation necessary for reaction
      c. provide proton donating or accepting groups as required

Carbohydrates

6⊠

A. Carbohydrates - a large class of polyhydroxylated aldehydes and ketones.
   1. Monosaccharides - carbohydrates that can't be broken down into smaller molecules by hydrolysis with aqueous acid.
   2. Polysaccharides - compounds that are made of many simple sugars linked together and that cleave into many molecules of simple sugars upon hydrolysis.

B. Monosaccharides.
   1. Aldose - contains an aldehyde carbonyl group.
   2. Ketose - contains a ketone carbonyl group.
   3. "ose" suffix - used to indicate a sugar.
   4. Use one of the Greek prefixes to indicate the number of carbon atoms in the sugar.

Handedness of Carbohydrates

7⊠

A. Compounds are chiral if they have a carbon atom bonded to four different atoms or groups of atoms.
   1. Lack a plane of symmetry.
   2. Can exist as a pair of enantiomers.
      a. right-handed - D form
      b. left-handed - L form

B. A compound with n chiral carbon atoms has a maximum of 2^n possible forms.

7⊠   **Cyclic Structures of Monosaccharides**
        A. Monosaccharides are shown as having open-chain structures.
        B. Exist as cyclic molecules.
            1. -OH group near the bottom of the chain adds to the -C=O group near the top of the chain.
            2. Cyclic α form - the C1 -OH group is on the bottom side of the ring.

Some Common Disaccharides and Polysaccharides
        A. Lactose - major carbohydrate present in mammalian milk.
            1. Disaccharide whose hydrolysis with aqueous acids yields one molecule of glucose and one molecule of galactose.
            2. Two sugars bonded together by a 1,4 link.
                a. bridging oxygen atom between C1 of β-galactose and C4 of β-glucose
        B. Sucrose - most common pure organic chemical in the world.
            1. Hydrolysis yields one molecule of glucose and one molecule of fructose.
            2. Invert sugar - 50:50 mixture of sugar produced from the hydrolysis of sucrose.
                a. commonly used as a food additive
        C. Cellulose - consists of several thousand β-glucose molecules joined together by 1,4 links to form an immense polysaccharide.
            1. Used as a structural material in plants.
        D. Starch - consists of several thousand α-glucose molecules joined together by 1,4 links.
            1. More structurally complex than cellulose.
            2. Two types.
                a. amylose - several hundred to 1000 α-glucose units joined together in a long chain by 1,4 links
                b. amylopectin - larger than amylose and has branches every 25 units along its chain
        E. Glycogen - a long polymer of α-glucose units with branch points in its chain.
            1. Breaks down into simple glucose units.
                a. some used as fuel
                b. some is stored for later use.

8⊠   **Lipids**
        A. Lipids - naturally occurring organic molecules that dissolve in nonpolar organic solvents.
            1. Defined by solubility rather than by chemical structure.
            2. Contain large hydrocarbon portions.
                a. accounts for their solubility behavior
        B. Fats and oils - most plentiful lipids in nature.
            1. Triacylglycerols (triglycerides) - triesters of glycerol with three long-chain carboxylic acids (fatty acids).
                a. fatty acids - unbranched and have an even number of carbon atoms
                b. 3 fatty acids can be different
            2. Monounsaturated - fatty acid with only one double bond.
            3. Polyunsaturated fatty acids - have more than one carbon-carbon double bond.
        C. Steroids - a lipid whose structure is based on the tetracyclic system.
            1. 3 rings are six-membered, while the fourth is five-membered.
        D. Cholesterol - most abundant animal steroid; serves two important functions.
            1. Minor component of cell membranes.
            2. Serves as the body's starting material for the synthesis of all other steroids.

Nucleic Acids
        A. Deoxyribonucleic acid (DNA) and ribonucleic acid (RNA) are the chemical carriers of an organism's genetic information.
9⊠   B. Nucleic acids - polymers made up of nucleotide units linked together to form a long chain.

388

    1. Nucleotide - composed of nucleoside plus phosphoric acid.
        a. nucleoside - composed of an aldopentose sugar plus an amine base.
            i. RNA - sugar = ribose
            ii. DNA - sugar = 2-deoxyribose

C. Four different cyclic amine bases in DNA: adenine, guanine, cytosine, and thymine.

D. Four different cyclic amine bases in RNA: adenine, guanine, cytosine and uracil.

E. Cyclic amine base is bonded to C1' of the sugar, and the phosphoric acid is bonded to the C5' sugar position.
    1. Numbers with a prime superscript refer to positions on the sugar component of a nucleotide.
    2. Numbers without a prime superscript refer to positions on the cyclic amine base.

F. Nucleotides link through a phosphate ester bond between the phosphate group at the 5' end of one nucleotide and the hydroxyl group on the sugar component at the 3' end of another nucleotide.

G. Structure of a nucleic acid depends on its sequence of individual nucleotides.
    1. An alternating sugar-phosphate backbone with different amine baseside chains attached.

H. Describe the sequence of nucleotides by starting at the 5' phosphate and identifying the bases in order.
    1. Use first letter of the base name as an abbreviation for each nucleotide.

10⊠ Base Pairing in DNA: The Watson-Crick Model

A. DNA consists of two polynucleotide strands coiled around each other in a double helix.
    1. Sugar-phosphate backbone is on the outside of the helix.
    2. Heterocyclic bases are on the inside of the helix.
    3. Base on 1 strand points directly at a base on the other strand.
    4. 2 strands run in opposite directions.
        a. held together by hydrogen bonds between pairs of bases
            i. A and T
            ii. G and C

B. Two strands are complementary.
    1. Whenever a G base occurs in one strand, a C base occurs opposite it in the other strand.

Nucleic Acids and Heredity

A. Chromosomes - threadlike strands that are coated with proteins and wound into complex assemblies.

B. Gene - segment of a DNA chain that contains the instructions necessary to make a specific protein.

C. DNA functions as a storage medium for an organism's genetic information.

D. RNA reads, decodes, and uses the information received from DNA to make proteins.

E. Three processes in the transfer and use of genetic information.
    1. Replication - the means by which identical copies of DNA are made.
        a. forms additional molecules
        b. preserves genetic information for passing on to offspring
    2. Transcription - means by which information in the DNA is transferred to and decoded by RNA.
    3. Translation - means by which RNA uses the information to build proteins.

F. Replication - an enzyme-catalyzed process that begins with a partial unwinding of the double helix.
    1. New nucleotides line up on each strand in a complementary manner.
    2. 2 new strands begin to grow.

11⊠ G. Transcription - similar to replication except that new ribonucleotides line up.
    1. Uracil lines up opposite adenine.

H. Translation - protein biosynthesis; directed by messenger RNA (mRNA).
   1. Occurs on ribosomes - knobby protuberances within a cell.
   2. Ribonucleotide sequence in mRNA specifies the order of different amino acid residues.
      a. 3 ribonucleotides that are specific for a given amino acid
   3. Read by transfer RNA (tRNA) - contains a complementary base sequence that allows it to recognize a three-letter word on mRNA.
      a. acts as a carrier to bring a specific amino acid into place for transfer to the peptide chain

Self-Test

This section is intended to test your knowledge of the material covered in this chapter. Think through these problems and make certain you understand what is going on. Ask yourself if your answer makes sense. Many of these questions are linked to the chapter learning goals. Therefore, successful completion of these problems indicates you have mastered the learning goals for this chapter. You will receive the greatest benefit from this section if you use it as a mock exam. You will then discover which topics you have mastered and which topics you need to study in more detail.

Matching

Metabolism	a. a large class of polyhydroxylated aldehydes and ketones
Protein	b. large proteins that act as catalysts for biological reactions.
Dipeptide	c. proteins that consist of polypeptide chains arranged side by side in long filaments
Residue	d. the means by which RNA uses the information to build proteins.
Fibrous proteins	e. molecules composed of an aldopentose sugar plus an amine base.
Globular proteins	f. threadlike DNA strands that are coated with proteins and wound into complex assemblies.
Enzyme	g. an enzyme-catalyzed process that begins with a partial unwinding of the double helix.
Cofactor	h. proteins that consist of polypeptide chains coiled into compact, nearly spherical shapes
Carbohydrates	i. the sum of the many organic reactions that go on in cells
Lipids	j. small, nonprotein portions of an enzyme, either a metal ion or a small organic molecule
Nucleoside	k. polypeptide chain with more than one hundred amino acids

Chromosomes

l. the molecule that results when two amino acids are linked together by formation of a peptide bond between the -NH$_2$ group of one and the -COOH group of the second.

Replication

m. naturally occurring organic molecules that dissolve in nonpolar organic solvents.

Translation

n. individual amino acids that are linked together by peptide bonds to form proteins

Fill-in-the-Blank

1. In the second stage of catabolism, small molecules are broken down into _____ _____ and attached to _____ .

2. The entire process of energy production revolves around the _____ interconversion.

3. The two functional groups in amino acids are _____ _____ .

4. α-amino acids differ in the nature of the _____ .

5. An acidic α-amino acid has an additional _____ on its side chain.

6. A protein is written with the _____ residue on the left and the _____ on the right.

7. The α-helix structure of a protein is stabilized by the formation of _____ between the _____ group of one amino acid and the _____ group of another amino acid _____ .

8. The number of substrate molecules acted on by one molecule of enzyme per unit time is the _____ .

9. The difference between an aldose and a ketose is that an aldose contains _____ _____ and a ketose contains a _____ .

10. _____ consists of several thousand β-glucose molecules joined together by 1,4 links to form an immense polysaccharide.

11. Lipids are defined by _____ rather than _____ .

12. _____ and _____ are the chemical carriers of an organism's genetic information.

13. The four different cyclic amine bases in DNA are _____ .

14. The base _____ replaces the base _____ in RNA.

15. Transcription is similar to replication except that _____ line up and _____ lines up opposite adenine.

Problems

1. State the amino acids represented by the following abbreviations.

   a. Trp   b. Glu   c. His   d. Arg

2. Draw the structures of amino acid that contains

   a. an isobutyl group   b. a *sec*-butyl group

3. Identify the amino acids present in the following tetrapeptide:

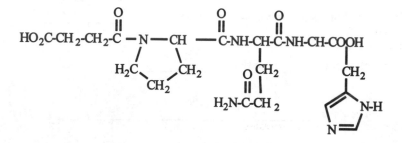

4. Which of the following amino acids are likely to be found on the outside and which are likely to be found on the inside of a globular protein?

   a. Tyr   b. Met   c. Lys   d. Glu

5. If the sequence A-G-T-A-A-T appeared on one strand of DNA, what sequence would appear opposite it on the other strand?

6. What RNA sequence would be complementary to the DNA sequence in #5?

7. Draw the structure of the dinucleotide A-G.

Solutions

Matching

    Metabolism - i
    Protein - k
    Dipeptide - l
    Residue - n
    Fibrous proteins - c
    Globular proteins - h
    Enzyme - b
    Cofactor - j
    Carbohydrates - a
    Lipids - m
    Nucleoside - e
    Chromosomes - f
    Replication - g

Translation - d

Fill-in-the Blank
1. 2-carbon acetyl groups; coenzyme A
2. ATP $\rightleftarrows$ ADP
3. the basic amino group, $-NH_2$, and the acidic group ($-OOH$)
4. side chain attached to the α-carbon
5. carboxylic acid group
6. N-terminal amino acid; C-terminal residue
7. hydrogen bonds; N-H group; C=O
8. turnover number
9. an aldehyde carbonyl; ketone carbonyl
10. Cellulose
11. solubility; chemical properties
12. deoxyribonucleic acid; ribonucleic acid
13. adenine, guanine, cytosine, thymine
14. uracil; thymine
15. ribonucleotides; uracil

Problems
1. a. tryptophan; b. glutamic acid; c. histidine; d. arginine

2. a.

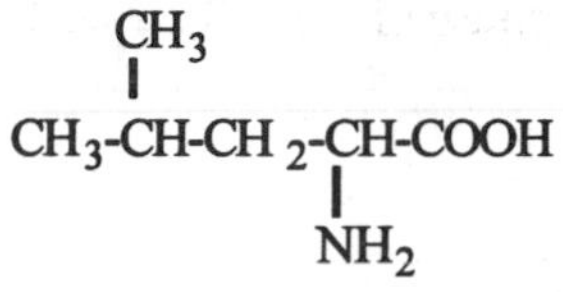

$$CH_3-CH-CH_2-CH-COOH \quad \text{Leucine}$$

with CH_3 attached above and NH_2 attached below

b.

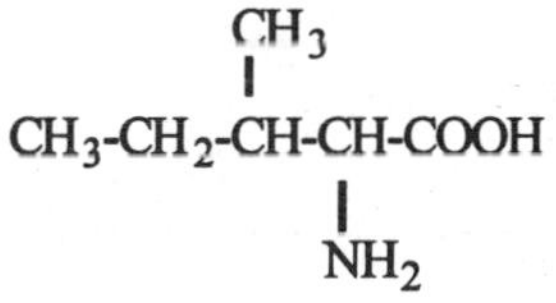

$$CH_3-CH_2-CH-CH-COOH \quad \text{Isoleucin}$$

with CH_3 attached above and NH_2 attached below

3. Asp-Pro-Gln-His

4. a. outside, hydrophilic; b. inside, hydrophobic; c. outside, hydrophilic d. outside, hydrophilic

5. T-C-A-T-T-A

6. U-C-A-U-U-A

7.

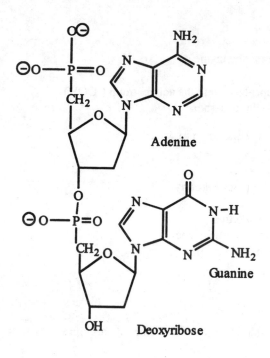

Adenine

Guanine

Deoxyribose

NOTES

NOTES

NOTES

NOTES

NOTES

NOTES

NOTES

NOTES

Useful Conversion Factors and Relationships

Length

SI unit: meter (m)

$$1 \text{ km} = 0.621\,37 \text{ mi}$$
$$1 \text{ mi} = 5280 \text{ ft}$$
$$= 1.6093 \text{ km}$$
$$1 \text{ m} = 1.0936 \text{ yd}$$
$$1 \text{ in.} = 2.54 \text{ cm (exactly)}$$
$$1 \text{ cm} = 0.393\,70 \text{ in.}$$
$$1 \text{ Å} = 10^{-10} \text{ m}$$

Mass

SI unit: kilogram (kg)

$$1 \text{ kg} = 10^3 \text{ g} = 2.2046 \text{ lb}$$
$$1 \text{ lb} = 16 \text{ oz} = 453.59 \text{ g}$$
$$1 \text{ amu} = 1.660\,54 \times 10^{-27} \text{ kg}$$

Temperature

SI unit: Kelvin (K)

$$0 \text{ K} = -273.15°C$$
$$= -459.67°F$$
$$\text{K} = °C + 273.15$$
$$°C = 5/9\,(°F - 32)$$
$$°F = 5/9\,(°C) + 32$$

Energy (derived)

SI unit: Joule (J)

$$1 \text{ J} = 1 \text{ (kg·m}^2)/s^2$$
$$1 \text{ J} = 0.239\,01 \text{ cal}$$
$$= 1 \text{ C} \times 1 \text{ V}$$
$$1 \text{ cal} = 4.184 \text{ J}$$
$$1 \text{ eV} = 1.602 \times 10^{-19} \text{ J}$$

Pressure (derived)

SI unit: Pascal (Pa)

$$1 \text{ Pa} = 1 \text{ N/m}^2$$
$$= 1 \text{ kg}/(\text{m·s}^2)$$
$$1 \text{ atm} = 101{,}325 \text{ Pa}$$
$$= 760 \text{ mm Hg (torr)}$$
$$= 14.70 \text{ lb/in}^2$$
$$1 \text{ bar} = 10^5 \text{ Pa}$$

Volume (derived)

SI unit: cubic meter (m³)

$$1 \text{ L} = 10^{-3} \text{ m}^3$$
$$= 1 \text{ dm}^3$$
$$= 10^3 \text{ cm}^3$$
$$= 1.0567 \text{ qt}$$
$$1 \text{ gal} = 4 \text{ qt}$$
$$= 3.7854 \text{ L}$$
$$1 \text{ cm}^3 = 1 \text{ mL}$$
$$1 \text{ in}^3 = 16.4 \text{ cm}^3$$

Fundamental Constants

Atomic mass unit	1 amu	$= 1.660\,540 \times 10^{-27}$ kg
	1 g	$= 6.022\,137 \times 10^{23}$ amu
Avogadro's number	N_A	$= 6.022\,137 \times 10^{23}$/mol
Boltzmann's constant	k	$= 1.380\,66 \times 10^{-23}$ J/K
Electron charge	e	$= 1.602\,177\,3 \times 10^{-19}$ C
Faraday's constant	$\mathfrak{F}$	$= 9.648\,531 \times 10^4$ C/mol
Gas constant	R	$= 8.314\,51$ J/(mol·K)
		$= 0.082\,057\,8$ (L·atm)/(mol·K)
Mass of electron	m_e	$= 5.485\,799 \times 10^{-4}$ amu
		$= 9.109\,390 \times 10^{-31}$ kg
Mass of neutron	m_n	$= 1.008\,664$ amu
		$= 1.674\,929 \times 10^{-27}$ kg
Mass of proton	m_p	$= 1.007\,276$ amu
		$= 1.672\,623 \times 10^{-27}$ kg
Pi	π	$= 3.141\,592\,653\,6$
Planck's constant	h	$= 6.626\,076 \times 10^{-34}$ J·s
Speed of light	c	$= 2.997\,924\,58 \times 10^8$ m/s

STUDY GUIDE
DonnaJean Fredeen

CHEMISTRY
Second Edition
McMURRY • FAY

This Study Guide was written specifically to assist the student using *Chemistry 2/e* by McMurry and Fay and presents, in outline form, the major concepts, theories, facts and applications found in the text. The second edition of the Study Guide has been expanded to include more examples within the outlines and more questions and problems in the Self Tests.

Every chapter is keyed to the main text, and is presented in four sections:

- Listed Learning Goals. These lists alert the student to key concepts that will be covered in each chapter.

- Chapter Overview. Overviews sumerize major material that will be covered in that chapter, placed in its context.

- Chapter Outline. Most students benefit from the use of outlines. For each chapter, the Study Guide provides a synopsis of major textual material in outline form. The Learning Goals are clearly marked and expounded upon within these chapter outlines. Each outline includes detailed examples showing applications of major chemical concepts. The outlines also provide the student with an initial set of notes that can then be tailored and annotated with lecture material.

- Self Tests and Solutions. Modeled on the text problems and linked to the Chapter Learning Goals, these questions and problems test the student's knowledge of the material covered in each chapter. If used as practice exams, the Self Tests allow the student to assess which topics have been mastered and which topics need more attention.

PRENTICE HALL Upper Saddle River, NJ 07458
http://www.prenhall.com

ISBN 0-13-757436-3
90000

9 780137 574360